MW01595786

Principles of Fire Protection

**By
Arthur Cote, P.E.
and Percy Bugbee**

National Fire Protection Association
Batterymarch Park
Quincy, MA 02269

J2 2140058

Sixth Printing, 1998

Project Manager: Gene A. Moulton
Editor: Hilary Davis
Art Coordinator: Hilary Davis
Interior Design: Frank Lucas
Composition: Sharon Summers
           Louise Grant
Production Coordinators: Elizabeth Carmichael
                     Debra Rose

NFPA No. ST-1
ISBN 0-87765-345-3
Library of Congress Catalog Card No. 88-060164
Printed in the United States of America

ABOUT THE AUTHORS

Arthur E. Cote, P.E., graduated from the University of Maryland with a B.S. degree in Fire Protection Engineering. Presently Assistant Vice President/Standards at the National Fire Protection Association, Mr. Cote administers NFPA's 175 Technical Committees which develop the more than 250 firesafety codes and standards that make up the *National Fire Codes.* In addition, he is Secretary to the 13-member NFPA Standards Council, Secretary to NFPA's Advisory Committee on the Toxicity of the Products of Combustion, and Editor-in-Chief of the 16th Edition of the *Fire Protection Handbook.*

Mr. Cote is a registered professional engineer in fire protection in the State of Pennsylvania and a member of the Society of Fire Protection Engineers, having been Past President of the New England Chapter. He is a Corporate Member of Underwriters Laboratories Inc., and belongs to UL's Standards Review Council. His other active memberships include the Executive Standards Council - American National Standards Institute, Consultative Council - National Institute of Building Sciences, and Charter Member of the World Organization of Building Officials.

Percy Bugbee graduated from MIT with a degree in Chemical Engineering, and soon began his career at the NFPA in Boston in 1921. Mr. Bugbee was the Association's General Manager for 30 years until his retirement in 1969. Since retirement, he continues to serve as Honorary Chairman of the Board of Directors.

Mr. Bugbee is Honorary President of the World Conference of Fire Protection Associations and is an Honorary Life Member of such organizations as the Australian Fire Protection Association, the British Institution of Fire Engineers, the Fire Marshals Association of North America, the French Federation of Fire Fighters, the International Association of Fire Chiefs, the International Association of Fire Fighters, and the Society of Fire Protection Engineers.

ACKNOWLEDGMENTS

The preparation and development of any technical book requires the cooperation of many individuals.

The authors wish to thank the following technical experts for their time and effort in contributing to this second edition: Wayne G. "Chip" Carson of Carson Associates; Robert P. Benedetti, Richard P. Bielen, Judy Comoletti, Ron Cote', Martin F. Henry, Dr. John Hall, Theodore C. Lemoff, Carl E. Peterson, Robert E. Solomon and Bruce W. Teele, all of the NFPA Staff.

Special thanks are also due to Richard P. Custer, Associate Director, Center for Firesafety Studies, Worcester Polytechnic Institute, who reviewed the first edition and suggested many of the changes incorporated into this second edition.

Much of the material in this second edition of *Principles of Fire Protection* is based on the first edition which was authored by Percy Bugbee, and on information contained in the latest (16th) edition of the *Fire Protection Handbook.*

Arthur E. Cote, P.E.

Fire is a fundamental force in nature. Without fire, life as we know it today would not exist. Friendly fires heat our homes, cook our food, and help to generate our energy. Like any force of nature, however, fire also carries with it the potential for great destruction. Fire poses a potential threat to our lives, property, and resources.

Civilian fire deaths in the United States have declined fairly steadily over the past decade from a high of over 7,700 per year in 1978 to an average of about 6,000 per year since 1982, except in 1984 when fire deaths fell to an all-time low of approximately 5,200.

Despite this sustained decline, however, the U.S. and Canada still have the highest fire death rates of all developed countries in the free world.

Annual estimates of civilian injuries indicate that, on average, approximately 30,000 civilians are injured annually by fire in the U.S. Since many injuries are not reported to the fire service, these estimates are probably on the low side. The number of fires in the United States has declined over the past decade from over 3,000,000 in 1977 to under 2,400,000 since 1983.

Despite these figures, the vast majority of people tend to think of a serious fire as they would a serious automobile accident or other tragedy—as something that happens to someone else. This is because the average person suffers a minor burn or experiences a small fire only once or twice in a lifetime, and thus the threat of a truly destructive fire seems improbable and remote.

The impressive decreases in the numbers of fires and civilian fire deaths in the U.S. over the past decade are probably due in large part to a greater awareness on the part of the general public as to the potentially destructive nature of fire. As a direct result of that awareness, there are more people in the fire protection field than ever before. People in many different fields and organizations, both private and public, are working to make fire protection technology and practices as sophisticated as possible. They include, among others, paid and volunteer fire fighters; federal, state, and local fire officials and fire investigators; fire insurance inspectors, raters, and agents; industrial safety and fire protection personnel; fire equipment manufacturers and salespersons; builders and building inspectors; electrical manufacturers, installers, and inspectors; and architects and engineers. New people enter these fields daily.

Principles of Fire Protection is intended to be a basic text on fire protection, a text whose objective it is to provide both fire science students and new fire protection personnel with an overview of the fundamental methods of fire protection, prevention, and suppression. It is intended to provide basic information on firesafety for people and property, the basic characteristics and behavior of fire, fire hazards of materials and buildings, fire protection equipment and systems, codes and standards for fire protection and prevention, and fire fighting forces and how they operate. Such a textbook should be a valuable resource guide in a world whose fire hazards increase proportionally with its complexity. The authors sincerely hope that *Principles of Fire Protection* will help to fulfill this objective.

Arthur E. Cote, P.E.
Percy Bugbee

CONTENTS

CONTENTS

CONTENTS

CONTENTS

CONTENTS

CONTENTS

Chapter 1

Fire—The Destroyer

Since prehistoric times, fire has been viewed as a force both beneficent and destructive. From the days of the Roman Empire to the present, many of the greatest losses of life and property can be attributed to fire. However, modern fire protection techniques are helping to guard people and their environment from fire's destructive potential.

FIRE: HISTORICALLY A TWO-SIDED GOD

Development of the human race's use of fire probably had four stages. First, people observed about them natural sources of fire, such as volcanoes and trees set afire by lightning. Second, they acquired fire from natural sources and used it for warmth, light, and protection from wild beasts. Third, they learned to make fire themselves. And, lastly, they learned to control fire to make life more comfortable and pleasant.

The keeping and use of fire probably had an influence in ending nomadism and consequently in the development of social and political institutions connected with living in fixed abodes.

Because of its beneficial nature, fire has been worshiped by many peoples as a divine or sacred element. In most mythology is an account of the method by which fire was brought to mankind. However, in addition to its beneficial nature, fire can be tremendously destructive. Uberto C. Crosby, second president of the National Fire Protection Association (NFPA), described this two-sided nature of fire in an address given in 1900 at NFPA's fourth Annual Meeting, in New York City.[1]

> A strange thing this fire from which we seek protection. No wonder the ancients worshiped it as a God, that primitive man guarded it jealously, keeping it constantly burning. A strange, inconsistent, two-sided god it has always been to man; now giving comfort and blessing in manifold ways, and then, without warning, turning and destroying the objects of its benefaction. It warms and lights our homes, builds and runs our workshops and factories, furnishes the life and power of our modern civilization. It seems a friendly, beneficent factor, and yet at the very time it is showering blessings, with persistency and cunning, it seeks to destroy, and while we rest in financial security, breaks through the barriers with which we seek to surround it, and like a mighty avalanche sweeps away homes, blocks, towns, and cities in one common destruction.

1

At the same time a friend and foe; in all ages and climes man has worked to obtain its blessings and, at the same time, to prevent its ravages.

Fire Protection During the Roman Empire

Historically, the first recorded attempts to control the ravages of fire took place about 300 B.C. in Rome when fire fighting duties and night watch services were delegated to a band of slaves, the *Familia Publica*, supervised by committees of citizens. During the reign of Augustus Caesar (Gaius Julius Caesar Octavianus) from 27 B.C. to 14 A.D., Rome developed what might be considered the first municipal-type fire department by organizing these slaves and citizens into a *Corps of Vigiles* (watch service). Decrees were issued stating the measures that all citizens should take to prevent and check fires.

The *Corps of Vigiles* represents the first organized form of fire protection. Night patrolling (performed by *Nocturnes*) and night watch forces were its principal services. In addition, each of the *Vigiles* was assigned a particular task during a fire. For example, some members (*Aquarii*) carried water to the fire scene in jars. Still later, aqueducts were built to carry water around the city, and handpumps were developed to help get the water on the fire. Pump supervisors were titled *Siponarii*, and the earliest recorded fire chief was the *Praefectus Vigilum*, who was charged with the overall responsibility for the *Corps of Vigiles*. Roman law decreed that the *Quarstionarius*, the Roman equivalent of today's state fire marshal, determines the cause of all fires. During the time of the Roman Empire, leather hose came into use, and large pillows were carried to the fire scene so that people trapped in taller buildings could jump onto them.

Marco Polo's account of the great Oriental civilization of the 13th century noted that a civil force of "watchmen" and "firemen," who had fire prevention duties, was maintained in Hangchow (referred to by Marco Polo as "The Celestial City"), and could turn out "one to two thousand men" to deal with a fire. The force was divided into companies of ten men each, five of whom were on watch by day and five by night.

Early Fire Protection Regulations

Following the fall of the Roman Empire, there was an extended period of time when apparently little or no organized effort was made to prevent or control fires. About the only public regulation for fire protection was the "curfew" (from the French word meaning "cover fire") requiring that fires be extinguished at a fixed hour in the evening.

Such a curfew was the subject of another early fire protection regulation, adopted in Oxford, England, in the year 872.

In 1189, London's first Lord Mayor issued an ordinance requiring that new buildings have stone walls and slate or tile roofs, thus banning the previously widespread use of thatched roofs. In 1566, an ordinance requiring the safe storage of fuel for bakers' ovens was put into effect in Manchester, establishing what was probably the first enactment on a fire prevention subject not related to buildings themselves. The first state action was England's Parliamentary Act of 1583 forbidding candlemakers from melting tallow in dwellings. Still later, in 1647, wooden chimneys were banned, and after the London fire in 1666 a complete code of building regulations was adopted; however, commissioners to enforce the regulations were not appointed until 1774.

It was not until Scotland's Edinburgh Fire Brigade came into being in 1824 that public fire services began to develop more modern fire protection regulations and standards of operation. A surveyor named James Braidwood, appointed chief of the Edinburgh Fire Brigade, in 1830 wrote the first comprehensive handbook on fire department operation. His handbook included some 396 standards and described the kind of service a good fire department should perform.

On a worldwide basis, the early development of public fire protection regulations and types of public fire department organizations closely paralleled developments in Great Britain and the American colonies.

Major Fires in Early Times

Because of the lack of adequate fire protection regulations, organizations, and equipment, early cities were fire prone. In 1752, Moscow suffered a major conflagration that destroyed 18,000 homes. Moscow was again demolished by fire in the War of 1812.

Constantinople (now Istanbul) was the greatest sufferer from conflagrations of any city on record, having experienced major fire disasters in 1729, 1745, 1750, 1756, 1782, 1791, 1798, 1816, and 1870. In more recent times, Constantinople suffered further major fires in 1908, 1911, 1915, and 1918.

Many of the large cities in India, China, and Japan were wiped out by great fires, and while almost every young student learns that "Nero fiddled while Rome burned" in 64 A.D., few remember that Rome burned again in 1764. Venice was destroyed by fire in 1106 and again in 1577. London was ravaged by fire in 798, 982, 1212, and again in 1666 when in the Great London Fire some 436 acres were burned, and more than 13,000 buildings and the entire heart of London were reduced to ashes.

Fire Protection in Colonial America

Night fire watches were instituted in the larger cities of America in colonial times. In Boston in 1654, a "bellman" was put to work from 10 p.m. to 5 a.m., and in 1657 in New York fire wardens were supplemented by a company of eight night fire watch volunteers. These volunteers were called the "rattle watch" because of the large rattles they used to sound alarms. A night fire watch of four town criers was established in New York in 1687.

The night fire watch service was a community institution before there were municipal police forces. The early night fire watch patrols were influenced by the necessity to control losses in insured properties; such patrols helped get the new institution of fire insurance accepted by the public.

As a result of a disastrous 1631 fire in Boston, the first fire ordinance in America was adopted. It prohibited thatched roofs and wooden chimneys, and was enforced by Boston's governing board of selectmen.

In 1647, New Amsterdam (now New York City) appointed surveyors of buildings to control the worst fire hazards, and the following year named five municipal fire wardens who had general fire prevention responsibilities. This is generally viewed as the origin of the first fire department in North America.

Following a major fire on January 14, 1653, Boston selectmen were given authority to buy a fire engine. Instead, they contracted for the service of an engine to be brought to fires. There is no record as to the nature of this engine or service performed. At the same time, additional fire protection regulations were adopted. A 1653 ordinance required all householders to keep a 12-foot swab for extinguishing roof fires and to maintain a ladder capable of reaching the ridge of the roof. At the same time, the town provided ladders, hooks, and chains to pull down houses in the path of fires; gunpowder also was used for this purpose occasionally. It was decreed that the owners of razed houses had no redress.

Establishment of the First Paid Fire Department: Another large fire in Boston, in 1679, led to the organization of the first paid fire department in North America, if not in the world. Boston selectmen imported a fire engine from England and employed a fire chief, Thomas Atkins, and 12 fire fighters to operate it. Massachusetts adopted the practice of using paid municipal fire fighters on a "call" basis, instead of unpaid volunteer fire companies such as those organized later in the southern colonies. (See Figure 1.1.)

In colonial American communities, each householder was required to keep two fire buckets on hand. When church bells rang to report a fire, people formed lines to pass water from wells or springs to the scene of the fire. Although, when fire engines were introduced, and companies were organized to operate the engines, citizens still were required to respond with buckets of water to fill the engines. In Boston in 1711, fire wardens were appointed to respond to fires with their staffs and to supervise the citizen bucket brigades. As late as 1810, Boston citizens were subject to a ten-dollar fine for failure to respond to alarms with their buckets. (The laws in a number of states still impose penalties upon citizens who refuse to obey the orders of fire officers to assist in fighting fires.) By 1715, Boston had six fire companies with engines of English manufacture. This was before either New York or Philadelphia had even a single fire engine in service.

Fig. 1.1   As a result of a fire in 1653, Boston selectmen contracted for the use of a fire engine and adopted fire protection regulations requiring householders to keep a swab and a ladder tall enough to reach the ridge of the roof. (From Library of Congress.)

Establishment of Mutual Fire Societies: Although fire fighting was now handled by fire companies, the first of a number of mutual fire societies was formed in Boston in 1718. Society members were the more affluent citizens who organized to assist each other in the salvage of goods exposed to fires in their homes or places of business. Their equipment consisted of a screw driver, a bed key, and a bag in which to collect valuables. The bed key was a very important tool, for with it beds—considered valuable possessions —could be taken apart and brought outside a burning building. Fire societies became inactive early in the 19th century when fire insurance was made available to the more prosperous citizens.

Formation of Fire Insurance Companies

In England, the Great London Fire in 1666 stimulated at least some constructive action: Fire insurance companies were started primarily as a result of this disaster. The fire insurance offices banned wooden chimneys, thatched roofs, and wooden roofs. In order to further protect their insured properties, the early fire insurance offices hired fire fighters and, in 1667, formed the first real fire brigades in England. These fire brigades were equipped with leather buckets, hooks, ladders, large syringes (the first fire extinguishers), and leather hose. Wooden water mains carried water around the city.

In *A History of the British Fire Service*,[2] G.V. Blackstone expresses the opinion that the history of English fire brigades should date from the formation of the insurance brigades in 1667. These brigades were without statutory authority or obligations, and the insurance company offices—not the government authorities—decided where the brigades would be located. The companies were based in the larger cities and in areas where insured values were concentrated. The brigades of the various insurance offices were maintained in part for advertising value, and were, therefore, competitive. The bad features of this competition eventually led to consolidation of the fire companies in London in 1833 into the London Fire Engine Establishment. The London Establishment was taken over by the Metropolitan Fire Brigade in 1865.

In 1736 Benjamin Franklin recommended formation of a volunteer fire fighting force called the Union Fire Company, and served on it as America's first volunteer fire chief. Franklin also organized the first fire insurance company in the United States, the Philadelphia Contributionship. However, the actual job of fire fighting was performed either by fire companies operating under the authority of the municipality or by independent volunteer companies that owned their own stations and apparatus. American insurance companies frequently contributed to the support of the volunteer fire companies.

In 1835, the Manufacturers' Mutual Fire Insurance Company was established in Providence, Rhode Island. This company would insure only those mills utilizing good fire prevention and protection standards of the time.

Progress of Fire Protection in 19th Century America

In America in the 1800s, fire protection and fire prevention regulations still required major disasters before they were enacted and enforced, as can be evidenced by the Great Chicago Fire. On October 9, 1871, a sweeping conflagration destroyed most of Chicago. (Traditionally, "Mrs. O'Leary's cow" has been blamed for the fire's start.) Following the Great Chicago Fire, the Chicago City Council decreed that the city be rebuilt—of brick and stone. Fire Prevention Week, established in 1922 to mark the anniversary of the Chicago disaster, is intended to serve as a reminder of the destructiveness of fire and the importance of its prevention.

In 1906, the San Francisco earthquake and resulting conflagration resulted in 674 fatalities and destroyed more than 28,000 buildings. It is considered the last of the truly huge urban conflagrations in the United States.

CONFLAGRATIONS: THE NEED FOR DEVELOPMENT OF REGULATIONS AND EQUIPMENT

Prevention of Conflagrations

Virtually all of the fire regulations established until the early 1900s dealt with conflagration potential, and methods for mitigating that potential. Typical fire prevention and protection regulations included requirements for stone or brick walls and noncombustible roof coverings for all buildings in the principal business districts of cities.

In addition, basic municipal fire fighting apparatus and equipment were developed and put into service in the latter part of the 19th century to help prevent the spread of fire from one building to another. The advent of suction engines created a need for the construction of cisterns, which were located in much the same manner as present-day hydrants.

Fire Hydrants: From 1830 through the 1840s, the first fire hydrants were installed on public mains. Before that time, some cities and towns depended on networks of wooden piping from which water could be obtained for fire fighting purposes. Because the primitive hydrants were unreliable and were supplied by three- or four-inch mains, cisterns long remained a principal source of water for fire engines. Boston Fire Chief John S. Damrell, in his testimony concerning Boston's 1872 conflagration, stated that the hydrants on four-inch mains were incapable of adequately supplying steamers, and that the cisterns were preferable and had been used to supply out-of-town engines. Damrell encouraged the construction of large mains which, following major fires in 1889 and 1893, were finally installed. Also installed in Boston were large pipes that ran from the waterfront area to the conflagration district, enabling fireboats to supply pumpers. In the 20th century, these mains were incorporated into the high-pressure fire main system.

Use of Fire Hose: In the American colonies, the use of fire hose was a comparatively late development in fire fighting. Boston imported a few short lengths of leather hose from England in 1799. This made it possible for the nozzle to be advanced close to the fire, whereas for more than a century hose nozzles had been mounted directly atop the pumps. Within a few years, fire hose and hose reels became a necessary part of fire department equipment.

In Boston in 1871, 1½-inch fire hose was placed into service, and woven-jacketed rubber-lined hose was introduced to replace leather hose. Expansion ring couplings with clear waterways replaced old-style couplings with restricted waterways. Increased interest in standardization of hose threads followed Chicago's fire of 1871 and Boston's fire of 1872. However, no significant progress was made until the fledgling NFPA was given the job of thread standardization, following the Baltimore fire of 1904. The dimensions of National Standard Thread (NST) for 2½-inch hose threads were selected because 70 percent of the existing threads could be recut to that specification. Although not universally accepted, NST is utilized in the majority of cities and towns throughout the U.S. today.

Advances in Fire Apparatus: By the 1870s, self-propelled steam fire engines were in service in New York City, Boston, and Detroit, and in 1873 steam-propelled fireboats were introduced. That same year, "Babcock" chemical engines were introduced to provide fast fire attack with ¾-inch hose, and the Babcock hose thread became the standard for booster hose. Also put into service in 1873 were the first aerial ladders. By 1882, water towers were in service for providing powerful streams of water to the upper floors of buildings. In 1905 spring-assisted aerial ladders came into use. And in the mid-1930s, power-operated 100-foot metal aerial ladders were introduced.

By 1910, introduction of automotive fire apparatus was well under way. This gradually eliminated the separate chemical and hose wagons in many fire departments, because each pumper could carry its own ancillary equipment. The first NFPA standard on automotive fire apparatus was adopted in 1914.

Major Hydraulic Studies: Much of the fundamental data used in hydraulic work in fire protection was developed in a series of extensive investigations by John R. Freeman in 1888 and 1889, and by Boston city engineer William Jackson in 1893. After measuring flows from nozzles and friction losses in fire hose, Freeman suggested the standard 250-gpm fire stream. His recommendations led to adoption of 3-inch fire hose by most fire departments. Jackson conducted detailed tests of pumper performance, water tower operation, and practical fireground layouts with both 2½-inch and 3-inch hose with various sizes of nozzles, resulting in improved fire fighting procedures.

Fire Alarm and Extinguishing Systems: The first municipal fire alarm system using a telegraph was installed in Boston in 1851. Such fire alarm systems were in widespread use by the time the telephone was introduced in 1877.

Conflagrations in the 20th Century

Since the San Francisco earthquake and fire of 1906, major conflagrations in the U.S. have been limited to those locations where, in addition to strong winds and a period of hot, dry weather, one of the following also exists:
1. Wood-shingle roofs (*not* fire-retardant treated);
2. Very closely spaced combustible buildings (typically, in very old sections of old cities); and

Fig. 1.2   Wood-shingle roofs have helped spread many fires, particularly in dry areas of the southern and western United States. (Courtesy of Los Angeles Times News Bureau.)

3.   Extensive urban/wildland interface where forest fires have led to conflagrations.

These conditions resulted in conflagrations which destroyed 3,500 buildings in Chelsea, Massachusetts in 1908; 1,600 buildings in Salem, Massachusetts in 1914; 1,440 buildings in Paris, Texas in 1916; 1,938 buildings in Atlanta in 1917; and more than 1,200 buildings in Southwestern Maine in 1947, to name just a few incidents.

FIRE IN 20th CENTURY AMERICA

Active Approaches to Fire Protection

The beginnings of modern industrial fire protection and prevention were established in the textile mills of New England in the 1830s. The automatic sprinkler system, one of the most important inventions for the control of fire, was conceived and put into use in the latter part of the 19th century.

According to Gorham Dana in *Automatic Sprinkler Protection*,[3] the first recognized patent for a sprinkler system was issued in 1723 to Ambrose Godfrey, a chemist. Godfrey's system consisted of a cask of fire extinguishing fluid, usually water, containing a pewter chamber of gunpowder. The chamber of gunpowder was connected to a system of fuses that were ignited by the flame of the fire. The ignition exploded the gunpowder and scattered the extinguishing liquid.

By the middle of the 19th century, additional fire-protection requirements were reflected in the automatic sprinkler systems developed in England. The first automatic sprinkler system patented in the United States was developed by Philip W. Pratt in 1872 in Abington, Massachusetts. From 1852 to 1885, perforated pipe systems were used extensively in textile mills throughout New England, and from 1874 to 1878 Henry S. Parmelee of New Haven, Connecticut, made continued design improvements on his invention: the first practical automatic sprinkler head. (See Figure 1.3.) Both the design and the installation of the Parmelee automatic sprinkler system involved some basic principles still being applied today.

Three important organizations created during the 19th century would prove to exert a profound effect on the standardization of fire protection and

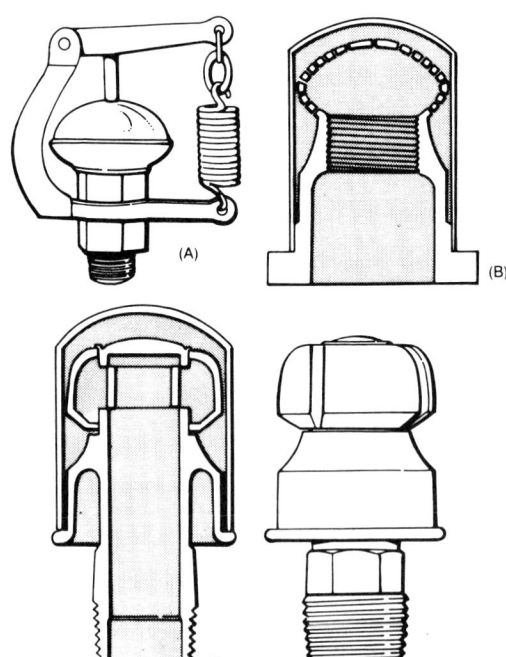

Fig. 1.3 (A) The 1874 Parmelee automatic sprinkler head; (B) the 1875 Parmelee automatic sprinkler head; (C) and (D) the 1878 Parmelee heads. (From Automatic Sprinkler Protection, by Gorham Dana.[3])

prevention systems and methods throughout the United States. These organizations are the Factory Mutual System (founded in 1835), Underwriters Laboratories Inc. (1894), and the National Fire Protection Association (1896).

Major Loss-of-Life Building Fires

By the end of the 19th century, active fire protection—in the form of automatic sprinkler systems and automatic detection and alarm systems—was becoming more common in industrial plants as the emphasis began to shift from prevention of city-wide conflagrations to control of fires within individual buildings. Until this shift was widespread, numerous fires in individual buildings resulted in tremendous losses of life as well as property.

Figure 1.4 presents the 20th century record in the United States for building fires claiming at least 25 lives. Although some fires in industrial and storage properties and a number of fires involving vehicles are included, the record is dominated by theatres, night clubs, schools, institutions, hotels, and similar properties.

Except for the Great Depression and World War II periods, which presented unique problems for code enforcement and compliance, the record shows considerable progress during the 20th century against major loss-of-life building fires. This progress is particularly impressive considering the fact that during the eight decades covered by Figure 1.4, the U.S. population nearly tripled, and great numbers of buildings addressed by the fire codes were constructed. These buildings usually include facilities that concentrate very large numbers of people.

Significant Loss-of-Life Fires

Historically, fires resulting in a major loss of life have brought about important changes in building and fire codes and in standard fire protection or prevention practices. In the early 1900s, four building fires—the Rhoades Opera House in Boyertown, Pennsylvania (1903), the Iroquois Theatre in Chicago (1903), the Lakeview Grammar School in Collinwood, Ohio (1908), and the Triangle Shirtwaist Factory in New York City (1911)—were largely responsible for the appointment in 1913 of the NFPA Committee on Safety to Life. The opening summary of the "Origin and Development of 101" in the current *Life Safety Code®* (NFPA *101*-1988)[4] states:

> For the first few years of its existence, the Committee devoted its attention to a study of the notable fires involving loss of life and in analyzing the causes of this loss of life. This work led to the preparation of standards for the construction of stairways, fire escapes, etc., for fire drills in various occupancies, and for the construction and arrangement of exit facilities for factories, schools, etc., which form the basis of the present *Code*.

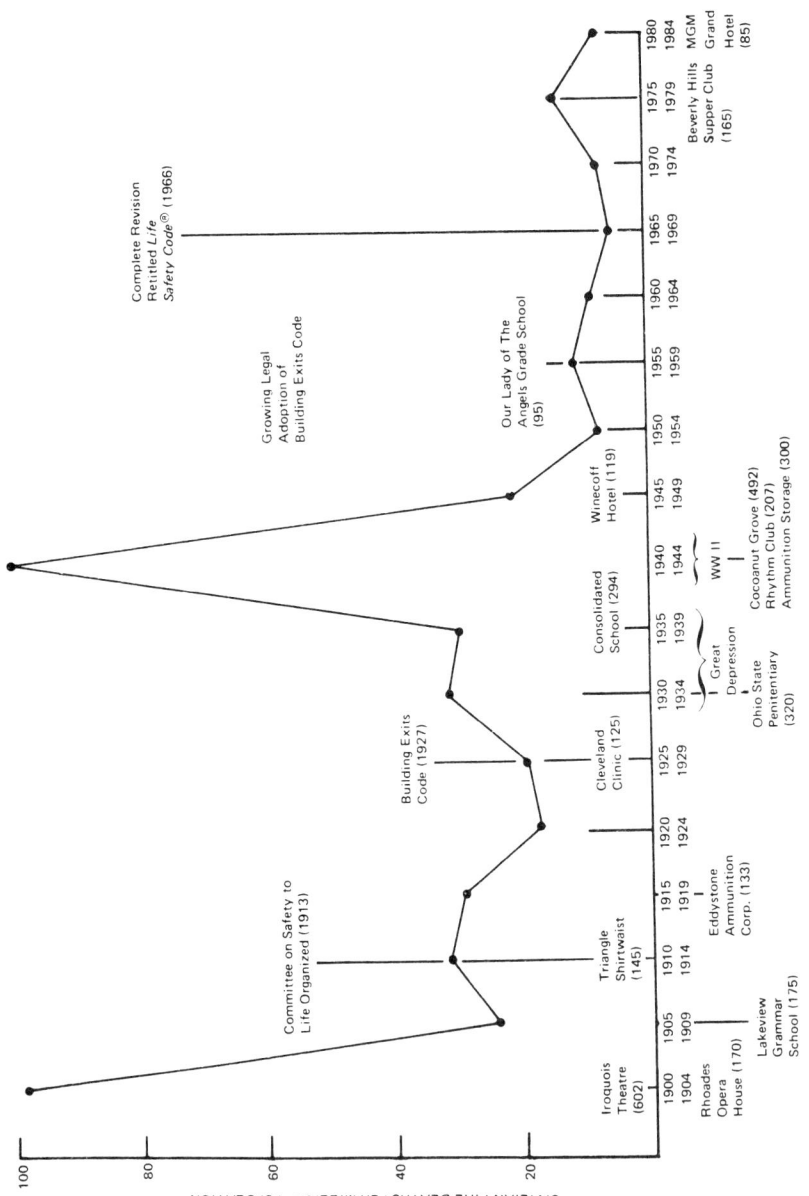

Fig. 1.4 Major loss of life in 20th century U.S. building fires claiming at least 25 lives.

The 1937 fire at the Consolidated School in New London, Texas, pointed out the need for state laws to protect public buildings not subject to municipal ordinance and inspection. Then, in the 1940s, a series of multiple-death fires—including those at The Rhythm Club, The Cocoanut Grove, and the La Salle, Canfield, and Winecoff Hotels—focused national attention upon the need for adequate exits and other firesafety features in hotels and public buildings. These fires resulted in major changes to the "Building Exits Code" (as the *Life Safety Code* was then known) over a period of almost two decades. NFPA 102, *Standard for Assembly Seating, Tents and Membrane Structures,*[5] was the result of still another multiple-death fire of the 1940s—the 1944 Hartford, Connecticut circus tent fire in which 168 people were killed.

Three hospital fires—St. Anthony's in Effingham, Illinois in 1949 (74 killed); Mercy Hospital in Davenport, Iowa in 1950 (41 killed); and Hartford Hospital in Connecticut in 1961 (16 killed)—prompted hospital administrators and fire prevention officials across the nation to assess the quality of construction and fire protection systems in health care facilities.

The Our Lady of Angels School fire in Chicago on December 1, 1958 probably resulted in the swiftest action in the wake of any major fire since World War II. (See Figure 1.5.) Within days of the fire, state and local officials throughout the nation ordered fire inspections of schools, and within one year it was reported that major improvements in life safety had been made in 16,500 schools across the country. Improvements in the frequency and quality of exit drills and inspections, in the storage of combustible supplies, and in the disposal of waste materials also were reported in almost every community where schools were surveyed.

The effects of more recent fires such as the Beverly Hills Supper Club (165 killed), the MGM Grand Hotel (85 killed), the Dupont Plaza Hotel (97 killed), and others are still being felt in the fire protection community.

FIRE PROTECTION IN THE SECOND HALF OF THE 20th CENTURY

Fire protection since the late 1950s has been marked by an increase in the active approach to fire protection coupled with increased emphasis on fire prevention.

Industrial fire protection has been based more and more on the installation of fire extinguishing equipment and systems such as portable fire extinguishers and automatic sprinklers; carbon dioxide, dry chemical, and halon extinguishing systems; and sophisticated fire detection and alarm systems.

This active approach to fire protection has been extended into the home environment in the 1970s and 1980s with the advent of residential smoke detectors, which now are installed in nearly three-forths of all U.S. households.

Fig. 1.5   A view of the fire scene at Our Lady of the Angels grade school in Chicago. The December 1, 1958 fire took 95 lives. (From *Men Against Fire*, by Percy Bugbee.[1])

Figure 1.6 shows the remarkable growth in detector installations from 1970 through 1984. A 1985 Louis Harris poll provides the latest benchmark: 74 percent of U.S. households now have detectors. Clearly, the detector explosion is unchallenged as the largest fire protection improvement of our time.

In spite of the decrease in home fire deaths in the U.S. in the last decade, these fires still accounted for 92 percent of all structure fire deaths, or an annual average of 5,193 fatalities from 1977 through 1984. This means that home firesafety remains the key to any major improvements in the overall fire death picture in the U.S.

Home firesafety education is vital. Topics such as the major causes of fatal home fires, the use and maintenance of smoke detectors, and escape planning must continue to be brought to the attention of dwelling and apartment residents of all ages.

Recent research has shown that "quick-response" residential sprinkler systems using domestic water supplies are effective and are becoming more

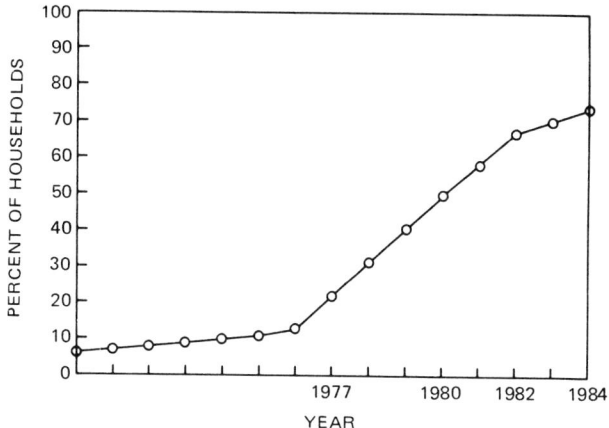

Fig. 1.6   Detector installation grew significantly from 1970 through 1984.

affordable. Some communities have begun to require residential sprinkler systems in new homes and, in some cases, in existing homes as well.

Economic Impact of Fire Protection

Fire suppression by public fire departments is a vital service. It is, however, a "last resort action." Prevention, detection, automatic extinguishment, and restraints against spread of fire are, in that order, the logical steps that should precede public fire service suppression.

The preventive and remedial actions taken by public authorities to mitigate the losses of life and property from fire are a major interactive component of the fire problem. Although the majority of such actions are necessitated by failure to effectively control the root cause of fire, several courses of action, available to public authorities, could eliminate much of the problem. The principal courses of action are:
1.   Fire prevention education and awareness;
2.   Firesafety code adoption and enforcement; and
3.   Fire suppression.

The cost of operating public fire protection services in the U.S. runs to several billion dollars per year, over 90 percent of which is expended on suppression activities. In addition, fire fighting is one of the most dangerous of all occupations. In the years ahead, the challenge of fire fighting safety will become even greater as fire departments confront new technologies and hazards, while simultaneously feeling the impact of reduced funds for personnel and/or the purchase and maintenance of equipment.

More emphasis by fire services on preventive measures, together with increased effectiveness of suppression techniques, can help eliminate many root causes of fire.

The heavy burden that reliance on public fire protection services places on the taxpayer points up the desirability of shifting the burden into known, more cost-effective lines of action. The balance between private protection and publicly provided service should constantly be reevaluated, and the emphasis changed to where beneficial results can be achieved.

A true balance must also be achieved in the firesafety code adoption and enforcement system, whereby communities are permitted to grow and expand, but with proper protection for the firesafety hazards encountered. It is not the purpose of the firesafety code adoption and enforcement system to provide, without exception, a risk-free, nonimposing, hazardless environment. Every event in each minute of each day involves risk in some form and to some degree. A firesafety code adoption and enforcement system must, therefore, protect adequately without undue imposition.

There is increasing demand for evidence of cost effectiveness and cost benefit of all firesafety codes and standards. This is brought about by the heavy expenditures that property owners sometimes must make in order to comply with the requirements of these codes and standards. More precise formulation of fire protection requirements, recognizing their economic implications, has been one of the major fire protection challenges of the 1980s.

Incendiarism, Arson, and Suspicious Fires

Incendiary fires are set fires. Motives for deliberately starting fires include fraud, vandalism, spite, politics, and crime coverup, as well as the compulsions of "pyros" and "psychos."

Suspicious fires have many of the characteristics of incendiary fires but are not conclusively judged to be incendiary. However, for most purposes suspicious fires are included in estimates of the size of the arson problem. Arson is the crime of setting an incendiary fire, and most incendiary fires are arson offenses. The rare exceptions are fires set by people incapable of criminal intent, such as those who are mentally ill persons and unable to understand the nature of their actions.

Incendiary and suspicious fires are of particular concern because they account for a major share of the worst multiple-death fires, and because they constitute the largest single cause of fire loss overall.

Public Apathy

Public apathy plays a large role in keeping U.S. fire statistics as high as they are. Many adults have the notion that "fire only happens to the other fellow." Until fire strikes their home and their family, they largely ignore fire prevention information, fail to install and maintain smoke detectors, do not practice a fire escape plan for use in emergencies, and generally omit the other steps necessary to keep their households safe from fire. For many, once

Fire Prevention Week comes and goes each October, not much thought is given to firesafety until the next campaign a year later.

Fire departments, schools, civic organizations, and many citizens work throughout the year to keep firesafety part of the normal routine of people of all ages, but efforts must continue if public apathy is to be overcome.

Summary

The total size and scope of fire waste from the earliest times to the present is phenomenal. Fire always has been and still is a destructive threat to all people and property. However, it should not be concluded that the situation is hopeless and that no progress has been made to prevent and combat fire. Although much more needs to be done to reduce the human suffering and heavy economic cost of fire, there has been a tremendous increase in knowledge about fire, fire fighting, fire science technology, and fire protection and prevention.

References

[1]Bugbee, Percy 1971. *Men Against Fire*, National Fire Protection Association, Boston.

[2]Blackstone, G.V. 1957. *A History of the British Fire Service*, Routledge and Kegan, London.

[3]Dana, Gorham 1919. *Automatic Sprinkler Protection*, Wiley, New York.

[4]NFPA *101*-1988. *Life Safety Code*, National Fire Protection Association, Quincy, MA.

[5]NFPA 102-1986. *Standard for Assembly Seating, Tents and Membrane Structures*, Quincy, MA.

Additional Reading

Braidwood, James 1830. *On the Construction of Fire Engines and Apparatus, the Training of Firemen and the Method of Proceeding in Cases of Fire*, Bell and Bradfute and Oliver and Boyd, Edinburgh.

Bryan, John L. 1976. *Automatic Sprinkler and Standpipe Systems*, National Fire Protection Association, Boston.

Fire Protection Handbook, 16th ed. 1986. National Fire Protection Association, Quincy, MA.

Freeman, J.R. 1883. *Transactions of the American Society of Civil Engineers*, Vol. XII.

Freeman, J.R. 1891. *Transactions of the American Society of Civil Engineers*, Vol. XXIV.

Hall, John R., Jr. 1985. "A Decade of Detectors: Measuring the Effect," *Fire Journal*, National Fire Protection Association, Quincy, MA.

2

Firesafety for People and Property

Life and property safety in a fire is dependent on many factors. Human characteristics, building construction, type of occupancy, and level of public awareness must all be considered to evaluate those fire protection plans and techniques that will yield the most effective response in a fire.

ASSESSING LIFE SAFETY

Fire is one of the most dangerous threats facing the occupants and owners of buildings. Fire is the second leading cause of deaths in buildings, after falls, according to analyses by the National Safety Council. In addition, fire accounts for nearly half the claims against fire-and-multiple-peril insurance policies and is one of the leading causes—if not the leading cause—of property loss to buildings and their contents.

In recent years, deaths due to fires in buildings have averaged more than 5,000 a year. While this represents a tremendous decline from the fire death rates of past decades, it still means an average of more than a dozen people killed every day of the year. The death toll is still so high that, in any group of 40 to 50 persons, there is a good chance that someone has had a friend or ·relative who died in a fire.

Direct property damage in fires in buildings totals billions of lost dollars each year, not counting the occasional large indirect losses when businesses are forced to shut down for prolonged periods. As a very rough rule of thumb, for every dollar lost each year to fire damage, a second dollar is spent on fire departments, a third dollar is spent on built-in fire protection for new construction (either for fire protection systems or for construction requirements), and another 50 cents to one dollar is spent on fire policies of the insurance industry (excluding claims paid for losses). This means that tens of billions of dollars are spent or lost every year to deal with fire.

Too many people believe fire happens only to others, not themselves. In fact, however, the average household will experience three fires a decade—whether reported to the fire department or not—and two fires per lifetime that will be serious enough to report to the fire department. Applications of fire protection principles are not simply engineering for what could happen; they are vital decisions that have important consequences for American lives every day.

Sources of Risk to Life, Limb, and Property

The contents and furnishings of modern buildings pose fire risks in several different ways. Items in buildings vary in the inherent riskiness of their mode of use. For example, portable heating devices carry an inherent risk when placed near an ignitable material. Items also vary in terms of ignitability. Once ignited, items vary in terms of their potential intensity—the peak rate at which they can release heat—as well as the speed at which they can reach that intensity and the length of time they might contribute to a fire. Also, the hazardous by-products of combustion—e.g., carbon monoxide, hydrogen cyanide, and carbon dioxide—vary from one item to another.

Depending on its design, a building might either contain or accelerate a growing fire. Interior finishes might slow or speed the spread of flame. Walls, doors, duct systems, and other barriers and routes in a building can either compartmentalize or channel a fire. Pathways for occupants might or might not be protected, lighted, of adequate capacity, sufficiently well marked, blocked, or locked. Fire detection and suppression systems might or might not be provided. And the list does not end here.

Every decision about a building has firesafety consequences. These consequences might be addressed explicitly, in which case the principles of fire protection contained in this text will show the way to proceed; or the consequences might be built in or brought in unrecognized, in which case the building stands a good chance of entering the fire records as a set of object lessons taught by yet another fire.

The essence of fire protection principles, then, is to see every building design decision as a choice among options which differ in the hazard or risk of fire they impose. Often, codes and standards indicate what range of choices are believed to eliminate the most severe hazards and risks. In recent years, increasingly sophisticated models permit the decision-maker to calculate the fire consequences of a decision and compare these to some baseline. Product design, code-equivalency building design, automatic system design, and many other kinds of choices can benefit from this rapidly growing array of tools and techniques.

Determining sources of hazards and risks can be difficult. The rapid growth in heating fires in the early 1980s, for example, resulted from the increased use of portable and area heating equipment. This use in turn represented a reaction to two other risks—freezing or poverty—posed by sudden and drastic increases in fuel prices. Sometimes there are direct tradeoffs in firesafety. In selecting materials for upholstered furniture, for example, some materials that reduce ignitability also increase intensity if ignition occurs. Or consider the home kitchen range: if controls are placed in front, there is a greater risk of fires started by children playing; if controls are placed in back, there is a greater risk of clothing (sleeve) ignitions during ordinary use.

It is important for the fire protection professional to understand the complex interrelationships that affect firesafety. It also is useful to know some of the key high-risk groups.

Proportionally, young children and the elderly die in numbers well above their share of the population. Compared with young adults, children under age 6 have three times the risk of dying in fire, while elderly persons over age 74 have four times the risk. Roughly one-third of all fire deaths occur in these age groups.

Rural areas suffer fire deaths at a higher rate than even the largest cities. Portable and area heating equipment are primarily responsible for the difference between rural and nonrural fire death rates. Rural areas lead in fire death rates due to every cause except arson and suspected arson, where the larger cities are highest.

In relation to the person-hours spent there, the most dangerous buildings for fire probably are homes; they account for more than 90 percent of all fire deaths in buildings. Car and truck fires kill twice as many people as all fires in buildings other than homes. Nevertheless, the potential for large-life-loss fires is greatest in properties holding very large numbers of people. Also, public buildings must meet a higher standard of firesafety than homes. People want the freedom to define their home environment, even if some risks are involved.

The public's fear of fire is partly based on the reactions of strangers in strange environments. However, the risk of death is far greater from a family member's discarded cigarette than from a cigarette discarded by another hotel guest, although fear of the latter scenario is far greater. Fire protection principles must address the objective, continuing hazards and risks as well as the public's greater concern over low-frequency, high-severity fires, such as in hotels.

The same is true, although to a lesser degree, of property loss. Most direct loss in fire does not occur in large-loss fires, and it generally is estimated now that indirect losses (e.g., lost productivity) are far smaller than direct losses, on a national average. One reason is that business lost to fire usually is not really lost; it merely switches to another supplier. National productivity therefore may be affected less than would appear based on an examination of the directly affected parties alone.

Unquestionably, the largest single source of fire risk is lack of knowledge. The overwhelming majority of fires begin with some type of human error—for example, careless disposal of hot items, such as cigarettes; overtaxing of portable heating devices and other appliances; or failure to properly monitor hazardous processes, including cooking. Even fires that begin with equipment often have human origins, such as flagrant disregard for code requirements in the modification of an electrical system. Once begun, many fires become more severe because of additional human errors—for example, the disabling of detection devices, the locking of fire exits, the failure to rehearse exit plans, or the determination to save property despite

clear risk to life. Accordingly, education and motivation in all their forms—
public firesafety education, fire-code-based inspections, activities of company
firesafety officers and their programs, etc.—must be included in any comprehensive treatment of the principles of fire protection.

HUMAN BEHAVIOR

How a person reacts during a fire emergency is related to: (1) the role
assumed, previous experience, education, and personality; (2) the perceived
threat of the fire situation; (3) the physical characteristics and means of egress
available within the structure; and (4) the actions of others who are sharing
the experience. Post-event analysis of behavior has described actions as
adaptive or nonadaptive, participative or inhibited, and altruistic or individu-
alistic. Detailed interview and questionnaire studies over the past 30 years
have established that instances of nonadaptive (panic) type of behavior are
rare events that occur under specific conditions. Most behavior in fire
incidents is determined by information analysis, resulting in cooperative and
altruistic actions.

Human Perception in Fires

Group and individual behavior characteristics during a fire have been
determined primarily by research studies in which individuals were interviewed by fire department personnel following the fire incident. It must be
recognized that an individual's behavior in a fire incident is affected by the
design variables of the building in which the fire incident occurs and the
appearance of the fire incident at the time of detection, e.g., if they perceive
an odor of smoke, rather than visible flames, or dark acrid smoke completely
obscuring a corridor. Variables within the fire protection provided for the
building also may be critical to the individual's perception of the threat
involved in the fire incident. Obviously, in life-threatening situations the most
important individual decisions and behavior occur in the early stages of the
fire incident, prior to the arrival of the fire department. Studies of fires in
health care facilities have indicated the importance of this early behavior.

> In the process of investigating these case studies we have come to
> believe that the period between detection of the fire and the arrival
> of the fire department is the most crucial life-saving period in terms
> of the first compartment (the area in direct contact with the room of
> origin and the fire).[3]

Thus, the behavior of the individuals intimately involved with the
initiation of the fire incident is critical not only for themselves but often for
the other occupants of the building. It should be recognized that the altruistic
behavior observed in most fire incidents (with the interaction of the occu-

pants and the fire environment in a deliberate, purposeful manner) appears to be the general mode of reaction. The nonadaptive flight or panic-type behavioral reaction apparently is an unusual behavior in fire incidents.

Awareness of the Fire Incident

The way in which an individual is alerted to the presence of a fire may determine the degree of threat perceived. With intercom-type alerting systems in buildings, variations in voice quality, pitch, or volume, as well as the content of the message, tend to provide "threat" cues.[4] In most fire incidents occupants are alerted initially to the fire by the odor of smoke. However, when the two categories "notified by family" and "notified by others" are compared, personal notification becomes the most common means of initial perception of fire, as indicated in Table 2.1. The category of noise includes sounds from people moving downstairs and through corridors, plus miscellaneous noise sources, including the breaking of glass and the arrival of fire department apparatus.

A study of the NFPA-recommended smoke detector noise level of 75 dBA indicates that individuals with hearing impairments, taking sleeping pills, or on medication may require a detector noise level exceeding 100 dBA.[5] (See NFPA 74, *Standard for the Installation, Maintenance, and Use of Household Fire Warning Equipment.*[25]) Flashing or activated lights are effective fire signals in occupancies populated primarily by hearing-impaired people.[6] The 1981 edition of NFPA *101*, the *Life Safety Code*,[26] first permitted the flashing of exit signs along with activation of an audible fire alarm system.

Decision Processes of the Individual

Seven processes that an individual may utilize in attempting to structure and evaluate situational threat cues have been identified.[7] Six of these

**Table 2.1   Means of Awareness
of the Fire Incident**

Means of awareness	Participants	Percent
Smelled smoke	148	26.0
Notified by others	121	21.3
Noise	106	18.6
Notified by family	076	13.4
Saw smoke	052	09.1
Saw fire	046	08.1
Explosion	006	01.1
Felt heat	004	00.7
Saw/heard fire department	004	00.7
Electricity went off	004	00.7
Pet	002	00.3
N = 11	569	100.0

processes are presented in Figure 2.1 as recognition, validation, definition, evaluation, commitment, and reassessment.

A detailed description of each process is included in the NFPA *Fire Protection Handbook*, 16th edition.[8]

Behavior Actions of Occupants

One study (which involved 952 fire incidents and 2,193 individuals interviewed by fire department personnel at the fire scene in Great Britain) found that the most frequent responses to fire involved evacuation of the building, fighting or containing the fire, and alerting of other individuals or the fire brigade.[2] An identical type of broad categorization of behavior was found in a similar study of 335 fire incidents in the U.S., involving a total of 584 people. Interviews were conducted by fire department personnel who used a structured questionnaire at the scene of the fire incidents.[1]

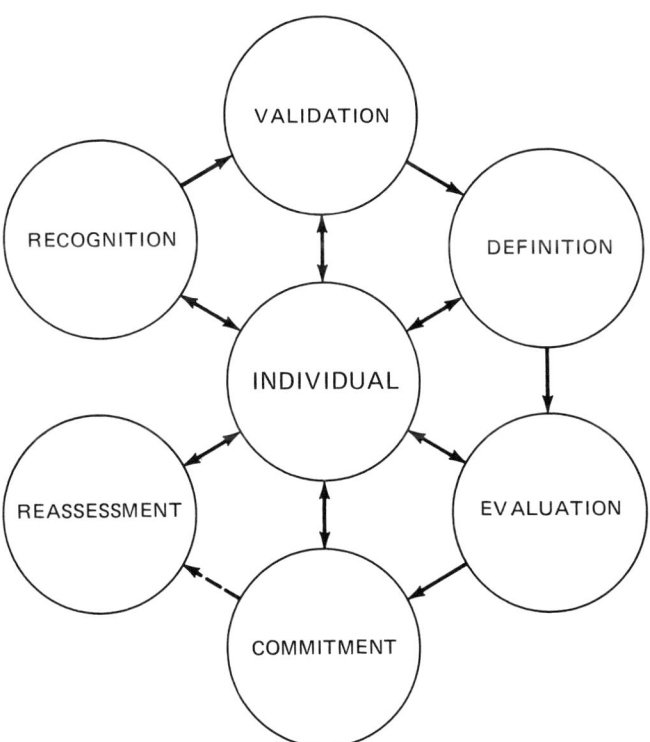

Fig. 2.1   Decision processes of the individual in a fire incident.

Panic Behavior

One concept almost always discussed following fire incidents[8] in which multiple fatalities occur is panic behavior. Panic can be defined as:

> A sudden and excessive feeling of alarm or fear, usually affecting a body of persons, originating in some real or supposed danger, vaguely apprehended, and leading to extravagant and injudicious efforts to secure safety.[9]

According to this definition, panic is a flight or fleeing type of behavior not likely to be limited to a single individual but rather to be transmitted and adopted by a group of people. From simulation experiments, a panic type of behavior reaction has been defined as:[10]

> A fear-induced flight behavior which is nonrational, nonadaptive, and nonsocial, which serves to reduce the escape possibilities of the group as a whole.

Panic is often used to explain the occurrence of multiple fatalities in fires, even when there is no physical, social, or psychological evidence showing that competitive, injudicious flight behavior actually took place. Media representatives and public officials often label various types of fire-incident behavior as panic. Yet, in the Beverly Hills Supper Club fire,[11] the evidence accumulated from interviews with participants and questionnaires completed by occupants provided no evidence of the classical group type of panic behavior with competitive flight for the exits.

The term "panic" must be separated from the terms "anxiety" and "fear." Self-destructive or animalistic panic responses to stimuli, such as in the presence of smoke, has not been supported by the research on human behavior in fire incidents. As reported, [1,2,4,12–14] it is rare to have panic behavior in which the flight is characterized by competition among the participants, with resultant personal injuries.

In an interview study of 100 participants in single-family dwelling fires, no instances of panic behavior were found. Primarily altruistic, helpful behavior was found instead.[14]

Reentry Behavior

The study of the 1956 Arundel Park Hall fire in Maryland documented the initial examination of the phenomenon of reentry behavior.[15] Some older codes and regulations affecting design of the means of egress appeared to be based on the assumption that pedestrian traffic only moves away from the fire area and away from the area or floor of the building involved. Conversely, the

Arundel Park Hall study indicated that approximately one-third of the survivors interviewed had reentered the building.

Thus, it has become apparent that doors, stairways, and corridors often will be subjected to two-way movement of occupants and others. The occupant who, after leaving the building safely, turns around and reenters is usually completely aware of the fire in the building and of the specific portions of the building involved in the area of fire origin and smoke propagation.

Reentry behavior often is used to assist or rescue persons remaining or believed to be remaining in the building, such as when parents seek children missing during a fire incident. The behavior often is undertaken in a rational, deliberate, and purposeful manner, without the anxiety characteristics usually associated with nonadaptive behavior. However, reentry behavior has been considered nonadaptive, since reentry of people into a burning building is often nonadaptive in relation to the efficient and effective evacuation by other people through the same means of egress selected for reentry.

Occupant Fire Fighting Behavior

Occupants who engaged in fire fighting behavior during fire incidents were predominately male; this behavior now appears to be a culturally determined and expected aspect of the male role. However, it should be noted that in the "Project People" study of 335 U.S. fire incidents, approximately 23 percent of the study population of 584 individuals were involved in occupant fire fighting behavior. Of these, 37.3 percent were females. Of the 134 individuals who participated in fire fighting behavior, 50 were female and 84 male.

Occupant fire fighting behavior appears most prevalent in occupancies in which the individuals are emotionally and economically involved, such as in their homes, or where such behavior is an assigned role as a result of training.

Occupant's Movement Through Smoke

Often related to fire fighting behavior, and a definite component of evacuation behavior in many fire incidents,[1,2] is the movement of occupants through smoke. The principal variables influencing the occupant's decision to move through smoke appear to be (1) recognition of the location of the exit and thus being able to estimate the travel distance required, (2) the appearance of the smoke, (3) the smoke density, and (4) the presence or absence of heat with the smoke.[16,17] However, in some cases occupants have attempted evacuation through smoke, even for extended distances under conditions of extremely limited visibility at personal risk, but were forced to turn back without completing the evacuation.[1,2,16,17]

Handicapped or Impaired Occupants

Handicapped people can have a variety of limitations that increase their risk in a fire situation, including: (1) sensory problems such as deafness and blindness, (2) mobility problems such as the need for a wheelchair, and (3) intellectual problems such as mental retardation.

Studies have indicated that many handicapped people with mobility problems are concerned about their personal risk in high-rise office and residential buildings where use of elevators is not allowed in a fire. In such situations, adequate areas of refuge must be provided for handicapped as well as nonhandicapped occupants.[18]

Fire problems involving occupancies such as nursing homes and hospitals, designed for permanently or temporarily disabled persons, have been lessened due to good building design, adequate staff training, and ability to protect the occupants in place until evacuation is possible. An extensive study of human behavior in health care facilities[19] indicated that the nursing staff performed their professional roles related to patient responsibility even in situations with a high degree of personal risk.

A study of a number of evacuation drills in high-rise office buildings in Canada has indicated that approximately 3 percent of the occupants will be unable to use the stairs due to permanent or temporary conditions limiting their mobility.[17] The study population included occupants with heart conditions and individuals recovering from surgery, other illnesses, and accidents.

The few studied fire incidents involving handicapped persons in occupancies other than health care facilities or high-rise buildings primarily have been in residential occupancies.

Human Behavior Summary

Behavior in fires can be understood as a logical attempt to deal with a complex, rapidly changing situation in which minimal information for action is available. It is suggested that the goals of codes should be "reoriented to increase the likelihood of informed decisions being made by people in fires."[21] Examination of behavior in the Beverly Hills Supper Club fire led to the recommendation that "firesafety education should consider and be based on people's erroneous conceptions about distance being related to safety and the time needed to escape from a fire emergency."[22] More than a decade of detailed systematic research on human behavior in fires has resulted in the following consensus[12] on the behavior of most persons: "Despite the highly stressful environment, people generally respond to emergencies in a rational, often altruistic manner, insofar as is possible within the constraints imposed on their knowledge, perceptions, and actions by the effects of the fire. In short, 'instinctive panic' type reactions are not the norm."

The relationship between the physical and social environment in which fire behavior occurs is complex. The situation is complicated by the individual's perception of ambiguous fire cues, which is influenced primarily by the person's relevant training and previous fire experience, if any. It must be recognized that fire cues are a product of a rapidly changing dynamic process which is constantly altering the decision choices of the building occupant. This decision dilemma has been summarized, "What is an appropriate action at one stage may be quite inappropriate a minute later."

FUNDAMENTALS OF BUILDING DESIGN

One of the major goals in designing a building is to provide a structure that fulfills both the needs of the owner and the purpose for which the building is intended. A well-designed building should make efficient use of space while presenting an appropriate aesthetic environment. It should be built as economically as possible, and should be efficiently planned to accommodate the activities of the people who will occupy it. Because buildings are designed for certain (and often specialized) types of activities and occupancy, careful consideration of inherent design features is of vital importance in assessing the life safety factor in case of fire.

Design features related to firesafety of the average one-family dwelling can be easily assessed. Conversely, planning for life safety from fire in multiple-purpose (mixed-occupancy) buildings such as those containing offices, stores, apartments, garages, and restaurants is complicated and difficult. Assessment of life safety from fire in multiple-purpose buildings must take into consideration many and varied design factors, so that if a fire develops the possibility of death or injury can be minimized.

A shopping mall building is a good example of the complexities involved in planning for life safety in the event of fire. In this type of occupancy many different stores of varying sizes and designs contain contents of varying combustibility, all under one roof—in effect, all in a single building. Each store opens onto the covered mall. A fire in one store may well vent itself into the entire mall building, thereby endangering the people in the covered mall as well as those in the other stores. If the store is small, as many stores in shopping malls are, it probably will have only one visible means of egress. This egress usually will be by way of the covered mall. This results in relatively long travel distances to exits from the mall—a situation which, in a fire emergency, will increase the hazard to people in the covered mall area. In addition, separation between stores may be difficult to achieve and maintain and must be extended to the roof space common to all the stores. Some malls are more than one story high, further complicating easy and speedy egress to a safe point outside the mall building.

Mixed-occupancy high-rise buildings present great problems in planning for life safety from fire. Some other examples of buildings whose design presents varying and complex problems when planning for life safety from fire are:

1. Large, undivided industrial buildings where compartmentation becomes impractical and travel distance to exits can become excessive.
2. School buildings of open and flexible plan layouts where, again, compartmentation is impractical and access to exits may be varied and obstructed.
3. Membrane-rooted (astrodome) structures from which large numbers of people must be quickly evacuated in fire emergencies.
4. Hotel exhibition halls that involve heavy fire loading and numerous ignition sources.
5. Windowless buildings that make entrance for fire fighting and rescue difficult.

Unfortunately, many buildings are in the advanced design stage before local firesafety codes are consulted. Code-complying firesafety features should be considered and included starting with the initial designing and planning stages of buildings. It must be remembered that local firesafety codes usually provide for only minimum requirements. The codes may not deal with the more complex design problems and considerations necessary to ensure life safety as well as property protection in more complex buildings.

Concepts of Egress Design

The degree of firesafety in a building is dependent on many components, including means of egress, fire detection and suppression systems, combustibility of interior finish and furnishings, presence of hazardous materials, training of occupants, occupant crowding, and the condition of the occupants, e.g., sleeping or physically incapacitated. The most important factor (1) in building design and construction is the provision of adequate exits; (2) in operating a building is the proper maintenance of the exits; and (3) in surviving a fire in a building is knowing how to reach the exits quickly.

Occupant Characteristics

When designing a system for transporting objects, it is important to know the characteristics of those "objects," including people. Since means of egress are used to transport people, the physical dimensions and characteristics of people must be understood.

The majority of adult males do not exceed a width of 20.7 inches (52.8 cm) at the shoulders. Additional width is added by clothing, particularly in colder climates. It is also important to recognize the tendency of people to avoid bodily contact. People generally prefer to establish an area around them—a buffer zone between themselves and others. These characteristics have brought forth the concept of a "body ellipse" (see Figure 2.2). For Americans, this body ellipse is given a major axis of 24 inches (60.9 cm) and a minor axis of 18 inches (45.7 cm). This ellipse has an area of 2.3 square feet (0.21 m^2) and represents the space occupied by a standing person.

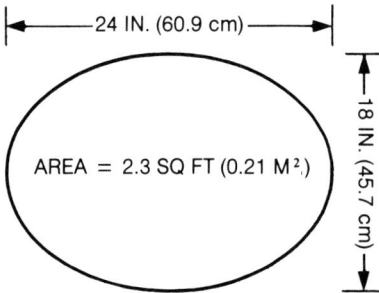

Fig. 2.2   Concept of a "body ellipse."

Yet, the body-ellipse space cannot be assumed to be the entire space that a person occupies when walking. As a person moves forward, side to side swaying occurs. This sway can vary from 1½ inches (3.8 cm) to each side during normal walking with relatively free movement to nearly 4 inches (10.2 cm) when movement is restricted such as when in a crowd. Body sway also is increased when moving on stairs. This increased space required for people movement indicates that a total width of nearly 30 inches (76.2 cm) is needed for an average single file of people moving up or down stairs.

Studies have shown that the average free-movement walking speed on a level surface, with approximately 25 square feet (2.32 m²) available per person, is approximately 250 feet (76.2 m) per minute. As crowding develops and people slow to a shuffle, the walking speed is reduced to approximately 145 feet (44.1 m) per minute. As crowding increases, movement slows further. At approximately 2¾ square feet (0.25 m²) per person, physical contact is unavoidable. At approximately 2 square feet (0.18 m²) per person, a "jam" point is reached and movement virtually stops. (See Figure 2.3.)

Ultimately, people's reaction to an emergency depends on their physical and mental conditions at the time of the emergency and their perception of the danger.

Evacuation vs. Defend-in-Place Behavior

The primary means of providing safety from fire in a building is to have the occupants leave the building *before* conditions become life-threatening. This requires that several parts of the firesafety system work properly, including:

- The fire is detected and the alarm is promptly activated
- The occupants recognize the alarm signal
- The occupants immediately proceed to evacuate the building
- The occupants proceed toward the exits in an orderly and efficient manner

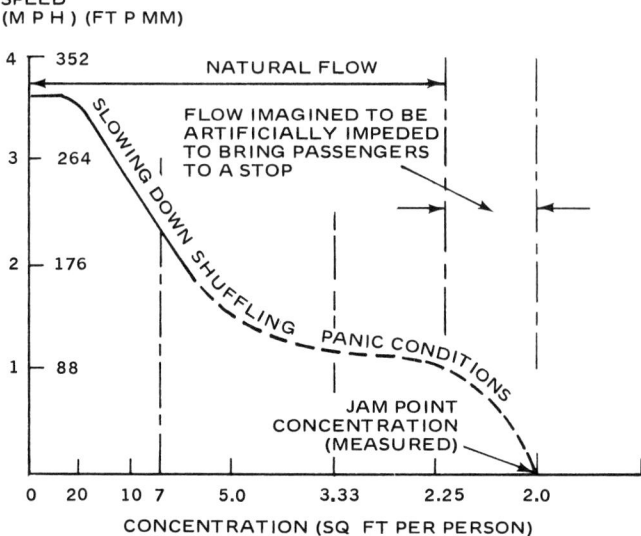

SPEED
(M P H) (FT P MM)

NATURAL FLOW

FLOW IMAGINED TO BE
ARTIFICIALLY IMPEDED
TO BRING PASSENGERS
TO A STOP

SLOWING DOWN SHUFFLING

PANIC CONDITIONS

JAM POINT
CONCENTRATION
(MEASURED)

CONCENTRATION (SQ FT PER PERSON)

Fig. 2.3 Speed in level passageways. (Adapted from Research Report No. 95, London Transport Board, "Second Report of the Operational Research Team on the Capacity of Footways.")

- The means of egress is adequate to accommodate the number of occupants
- The exits have been properly designed, constructed, and maintained to provide a safe environment
- The last occupant is out of the building before the fire develops life-threatening conditions.

This system of firesafety generally is evident in buildings where the occupants are physically capable of taking self-preservation actions. Examples include schools, businesses, stores, places of assembly, and industrial and storage occupancies.

Other means of providing protection is needed where evacuation of the occupants is difficult, such as in hospitals and jails or prisons. Hospital patients often are physically incapable of leaving the building by themselves, and their evacuation by staff may be extremely difficult due to life-support equipment. Occupants of jails and prisons generally are capable of evacuating themselves, but they are locked in and require the aid of staff to unlock the doors for means of egress.

In both these cases, the protection of occupants may be accomplished by applying the concept "defend in place." This involves a firesafety system where:

1. **The size of any expected fire is controlled.** This is accomplished by controlling the combustibility of interior finish and furnishings to limit

the speed with which a fire would develop and spread. Automatic sprinklers normally are considered necessary to assure that any fires that do start are quickly controlled.

2. **The building is compartmented.** Dividing the building with fire barriers limits the number of people exposed to a fire. Those in the area exposed to the fire can move horizontally through a fire barrier to an area of refuge.

3. **Exits are provided.** Adequate exits must be maintained in case building evacuation is required. However, evacuation generally is a last resort.

THE *LIFE SAFETY CODE*

NFPA *101, Code for Safety to Life from Fire in Buildings and Structures*,[26] commonly known as the *Life Safety Code*, addresses those items in building design and operation which affect safe egress from a building. This code does not address protection of property nor other building safety measures such as are included in a typical building code. Its principal objective is to specify those firesafety and egress characteristics in a building needed to safely evacuate the occupants.

NFPA began work on the *Code*[26] in 1913 when it appointed a Safety to Life Committee. The new committee studied several noteworthy fires and prepared standards for construction of exits for occupancies, such as factories, schools, and department stores. Over the years, the committee was enlarged to include a more diverse membership. In 1927 the first edition of the "Building Exits Code" was published. In 1963 the committee was restructured and subcommittees established for the various occupancies. In 1966 the *Code* title was changed to its present-day title and the text was put in "code language," with all explanatory material placed in the appendices.

The objectives of the *Life Safety Code* are based on the following 12 principles:

1. A sufficient number of unobstructed exits of adequate capacity and arrangement are provided.
2. The means of egress is protected against fire, heat, and smoke for the time it requires the occupants to leave the building.
3. An alternate means of egress is provided in the event one of the exits is blocked by fire, heat, and/or smoke.
4. Compartmenting the building with adequate fire barriers creates areas of refuge where evacuation is not the primary means of providing occupant safety.
5. Protection of vertical openings limits the movement of fire and smoke to multiple floors.
6. Fire detection and/or alarm systems alert occupants and, in some cases, notify the fire department of a fire condition.
7. Adequate illumination is provided for the means of egress.

8. Signs indicate the paths of travel to reach the exits.
9. Adequate protection is provided from unusual hazards which could impinge upon the means of egress.
10. Adequate evacuation plans and exit drills are provided.
11. Adequate instructions to occupants facilitate their effective movement, particularly in those occupancies where crowding or severe fire hazards are present.
12. Interior finish materials are controlled to prevent fast-spreading fires and dense smoke production that could endanger occupants.

The *Life Safety Code*[26] recognizes that many factors are involved in providing safety to the occupants of a building. In other words, the safety of the building occupants cannot depend upon a single safeguard, but is based on a system made up of several components. Any one component can fail due to human error or mechanical malfunction, so multiple safeguards are necessary to assure a reasonable degree of life safety.

It has been argued that existing buildings, which were constructed under earlier codes and which complied with all legal requirements at that time, should not be required to make changes due to changes in the *Code*.[26] The *Life Safety Code*[26] addresses this problem of firesafety in existing buildings by providing separate chapters for new and existing structures. The *Code* notes that age is not an acceptable excuse for maintaining a building in an unsafe condition. In those occupancies where there is little difference between the requirements for new and existing buildings, e.g., industrial occupancies, the *Life Safety Code* has only one chapter. In this case, the differences in requirements are handled by exceptions to *Code* provisions.

Fire hazards vary depending on the use of the building. The *Life Safety Code*[26] addresses the requirements for the means of egress based on the occupancy or building use. The *Code* recognizes that the occupants of any given building may have different situations that could decrease personal safety, such as:

*   *Physical ability to safely evacuate.* People in health care occupancies often are not capable of self-preservation. Also, young children may need adult supervision to safely evacuate.
*   *Physical restraint.* People in jails and prisons are not able to evacuate until locks are released.
*   *Crowding in unfamiliar surroundings.* People in places of assembly, e.g., hotel ballrooms, generally are unfamiliar with the means of egress. Also, the huge numbers of people who will use the exits can put a strain on the egress system.
*   *Sleeping.* People who are asleep generally require extra protection due to the time required to awake, react, and then evacuate a building.

The *Life Safety Code*[26] defines and addresses different occupancy classifications, including the following.

Places of Assembly

Assembly occupancies are buildings in which 50 or more persons gather for such purposes as deliberation, worship, entertainment, eating, drinking, amusement, or awaiting transportation.

Generally, people in an assembly occupancy are not familiar with the building, but they are mobile and capable of self-preservation. The firesafety concerns are that the occupants be warned of a fire early, that adequate aisle and exit capacity be available at all times, and that the means of egress be well lighted and identified. One of the primary concerns in this occupancy is that the means of egress be continually well maintained and kept unblocked.

The density of people in assembly occupancies can vary greatly depending upon the function or use of the space. The *Life Safety Code*[26] defines different densities by specifying the square feet (m²) per person expected, based on the use. This will vary from 15 square feet (1.4 m²) per person in areas where tables and chairs are used to 3 square feet (0.28 m²) per person for standing/waiting spaces. (See Table 2.2.)

Table 2.2   Occupant Load Factors

Use	Sq Ft	Sq M
Assembly		
Less concentrated use without fixed seating	15 net	1.4
Concentrated use without fixed seating	7 net	0.65
Waiting space	3 net	0.28
Library—stack areas	100 net	9.3
Library—reading areas	50 net	4.6
Mercantile		
Street floor and sales basement	30 gross	2.8
Multiple street floors—each	40 gross	3.7
Other floors	60 gross	5.6
Storage, shipping	300 gross	27.9
Malls	See Section 24-1.7.1, NFPA *101*	
Educational		
Classroom area	20 net	1.9
Shops and other vocational areas	50 net	4.6
Day-care centers	35 net	3.3
Business (offices), industrial	100 gross	9.3
Hotel and apartment	200 gross	18.6
Health care		
Sleeping departments	120 gross	11.1
Inpatient treatment departments	240 gross	22.3
Detention and correctional	120 gross	11.1

Educational

Educational occupancies are buildings used for educational purposes through the 12th grade by six or more persons for four or more hours per day or more than 12 hours per week.

Educational occupancies generally have a good life safety record. This can be attributed to the strong *Code* provisions requiring adequate exits and exit drills. A total of ten fire exit drills are required during the school year, with at least two drills taking place during the first two weeks of that period.

Health Care

Health care facilities are buildings in which occupants are not capable of self-preservation because of age and/or physical or mental disability.

Occupants of health care facilities are divided into two groups: (1) those who are mobile and can evacuate rapidly, such as outpatients, and (2) those who are physically or mentally incapacitated and cannot evacuate without assistance. These latter occupants must be protected using the defend-in-place concept.

Detention and Correctional

Detention and correctional occupancies are buildings which house persons under some degree of restraint or security.

Detention and correctional occupancies present a unique problem due to the security measures imposed. Generally, the tighter the security the greater the concern for firesafety. In jails and prisons, the occupants' total environment is controlled by management, including their ability to walk from one room to another. Since evacuation to the outside of the building generally is not desirable, the defend-in-place concept is applied.

Residential

Residential occupancies are buildings in which sleeping accommodations, as well as other living functions, are provided for "normal residential purposes." Residential occupancies include:

- Hotels
- Apartment buildings
- Dormitories
- Lodging and rooming houses
- Board-and-care homes
- One- and two-family dwellings.

One of the key factors that distinguishes residential occupancies is that the occupants are asleep a good portion of the time they are in the building. The *Code*[26] also recognizes that the degree of protection should increase as the concentration of occupants increases and the degree of occupant familiarity with the building decreases. For example, the requirements for hotels are more stringent than those for one- and two-family dwellings.

Mercantile

Mercantile occupancies are buildings displaying and offering merchandise for sale.

People in mercantile occupancies generally are assumed to be somewhat —if not totally—unfamiliar with the building arrangement. The occupants are expected to be mobile and capable of self-preservation but, due to crowding and lack of familiarity, might have some difficulty in locating and walking to exits in an emergency. Also, the display of merchandise can present a faster fire development potential than that present in other occupancies.

Business

Business occupancies are buildings in which financial, managerial, and technical work is performed related to the operation of a business.

People in a business occupancy generally are considered to be awake, mobile, and familiar with their surroundings. Also, the contents usually do not present severe fire hazards such as rapid fire development or explosions.

Industrial

Industrial occupancies are buildings in which processing, assembling, mixing, packaging, finishing, decorating, and/or repair operations are conducted.

Life loss as a result of fires in industrial occupancies is usually low. The majority of life loss here generally is related to flash fires or explosions.

Storage

Storage occupancies are buildings or structures used for the storage or sheltering of goods, merchandise, products, vehicles, or animals.

Storage occupancies have a very low life-loss record. The number of people within a storage building generally is low, and those occupants are usually familiar with the building arrangement. Wide aisles for the use of material-moving machinery facilitate egress.

Unusual Occupancies

Unusual occupancies are those buildings or structures which cannot be classified in one of the preceding groups due to some function or arrangement; these include piers, towers, and underground occupancies. Facilities within an unusual structure, such as a restaurant on a pier, should comply with the appropriate occupancy classification.

The *Life Safety Code*[26] establishes basic requirements for means of egress. These requirements are specified in the first seven chapters of the *Code*, referred to as the "building block" chapters. Chapters 8 through 30 then provide requirements for individual occupancy types. These occupancy chapters may place additional requirements on the means of egress or may take exception to the requirements specified in the "building block" chapters. Chapter 31 addresses the need for evacuation plans, evacuation drills, and the control of special indoor hazards, such as open-flame devices.

The *Life Safety Code*[26] is widely used and referenced throughout the world to establish minimum requirements for means of egress. While it is not possible to provide 100 percent life safety from fire to building occupants, the *Life Safety Code* does provide requirements to establish reasonable safety for egress from buildings. Preservation of property, protection of individuals from their own negligence, and prevention of personal injuries are not covered by the *Code*.

INDUSTRIAL FIRE BRIGADES

Many industrial facilities find it necessary, for a variety of reasons, to establish their own private or industrial fire brigades. (See Figure 2.4.) Often industrial facilities are beyond the jurisdiction of the local fire department or are involved in producing materials in which any fire requires immediate or specialized attention to contain. In instances such as these, industrial fire brigades perform a vital function until the fire department arrives. The life safety factor is heightened considerably because of industrial fire brigades. Because of the specialized materials involved in many industrial facilities, members of these industrial fire brigades often work in conjunction with the fire department once it arrives at the fire scene.

Components of Industrial Fire Brigades

The Industrial Fire-Loss Prevention Manager: The fire-loss prevention manager (usually the plant manager or the shift superintendent) has the primary responsibility for correlating activities with the public fire department so that agreement is reached on the mutual responsibilities in a fire emergency. Once the public fire department arrives, it is the responsibility of the fire-loss prevention manager to make sure the fire chief understands the situation and is made aware of steps already taken. If more than one public fire department is involved, the fire-loss prevention manager should communicate with the chief of the fire department having jurisdiction over the

Fig. 2.4 Members of industrial fire brigades should receive in-plant training by practicing fire suppression techniques under controlled conditions while being supervised by skilled instructors. (Los Angeles Times photo by Al Seib.)

property. That chief then will have the responsibility of communicating with the other fire departments.

Industrial Fire Brigade Organization: The fire brigade is composed of selected building, operating, maintenance, and security personnel. The brigade responds to all fire alarms, reports to the scene of the alarm, establishes communication with the appropriate personnel, takes action consistent with existing conditions to evacuate the area, and fights the fire with available equipment. Generally, upon the arrival of the fire department, the fire brigade will let the fire department take charge; members of the fire brigade then stand ready to assist as requested.

The organization of a fire brigade varies to suit the needs of the industrial facility it serves. The fire-loss prevention manager usually is the one who determines the nature of the brigade and its organizational structure. The fire brigade can consist of selected individual personnel or groups of people functioning as teams.

The availability and proximity of the public fire department having jurisdiction also can affect decisions on the nature and organization of an industrial fire brigade. No matter how the brigade is organized, provisions must be made in advance so that a sufficient number of members of the fire brigade are on duty during each work shift and during the periods when the plant is shut down.

Members of fire brigades should meet minimum physical requirements, should be available to answer alarms and attend meetings, should complete a required course of instruction, and should be prepared to attend monthly training sessions.

Traffic and Exit Drills

When an industrial fire emergency occurs, both ingress and egress traffic problems arise. Fire fighting personnel must be able to get to the location of the fire. At the same time, people not involved in fighting the fire must be evacuated. Traffic control is divided into two categories: (1) external traffic control on public streets and highways that surround and/or lead to the industrial structure, and (2) internal traffic control inside the property involved in the fire situation.

External Traffic Control: Preplanning for external traffic control in the event of an industrial fire involves working with the law enforcement agency having jurisdiction. If law enforcement staffing cannot adequately handle external traffic control, a facility might need to provide its own security guards. This type of traffic control plan should be designed in cooperation with the appropriate law enforcement agency.

It is wise to provide personnel involved in emergency fire control with clear identification. IDs approved by the proper law enforcement agency and Civil Defense will enable fire fighting personnel to get to the scene of the emergency without delay. In addition, alternate routes should be planned so that, in the event of an emergency, fire fighting equipment and other emergency vehicles can get to the scene quickly.

Internal Traffic Control: Internal traffic control responsibilities for both pedestrians and vehicles generally are assigned to private security personnel. Their major responsibilities are to clear the involved area of unnecessary personnel and to direct the fire department and other emergency forces to the emergency scene.

Federal law requires that industrial facilities properly identify their emergency exits and access routes. Exit drills are essential in preplanning for a possible fire emergency in an industrial facility.

The fire-loss manager or plant manager generally has the responsibility for evacuation drills. A recommended procedure for such drills is:

1. Employees should become familiar with the evacuation signal,
2. Employees should immediately turn off all equipment,
3. Employees should immediately proceed along the predetermined exit route (they should be trained to use alternate routes as well), and
4. Employees should report to a predetermined assembly point so that it can be determined if all personnel are present.

In addition, daily inspection of emergency routes and exits should be conducted. The frequency of such evacuation drills should be dependent upon both the degree of potential hazard in the facility and the complexity of the facility's layout.

PUBLIC EDUCATION AND COMMUNITY RELATIONS

Public firesafety education is one of the major fire prevention methods that can be used to reduce life loss from fire. Teaching people to prevent fires and the appropriate actions to take in case a fire does occur are the basic goals of a worthwhile firesafety education program.

The majority of fires and civilian fire deaths take place in residential properties. Most of these residential fire deaths are preventable. The key to survival is knowledge through education.

Fire Prevention Programs

Various projects in fire prevention education have demonstrated repeatedly the role that increased public awareness plays in reducing fire hazards. Awareness can be a significant factor in reducing loss of both life and property. Benefits of a firesafety education program far exceed its cost.

In a well-planned firesafety education program, the overall message must be concise, positive, relevant, and aimed at enlightening someone's attitude and behavior. Getting this message across is a team effort involving fire departments, families, teachers, media, businesses, and industries. The fire department's leadership role in this team effort exemplifies its position as a vital segment of the community. The fire department works in partnership with the public for the public's benefit and safety.

Methods of Public Education

Major goals of public education programs are to increase public awareness, support, and involvement in fire prevention. Local fire departments can make significant contributions to public education by periodically inspecting residential and commercial properties, distributing materials on firesafety, conducting firesafety programs for the public, and presenting educational programs in the schools.

Periodic inspections of homes and other buildings increase the public's awareness of fire hazards. Inspections also can provide good opportunities for the fire inspector to suggest actions occupants can take to prevent fires, as well as plans to be implemented in case of fire. A good educational program can motivate home and other property owners to plan and practice evacuation procedures, install and maintain smoke detectors, inspect for fire hazards, and practice vital self-protection behaviors. Some such behaviors are "stop, drop, and roll" to extinguish a clothing fire and "crawl low in smoke" to leave a smoke-filled building.

Media campaigns such as NFPA's *Learn Not to Burn®* program help make the public aware of fire hazards, good fire prevention techniques, and the services provided by the fire department. Media campaigns work, as evidenced by the number of lives saved through the *Learn Not to Burn* public service announcements on television. An effective media campaign utilizes all forms of media—print, television, and radio. Timely information should be disseminated at a continuous rate. Special events or seasons of the year present good opportunities for public education regarding specific fire hazards, such as portable heating devices. Spring clean-up (with emphasis on discarding buildups of combustible items, such as old newspapers), Fourth of July (for fireworks hazards), and Fire Prevention Week in October are major events around which to build educational efforts. Newspaper articles that give key firesafety tips, and public service announcements on local cable television or radio, provide excellent opportunities for reaching the public.

Firesafety education curricula are now available for schools. NFPA's *Learn Not to Burn Curriculum*,[23] first released in 1978, now is used in more than 43,000 classrooms across the nation. It is in its third edition with updates which recognize lifestyle changes and differing community needs. Designed for grades kindergarten through eight, the *Curriculum* is structured around 22 key firesafety behaviors, with instructions for developing three behaviors to meet local needs. The *Curriculum* provides classroom teachers with firesafety information, lesson plans, and evaluation instruments to measure the program's effectiveness. In conjunction, the fire department provides the technical support and motivation that help the teacher present firesafety education lessons in the classroom.

Firesafety for the Rest of Your Life[24] was developed by the Foundation for Firesafety and Office of Fire Prevention and Control, State of New York, and is distributed by NFPA. This program uses 12 firesafety modules to implement firesafety education in the high school classroom.

To assist in the implementation of these programs nationwide, NFPA developed the Learn Not to Burn Regional Representative Network. "Regional Reps" provide technical assistance and in-service training workshops to help educators and community leaders learn implementation strategies. The Regional Rep program is funded through a cost-sharing arrangement between NFPA and the Federal Emergency Management Agency (FEMA).

The number of firesafety educators and of others interested in firesafety education has increased dramatically over the last several years. Fire service organizations have responded by establishing specialized groups which enable educators to address issues and learn the most up-to-date techniques in firesafety instruction. The education section of NFPA, which provides a forum for educators from schools, the fire service, the health care field, and other areas of concern, is one such organizational opportunity.

Firesafety educators are growing in skills, such as organization and planning, and are using sound management of financial, material, and human resources to reach designated goals and objectives. They also are promoting firesafety by addressing the demonstrated firesafety needs of their communities and enlisting the cooperation of many local groups and individuals.

A good relationship with the community it serves is an important aspect of fire department public relations. Because fire departments are supported by public funds, an understanding by the public of the fire department's services is vital. Good community relations also result in a high level of understanding and cooperation by members of the community who can help achieve and support effective fire prevention programs. Establishing and maintaining good community relations are vital fire department activities and, as such, should be continual and effective.

Summary

Loss of life and property through fire is dependent on many factors. The human characteristics of the people involved in terms of age, agility, decision-making abilities, previous firesafety training, and so forth are a vital life safety factor in assessing fire protection planning and techniques. Life safety considerations are also dependent on hazards that may be involved in buildings and areas designed for specific activities, such as stores, restaurants, dwellings, and theaters.

The design of the building and the type of occupancy are also important factors to be considered. The requirements of the *Life Safety Code*[26] provide specific recommendations to enhance the probability of life safety in all types of occupancies. Most, if not all, of these recommendations have been enacted into law at both the federal and state levels.

Public education in firesafety is a major factor in fire prevention. Fire prevention programs and various methods of public education in fire prevention have proven to be effective deterrents to fire and fire loss. Good community relations between the fire service and the public is an important factor in public support for fire prevention and for the fire service itself.

References

[1]Bryan, John L., 1977. *Smoke As a Determinant of Human Behavior in Fire Situations*, Department of Fire Protection Engineering, University of Maryland, College Park, MD.

[2]Wood, Peter G. 1972. *The Behavior of People in Fires*, Building Research Establishment, Fire Research Note 953, Fire Research Station, Borehamwood, Hertfordshire, England.

[3]Lerup, Lars, *et al* 1978. "Human Behavior in Institutional Fires and Its Design Complications," *NBS-GCR-77-93*, Center for Fire Research, National Bureau of Standards, Washington, DC.

[4]Keating, John P., and Loftus, Elizabeth F. 1981. "The Logic of Fire Escape," *Psychology Today.*

[5]Berry, Charles H. 1978. "Will Your Smoke Detector Wake You?" *Fire Journal,* Vol. 72, No. 4, National Fire Protection Association, Boston.

[6]Cohen, Hal C. 1982. "Fire Safety for the Hearing Impaired," *Fire Journal,* Vol. 76, No. 1, National Fire Protection Association, Quincy, MA.

[7]Withey, Stephen B. 1962. "Reaction to Uncertain Threat," in *Man and Society in Disaster,* Edited by G.W. Baker and Dwight W. Chapman, Basic Books, NY.

[8]*Fire Protection Handbook,* 16th ed. 1986. National Fire Protection Association, Quincy, MA.

[9]Best, Richard L. 1978. "Tragedy in Kentucky," *Fire Journal,* Vol. 72, No. 1, National Fire Protection Association, Boston.

[10]Schultz, D.P. 1966. *An Experimental Approach to Panic Behavior,* Office of Naval Research, Department of the Navy, Arlington, VA.

[11]Kentucky State Police 1977. *Investigative Report to the Governor, Beverly Hills Supper Club Fire,* Kentucky State Police, Frankfort, KY.

[12]Sime, Jonathan D. 1980. "The Concept of Panic," in Canter, David, ed. *Fire and Human Behavior,* John Wiley & Sons, NY.

[13]Quanrantelli, E.L. 1979. *Panic Behavior in Fire Situations: Findings and a Model from the English Language Research Literature,* Disaster Research Center, Ohio State University, Columbus, OH.

[14]Keating, John P. 1982. "The Myth of Panic," *Fire Journal,* Vol. 76, No. 3, National Fire Protection Association, Quincy, MA.

[15]Bryan, John L. 1957. *A Study of the Survivors' Reports on the Panic in the Fire at Arundel Park Hall, Brooklyn, Maryland, on January 29, 1956,* Fire Protection Curriculum, University of Maryland, College Park, MD.

[16]Bryan, John L. 1983a. *An Examination and Analysis of the Dynamics of the Human Behavior in the MGM Grand Hotel Fire,* Revised Edition, National Fire Protection Association, Quincy, MA.

[17]Bryan, John L. 1983b. *An Examination and Analysis of the Dynamics of the Human Behavior in the Westchase Hilton Hotel Fire,* National Fire Protection Association, Quincy, MA.

[18]Levin, Bernard N., and Nelson, Harold E. 1981. "Firesafety and Disabled Persons," *Fire Journal,* Vol. 75, No. 5, National Fire Protection Association, Quincy, MA.

[19]Bryan, John L., *et al* 1980. "The Determination of Behavior Response Patterns in Fire Situations, Project People II, Final Report—Incident Reports, Aug. 1977 to June 1980," *NBS-GCR-80-297,* Center for Fire Research, Washington, DC.

[20]Pauls, J.L. 1977. "Movement of People in Building Evacuations," in Conway, D.J., ed. *Human Response to Tall Buildings,* Dowden, Hutchinson and Ross, Stroudsburg, PA

[21]Canter, David, *et al* 1978. *Human Behavior in Fires.* Department of Psychology, University of Surrey, Guilford, England.

[22]Pauls, J.L., and Jones, B.K. 1980. "Research in Human Behavior," *Fire Journal,* Vol. 74, No. 3, National Fire Protection Association, Quincy, MA.

[23]*Learn Not to Burn Curriculum* 1981. National Fire Protection Association, Quincy, MA.

[24]*Fire Safety for the Rest of Your Life* 1983. Foundation for Firesafety and Office of Fire Prevention and Control, State of New York.

[25]NFPA 74-1984. *Standard for the Installation, Maintenance, and Use of Household Fire Warning Equipment,* National Fire Protection Association, Quincy, MA.

[26]NFPA *101*-1988. *Life Safety Code,* National Fire Protection Association, Quincy, MA.

Chapter 3

Characteristics and Behavior of Fire

It is virtually impossible to predict exactly when a fire will occur and, upon its inception, the extent of its destructive potential. However, through scientific knowledge of ignition, the combustibility of solids, liquids, and gases, and the products of combustion, effective ways to control the dangers of fire and explosion can be determined.

THE UNPREDICTABILITY OF FIRE

It is foolhardy to attempt to completely describe in one chapter the chemical and physical reactions that take place during a fire. Such an attempt is hindered by the fact that the variables involving the essentials for fire—something to burn (fuel), a source of ignition, and the oxygen necessary to maintain combustion—are infinite. Although technical knowledge about flame, heat, and smoke continues to grow, and although additional information continues to be acquired concerning the ignition, combustibility, and flame propagation of various solids, liquids, and gases, it still is not possible to predict with any degree of accuracy the probability of fire initiation or the consequences of such initiation. Thus, while the study of controlled fires in laboratory situations provides much useful information, most unwanted fires happen and develop under widely varying conditions, making it virtually impossible to compile complete bodies of information from actual unwanted fire situations. This fact is further complicated because the progress of any unwanted fire varies from the time of discovery to the time when control measures are applied.

The combustion process is a very complicated chemical reaction affected by many variables. For example, the size and shape of a solid will influence the way it burns. A solid block of wood is hard to ignite. However, if the wood is in the form of shavings, it will ignite readily; if the wood is finely divided into dust and the dust is suspended in a cloud, it may explode violently when a source of ignition is present.

PRINCIPLES OF FIRE

Dr. Richard L. Tuve, in *Principles of Fire Protection Chemistry*, provides the following simple definition of fire:[1]

> Fire is a rapid, self-sustaining oxidation process accompanied by the evolution of heat and light of varying intensities.

Although this definition is complete and almost poetic in its simplicity, some of the basic principles underlying this definition should be considered. As previously mentioned in this chapter, three elements are essential for ignition—i.e., the initiation—of a fire: something to burn (fuel), a source of ignition (heat or thermal energy), and oxygen. Ignitability of fuels is presented in Chapter 4, "Fire Hazards of Materials." Sources of ignition will be described in more detail later in this chapter. The most common sources of ignition are the heat produced by a chemical reaction (such as striking a match), by electrical energy (such as an arc), or by mechanical energy (such as friction). Oxygen, the third element necessary for ignition, usually is supplied by the air. Together, these three elements customarily have been depicted as a *fire triangle*, as shown in Figure 3.1.

Dr. Tuve's definition of fire, otherwise known as combustion, contains three key words: rapid, self-sustaining, and oxidation. Working in reverse order, "oxidation" is a chemical reaction in which two materials—an oxidizing agent and a reducing agent—combine to form a product less reactive than the parent materials. Combustion is a particular type of oxidation reaction in which oxygen almost always is the oxidizing agent (there are specific exceptions) and a fuel, or something that burns, is the reducing agent. The most common reducing agents, or fuels, are materials containing great percentages of carbon and hydrogen.

The next key word is "self-sustaining." If the oxidation process is to be self-sustaining, it must proceed so it continues to grow, or at least to maintain

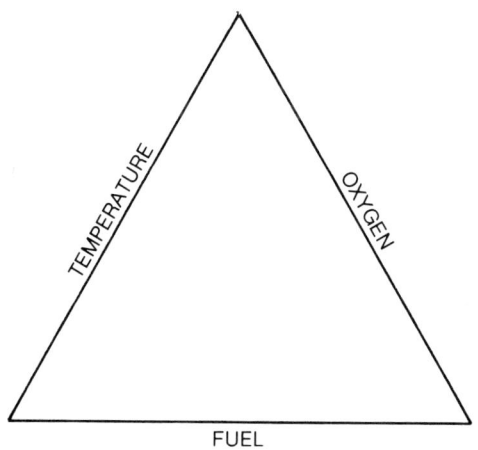

Fig. 3.1   Fire triangle.

itself. The ignition which triggers the oxidation reaction between one tiny bundle of oxygen and one tiny bundle of fuel must produce enough excess energy to trigger oxidation between two more tiny bundles of oxygen and fuel. In this manner, the combustion reaction continues in chain-like fashion.

The third key word is "rapid." This combustion reaction must proceed rapidly enough to produce excess energy, to evolve heat and light, and to grow. Combustion as a rapid oxidation process is vividly explained by comparison with other, slower forms of oxidation. The rusting of iron is a very slow oxidation process. The basic metabolism of the human body is a faster, but also very complex oxidation process. Simply stated, we breathe in oxygen that "burns up" our food, thus releasing energy for activity, both physical and mental.

The combined concept of a rapid, self-sustaining oxidation reaction is covered by the term "chain reaction." The concept is depicted by taking the fire triangle and expanding it to a fire tetrahedron, shown in Figure 3.2. As can be seen, each side of the tetrahedron shares an edge with the other three sides. This vividly, yet very simply, demonstrates the interdependence of fuel, heat or thermal energy, oxygen, and the chain reaction. The fire triangle shows the ignition sequence; the fire tetrahedron shows the growth of ignition to a fire.

It is helpful to examine the entire combustion sequence more slowly. Take a block of wood and place it very close to the glowing coil of an electric heating element. All three elements of ignition are present: fuel, heat, and oxygen. There is no self-sustaining reaction evident, so the chain reaction is not yet a factor. As the wood block is heated, its temperature increases. It begins to dry. A decomposition reaction called "pyrolysis" begins to occur, as

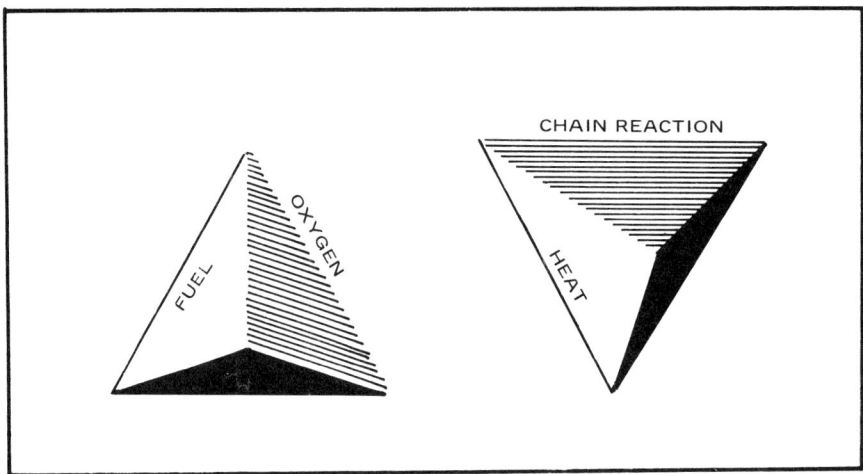

Fig. 3.2   The fire tetrahedron, which expands upon the one-dimensional fire triangle by the addition of the fourth element of fire, the chain reaction.

evidenced by evolution of smoke vapors and browning of the surface. The reaction, however, is not proceeding fast enough to be self-sustaining.

Move the block of wood a bit closer to the coil. Now, gases are evolved at a more rapid rate and are chemically different than they were at the beginning of the experiment. The discoloration gets darker, blacker; charring occurs as the reaction moves deeper into the wood. The pyrolysis is nearing the point where it will become self-sustaining, thereby instantly changing to combustion. If a spark or tiny flame is introduced periodically, close to the surface of the block where the gases are being driven off, eventually a flicker of fire will leap from the spark or flame. This is ignition.

Upon continued heating of the block, the reaction eventually proceeds fast enough so that an ignition source provides the final extra bit of energy to trigger the chain reaction, and gases released by the wood burst into flame. This is combustion. The concentration of the gases at this point is known as the "limit of flammability," and the temperature at this time is the "ignition temperature."

Chapter 4 will discuss in greater detail the complex reactions involved in burning and the concept of limits of flammability.

To summarize:

1. Ignition requires three elements: fuel, oxygen, and energy.
2. Propagation of ignition to self-sustained combustion requires a fourth element: the chain reaction.
3. The fuel must be in the proper concentration of oxygen, i.e., at least at its limit of flammability and at its ignition temperature.
4. Combustion will continue until:
    a. The fuel is consumed or removed.
    b. The oxidizing agent concentration is decreased below that necessary for combustion.
    c. The fuel is cooled to below its ignition temperature.
    d. The chain reaction is successfully interrupted.

All of the material presented in NFPA's *Fire Protection Handbook*[2] for the prevention, control, or extinguishment of fire is based on these principles.

HEAT MEASUREMENT

All combustion reactions are exothermic; that is, they give off heat. One of the major determinations a fire fighter must make in a fire situation is to estimate the intensity of the heat being emitted from the fire. This is vital in determining the potential danger of nearby combustibles reaching their ignition point from the heat being emitted by materials already involved in the fire.

In *Principles of Fire Protection Chemistry*, Dr. Tuve describes the importance of determining the intensity of heat as follows:[1]

When a fire fighter senses the temperature of a fire situation by sight, touch, or by the amount and type of smoke or gas odor smelled, such information must be translated into the "extent of heat" that has been evolved by the fire. High temperatures mean that ordinary combustibles not already burning are rapidly reaching their ignition points; in order to slow up the evolution of heat, the prompt preventive efforts of cooling are necessary. Ordinary carbon-hydrogen substances produce temperatures during burning of 1,100°F to 1,800°F (593°C to 982°C), whereas their ignition points may be only 350°F to 1,000°F (177°C to 538°C).

"Heat" and "temperature" are not the same. "Heat" is a measure of the quantity of energy contained by a material. "Temperature" is an indication of the level at which the energy exists or, more simply, an indication of the difference in energy levels between one material and another.

Temperature Units

Celsius Degree (also Centigrade): A Celsius (or Centigrade) degree (°C) is 1/100 the difference between the temperature of melting ice and boiling water at 1 atm pressure. On the Celsius scale, zero is the melting point of ice; 100 is the boiling point of water. This unit is approved by the International Systems of Units (SI).

Fahrenheit Degree: A Fahrenheit degree (°F) is 1/180 the difference between the temperature of melting ice and boiling water at 1 atm pressure. On the Fahrenheit scale, 32 is the melting point of ice; 212 is the boiling point of water.

Kelvin Degree: A Kelvin degree (K) is the same as the Celsius degree. On the Kelvin scale (sometimes called Celsius Absolute), zero is minus 273.15°C. The Kelvin degree is also an approved SI unit.

Rankine Degree: A Rankine degree (R) is the same as the Fahrenheit degree. On the Rankine scale (sometimes called Fahrenheit Absolute), zero is minus 459.67°F.

Zero K and zero R are the same temperature: absolute zero. At absolute zero, molecular vibration is considered to have ceased.

Heat Units

Joule (J): The amount of heat energy provided by one watt flowing for 1 second is called a joule. This is an approved SI unit.

Calorie (cal. or c.): The amount of heat required to raise the temperature of 1 gram of water 1°C (measured at 15°C) is called a calorie. One calorie equals 4.183 joules.

British Thermal Unit (Btu): The amount of heat required to raise the temperature of 1 lb of water 1°F (measured at 60°F) is called the British thermal unit. One Btu equals 1,054 joules (252 calories).

To illustrate the difference between heat and temperature, consider two blocks of ice at 32°F; one weighs 5 lbs, the other 15 lbs. To convert the smaller block of ice completely to steam (at 212°F) requires 900 Btu of energy:

$$5 \text{ lbs} \times 180°\text{F} \times 1 \ \frac{\text{BTU}}{16 - °\text{F}}$$

To convert the larger block completely to steam requires 2,700 Btu of energy. The heat required differs by a factor of 3 because the initial weights also differed by a factor of 3, yet the temperatures are the same for both blocks of ice and for both clouds of steam.

Heat Transfer

When heat causes a material to approach its ignition temperature, preventive cooling measures are necessary. The heat must be transferred to another material. For example, when water is poured on a material that is almost at its ignition temperature, the heat contained by the combustible material is transferred to the water, thus lowering the temperature of the endangered material. The transfer of heat is an important and, in many cases, the determining factor in the ignition and extinguishment of most fires. Heat can be transferred by one or more of the following three methods: (1) conduction, (2) radiation, or (3) convection.

Conduction: Heat may be transferred from one material to another by means of physical contact between the two. The ease with which heat flows from the hotter to the cooler material depends on the latter's thermal conductivity, i.e., its ability to conduct heat. For example, wood, when in direct contact with a steam pipe, eventually will char and burn. But the ignition time depends on the degree and constancy of the heat in the pipe, and it may be as long as weeks or months before burning occurs.

Thermal conductivities vary widely. Metals conduct heat very readily, which explains why they normally feel cool to the touch, even though they may be at room temperature. By contrast, a block of wood at the same temperature will feel much warmer, due to its much lower thermal conductivity. Gases such as air conduct heat very slowly. As a result, an air space between a hot object and a nearby combustible object will greatly reduce the chances of ignition.

Insulating materials cannot stop the flow of heat entirely. Their insulating ability depends to a great extent on air trapped within the interstices of the material. Fiberglass building insulation is a good example.

Radiation: Heat is energy and, as such, has the ability to travel in a straight line through space until it is stopped by some object. In so doing, the heat radiated by a fire can, if intense enough, ignite combustibles some distance away. This phenomenon is well known to fire fighters; that is why one of their first tasks at a fire is to protect exposures.

The quality and quantity of heat radiation depend on the temperature of the radiating body and the size of the radiating surface. When two bodies face each other and one body is hotter than the other, radiant energy travels from the hotter body to the cooler body until both attain the same temperature. The ability to absorb the radiated heat depends on the kind of surface of the cooler body and the area of the radiating surface of the hotter body. If the receiving surface is black or dark in color, it will absorb heat readily. If the surface is light in color, or shiny and polished, it will reflect much of the heat. This property of reflection is utilized in the design of the clothing worn by crews on crash trucks fighting aircraft fires; the shiny metal-coated fabrics help reflect the heat of the fire away from the fire fighters.

Radiant heat travels in a straight line, which is why the extent of the radiating surface of the hot body is important. Heat radiation will pass through air, glass, water, and transparent plastics, with merely a small absorption of energy that only slightly heats the transparent substance. Radiated heat can be reflected and can be concentrated in much the same way that light rays are reflected and concentrated, such as by a mirror or through a magnifying glass.

When large amounts of radiant heat are produced in large fires, the fire will spread rapidly to nearby combustible materials unless water in the form of hose streams conducts away the heat. The development of fog nozzles for fire fighting is most useful in blocking heat radiation. The water fog reflects the heat rays and breaks up the straight-line path of the heat radiation.

Convection: Heat energy also can be transferred by circulation of a fluid, such as air or a liquid. Hot gases, vapors, and liquids rise and progressively increase the temperature of the area and of the material in the area. Most homes are heated by a furnace or heater supplying warm air near the floor level. This air circulates upward by convection and mixes with the cooler air that travels downward. Likewise, in a large fire the hot burning gases from the fire flow upward and transfer heat energy up through the building.

Sources of Ignition (Sources of Heat Energy)

As previously mentioned, a source of ignition (heat) is needed to begin the combustion process. Heat energy may be produced from many sources and under many circumstances, as described by Dr. Tuve:[1]

> Unwanted or accidental fires are caused by many circumstances, any one of which might be singled out as the source from which runaway combustion begins. It is lengthy and difficult to catalog all such possible situations; such cataloging can usually be found in

studies of the science and art of arson investigation. However, it is important that certain basic categories of energy sources should be recognized as primary origins by which heat is developed—heat which, if it is not controlled in some way, can result in fires of a disastrous nature.

There are four major sources of heat energy: (1) chemical, (2) electrical, (3) mechanical, and (4) nuclear.

Chemical: Fire is, essentially, a chemical reaction, one of the basic components being oxidation—the chemical reaction caused by oxygen. Oxidation produces heat. The oxidation process will develop heat whether the oxidation is incomplete or total.

Air is the primary source of oxygen. Oxidation usually is limited by the air supply, which normally affects the amount of heat produced. Spontaneous heating occurs when the temperature of a material increases without drawing heat from its surroundings. If exposed to air, practically all organic substances capable of combination with oxygen will oxidize at some critical temperature, with a resultant evolution of heat. The rate of oxidation at normal temperatures usually is so slow that the released heat is transferred to surroundings as rapidly as it is formed. The result is that there is no temperature increase in the combustible material being oxidized. However, when the heat developed cannot be rapidly released, spontaneous ignition may occur. Common examples of spontaneous heating situations are oily rags in a confined space, and wet hay in a barn loft.

Electrical: Another common source of ignition is the heat produced by electrical energy. The energy required to move electric current through a wire or piece of electrical equipment appears in the form of heat, due to the resistance of the conductors. If the conductor is a good material, such as copper or silver, the resistance is low and little heat will be produced.

Dr. Tuve describes the effect of electrical energy on everyday lives, and the relationship between electrical energy and fire, in the following manner:[1]

> Because it is used in many ways, electrical energy is a common cause of unwanted fires. In home or industrial oil burners, electrical sparks are used to start and to sustain the ignition of the heating oil. In the kitchen electrical resistance heating elements or other forms of electric heating energy are needed to supply cooking temperatures. In certain living or working areas, electric energy is used to provide comfortable temperatures by means of resistance heaters. In our various appliances electricity is used in many ways, each of which is a potential fire hazard to nearby combustible materials. Unless electrical energy is used in an efficient manner, it almost always will produce heat as an unwanted by-product. Another factor that must continually be remembered is that the fire hazards of electrical energy almost always are concealed. In efforts to beautify

and hide electrical wires and equipment, they are often installed in walls and partitions or enclosures so that malfunctions or dangerous conditions that can initiate fires are difficult to detect. In addition, rather than installing additional wall outlets when and where they are needed, the tendency is to overload existing ones.

Five forms of heating due to electrical energy can result in a fire. These forms are: (1) resistance, (2) arcing, (3) sparking, (4) static electric charge, and (5) lightning.

Resistance: Overloading electrical circuits can result in resistance heating of electrical wiring as the wiring attempts to carry a heavier current than intended. This is a common cause of fire in the home. Portable electric room heaters produce substantial heat and will ignite combustible materials placed too close to them. Most electrical circuit wires are concealed in walls and partitions, which increases the chances of undetected ignition. Bare wires can carry more current than insulated wires, and the heat generated by bare wires is dissipated more readily. However, most wires are insulated; if a breakdown occurs, the insulation will contain the heat, but the insulation itself may then burn.

Arcing: Ignition also can be affected by arcing. When an electrical connection at a switch or fuse block or in an appliance is not good, the electrical energy may jump or arc across the gap. The spark can be hot enough to ignite insulation or nearby combustible material.

Sparking: Sparking can occur only once. This is in contrast to arcing, which can be continual or intermittent. Sparking may take place when a voltage discharge is too high for a low energy output.

Heating through sparking does not create a large or sustained heating effect. Danger can result, however, when sparking occurs in a flammable atmosphere.

Static Electric Charge: Static sparks can occur when an electrical charge accumulates on the surfaces of two materials that have been brought together and then separated. One surface becomes charged positively, the other surface becomes charged negatively. If the materials are not bonded or grounded, they eventually will accumulate sufficient electrical charge to enable a static electric spark to occur.

Static sparks do not produce enough heat to ignite ordinary combustibles such as paper, but they can ignite flammable vapors, gases, and dusts. Gasoline or oil flowing through a pipe can generate enough static electricity to ignite flammable vapor.

Lightning: A powerful, but infrequent, source of electrical energy is lightning. A properly installed metallic rod that will conduct the lightning into the ground is the safeguard for preventing fires caused by lightning.

Mechanical: Mechanical heat energy in the form of friction is a frequent cause of fire. Friction occurs when two solids are rubbed together. The friction transforms the energy into heat; unless this heat is rapidly dispersed, fire can result. The ageless procedure for starting a fire by rubbing sticks together is an example of ignition by mechanical heating. Friction is the cause of many fires. For example, in a cotton mill the friction heat of a slipping belt against a pully, or the friction sparks generated when a piece of metal gets in the cotton shredder, will readily ignite the cotton fibers. Sparks can be generated from the impact of shoe nails on a concrete floor, or from the contact of a steel tool with a concrete floor. The size of the particle involved has a pronounced effect on spark ignition. Small particles of metal can develop sparks capable of igniting dust or explosive materials. However, larger pieces of metal usually will not develop enough friction heat to reach dangerous temperatures.

Mechanical heat energy also can develop when a gas is compressed. For example, in diesel engines the heat of compression of the fuel-air mixture in the cylinder eliminates the need for a spark ignition system. First, air is compressed in the cylinder of the diesel engine, after which an oil spray is injected into the compressed air. The heat released when the air is compressed is sufficient to ignite the oil spray.

Nuclear: Nuclear energy is heat energy released from the nucleus of an atom through nuclear fission. Nuclear material, such as uranium or plutonium, is composed of atoms held together by tremendous forces. These forces can be released when the nucleus is bombarded by energized particles. This energy is released in the form of heat, pressure, and/or nuclear radiation. Nuclear energy may be a million times greater than the energy released by an ordinary chemical reaction. Controlled nuclear energy is used to generate electric power.

EXPLOSIONS

The basic difference between an explosion and a fire is the rate at which energy is released. In its broadest sense, an explosion can be defined as the result of a sudden and violent expansion of gases. These gases might already exist, or may be formed at the time of the explosion. The rapid expansion of energy in an explosion also may be accompanied by shock waves and/or the disruption of enclosing materials or structures. Explosions can be chemical, mechanical, atomic, or thermal.

Chemical Explosions

Chemical explosions are very rapid combustion reactions which are classified as detonations or deflagrations, depending on the rate of propagation of the flame front through the fuel-air mixture. Detonations propagate at the speed of sound (based on the temperature and pressure of the unburned mixture), producing a characteristic shock wave. Deflagrations propagate at less than the speed of sound, but nonetheless produce a sizable pressure effect. An example of a detonation is the explosion of a blasting agent or dynamite. Deflagrations most often occur in clouds of combustible gas or combustible dust.

In deflagrations of a combustible gas or vapor, the overpressure produced can vary greatly, depending on the amount of fuel, how well the fuel is mixed with air, and, mostly, on how much turbulence is present. Very turbulent gas-air mixtures can transit to detonation.

When combustible solids are pulverized or otherwise reduced to a fine particle size, then mixed with air, conditions are ripe for a dust explosion. Dust explosions can be just as devastating as vapor cloud explosions. The finer the dust particles, the more completely they will mix with the air and remain suspended there. A large variety of combustible solids are capable of yielding explosive dust particles. Table 3.1 lists some common substances with potential explosive properties.

The following excerpt from Dr. Tuve's book contains a further warning concerning explosive dust particles:[1]

> Obviously, dust explosions take place only when a source of ignition is present in, or at the edge of, a mixture of combustible dust and air. Open flames, electric arcs, suddenly broken electric light bulbs, friction sparks, and even static sparks can initiate dust explosions.

Table 3.1*   Common Combustible Solid Dusts Generating Severe Explosions**

Type of Dust	Maximum Explosion Pressure (psig)	(bar)	Maximum Rate of Pressure Rise (psig/sec)	(bar/sec)
Corn (processing)	95	(6.55)	6,000	(413.7)
Cornstarch	115	(7.93)	9,000	(620.5)
Potato starch	97	(6.89)	8,000	(551.6)
Sugar (processing)	91	(6.27)	5,000	(344.7)
Wheat starch	105	(7.24)	8,500	(586.0)
Ethyl cellulose plastic molding compound	102	(7.03)	6,000	(413.7)
Wood flour filler	110	(7.58)	5,500	(379.2)
Natural resin	87	(6.0)	10,000	(689.5)
Aluminum (powder)	100	(6.9)	10,000	(689.5)
Magnesium (powder)	94	(6.48)	10,000	(689.5)
Silicon (powder)	106	(7.31)	10,000	(689.5)
Titanium (powder)	80	(5.52)	10,000	(689.5)
Aluminum magnesium alloy (powder)	90	(6.20)	10,000	(689.5)

*From *Principles of Fire Protection Chemistry*, by Richard L. Tuve.[1]
**Extracted from Bureau of Mines Investigations and Reports, Nos. *RI 5753, RI 5971, RI 6516.*[3,4,5]

Mechanical Explosions

Mechanical or physical explosions, such as the bursting of a boiler, account for many explosions and resultant loss of life and property. Because of their large size, when boilers used in industry explode the results can be particularly devastating. Trapped-steam explosions, such as those that occur at foundries, and ruptures of pressurized equipment, such as pressurized tanks and processing machinery, are other fairly common sources of mechanical explosions. Adequate precautions and proper maintenance, as well as installation of effective pressure-relief devices, can reduce the danger of loss from explosions of this type. One type of mechanical explosion is a boiling liquid expanding vapor explosion (BLEVE), as described in Chapter 4.

Atomic Explosions

A nuclear or atomic explosion is the result of the redistribution of protons and neutrons within the interacting nuclei. Atomic explosions are produced by two processes: fission and fusion. Fission involves the use of uranium 235 and plutonium 239, while fusion uses deuterium. The resulting energy yield generally is divided as follows: blast and shock, 50 percent; thermal energy, 35 percent; and nuclear radiation, 15 percent.

Thermal Explosions

A thermal explosion occurs when an unstable material decomposes, with the resulting rapid production of gases and heat. Once decomposition begins, the heat generated by the reaction might not readily escape, thus increasing the temperature. As the decomposition accelerates, an explosion occurs. One factor that greatly affects the thermal explosion potential of a material is the shape of the vessel or container in which the material is confined or stored.

PRODUCTS OF COMBUSTION

There are four major products of combustion: (1) fire gases, (2) flame, (3) heat, and (4) visible smoke. All of these are produced in varying degrees by fire. However, the material or materials involved in the fire and the resulting chemical reactions produced by the fire create many variables that must be considered in fire protection.

Fire Gases

Most people think a fire death or injury results from contact with flame or heat. Actually, the primary cause of loss of life in fires is the inhalation of heated, toxic, and oxygen-deficient gases and smoke. The amount and kind of fire gases present during and after a fire vary widely with the chemical composition of the material burning, the amount of available oxygen, and the temperature. The effect of toxic gas and smoke on a person will depend on the length of time of exposure, the concentration of the gases in air, and also—to a large degree—on the physical condition of the individual.

Usually, several gases are present during a fire. Those most lethal are carbon monoxide, carbon dioxide, hydrogen cyanide, hydrogen chloride, acrolein, and insufficient oxygen.

Carbon Monoxide: Although carbon monoxide is not the most toxic of fire gases, it is the one produced in greatest quantity. If a fire burns with a good air supply, the carbon in most organic combustible materials is converted to carbon dioxide. However, most fires develop under conditions where the supply of air is not sufficient for complete combustion, so carbon monoxide is produced.

In a confined smoldering fire, the ratio of carbon monoxide (CO) to carbon dioxide (CO_2) is usually greater than in a well-ventilated, free-burning fire.

The toxicity of CO is primarily due to its affinity for the hemoglobin in blood. The carbon monoxide content of blood can readily be measured and is expressed as percent carboxyhemoglobin (COHb) saturation. Even partial conversion of hemoglobin to (OHb) results in a decreased supply of oxygen to body tissues (hypoxia).

There exists no minimum blood COHb saturation associated with death, below which it can be certain that a victim died from other causes or toxicants. The actual blood COHb saturation levels associated with both incapacitation and death vary quite widely over the general population and depend upon many factors. With some persons having preexisting functional impairments, even quite low COHb saturations are likely to be dangerous. The very young, the elderly, the physically disabled, those under the influence of alcohol, drugs, or medication, and those with heart diseases are particularly susceptible.

It is felt that any COHb saturation above approximately 30 percent would be potentially hazardous to most humans. A saturation of about 50 percent is likely to be lethal to many individuals.

In terms of CO concentrations required to reach hazardous COHb levels, a simple rule of thumb may be used. Any exposure in which the product of concentration (ppm) × time (minutes) exceeds approximately 35,000 ppm-min is likely to be dangerous. Thus, a 10-minute exposure to 3,500 ppm of carbon monoxide would be expected to be hazardous and possibly incapacitating to many people. The concentration × time rule of thumb must be applied with caution at high concentrations, since progressively lower doses can be tolerated as the concentration is increased. It is, however, reasonably applicable for the range of CO concentrations normally generated in fires.

Carbon Dioxide: Carbon dioxide is not toxic in the same way as carbon monoxide. However, large quantities of carbon dioxide usually are produced in a fire. Inhaling above-average amounts of this gas increases the speed and depth of breathing. Carbon dioxide at 2 percent concentration in air can increase breathing by about 50 percent. If the concentration of this gas nears 10 percent, carbon dioxide can cause death in a few minutes. Since high concentrations of carbon dioxide increase the breathing rate, danger to life is further heightened because other toxic gases developed by the fire might be inhaled more readily.

Hydrogen Cyanide: Hydrogen cyanide (HCN) is produced from the burning of materials which contain nitrogen (N). Natural and synthetic materials, such as wool, silk, acrylonitrile polymers, nylons, polyurethanes, and urea-containing resins, are included.

Hydrogen cyanide is a very rapidly acting toxicant which is approximately 20 times more toxic than carbon monoxide. It does not combine appreciably with hemoglobin, but the toxicant inhibits the use of oxygen by cells (histotoxic hypoxia).

Data relating symptoms in humans to various concentrations of HCN are very limited. One widely used descriptive account of hydrogen cyanide intoxication of humans reports that 50 ppm may be tolerated for 30 to 60 minutes without difficulty, 100 ppm for that same period is likely to be fatal, 135 ppm may be fatal after 30 minutes, and 181 ppm may be fatal after 10 minutes. Since incapacitation normally occurs at one-third to one-half the lethal dose, the data suggest that doses for incapacitation by HCN range from approximately 2,500 ppm-min at 100 ppm to 750 ppm-min at 200 ppm. Using a rule of thumb analogous to that for carbon monoxide, it would appear that a product of hydrogen cyanide concentration (ppm) × time (minutes) in the range of about 1,500 ppm-min would likely be hazardous to humans. With hydrogen cyanide, in particular, progressively lower doses can be tolerated as the concentration is increased. Therefore, the concentration × time rule of thumb must be applied with caution at high concentrations. It is, however, reasonably applicable for concentrations of hydrogen cyanide generally found in fire atmospheres.

Hydrogen Chloride: Hydrogen chloride is formed from the combustion of chlorine-containing materials, the most notable of which is polyvinyl chloride (PVC). It is both a potent sensory irritant and also a strong pulmonary irritant. Concentrations as low as 75 ppm are extremely irritating to the eyes and the upper respiratory tract, and impairment from a behavorial standpoint has also been suggested. However, hydrogen chloride gas has been found not to be physically incapacitating to nonhuman primates subjected to concentrations up to 17,000 ppm for 5 minutes. The toxicant was reported to cause post-exposure death at doses which did not appear to incapacitate. Comparable studies have not been conducted using actual PVC smoke, however, and it is claimed that other irritants also may be present from PVC in a real fire atmosphere. Furthermore, the question remains as to the extent of respiratory dysfunction and susceptibility to infection caused by exposure to hydrogen chloride and PVC smoke.

Acrolein: Acrolein is a particularly potent irritant, both sensory and pulmonary, which has been demonstrated to be present in many fire atmospheres. It is formed from the smoldering of all cellulosic materials and also from pyrolysis of polyethylene. Acrolein is extremely irritating, with concentrations as low as a few parts per million irritating to the eyes and possibly psychologically incapacitating. Surprisingly, studies with nonhuman primates have shown that concentrations up to 2,780 ppm for 5 minutes were not physically incapacitating during exposure. However, pulmonary complications caused by even lower concentrations did result in death within hours after the exposure.

Oxygen Deficiency: Another life-threatening effect of the combustion process is a drop in the oxygen level. The usual concentration of oxygen in air is about 21 percent. If the level drops to 17 percent, anoxia (diminished muscular control) develops. If the oxygen drops lower (10 to 14 percent), a person can remain conscious but judgment will be impaired and the person will tire easily. At an oxygen content range of 6 to 10 percent, a person will collapse and must be revived with fresh air or oxygen within a few minutes to prevent death.

Flame

The burning of combustible materials in air almost always is accompanied by visible flame. Serious burns can be caused by direct contact with flames, as well as by direct radiation of heat from flames.

Heat

In *Principles of Fire Protection Chemistry*, the characteristic of heat as a product of fire is explained by Dr. Tuve:[1]

> One of the basic characteristics of a fire is the emission of heat. All combustion reactions, or oxidation reactions, are exothermic, meaning they give off heat. The rate and the extent to which this heat is given off is highly variable and depends upon many factors. In the case of most fires, it is difficult—if not impossible—to identify the fire's rate and extent. Because of this, these factors can only be determined in a general way.

Heat from a fire will affect people exposed to it in proportion to the length of exposure and the temperature of the heat. The dangers of exposure to heat from fire range from minor injury to death. Exposure to heated air increases the heart rate and causes dehydration, exhaustion, blockage of the respiratory tract, and burns. Fire fighters should not enter atmospheres exceeding 120°F to 130°F (48.8°C to 54.4°C) without special protective clothing and self-contained breathing apparatus. The maximum survivable breathing level of heat from fire in a dry atmosphere for a short period has been estimated at 300°F (148.8°C). Any moisture in the air greatly increases the danger and sharply reduces the time of survival.

Burns: Burns usually are classified as first, second, or third degree. First-degree burns affect the outer layer of skin. They are painful but not as serious as second- or third-degree burns. Second-degree burns penetrate more deeply into the skin. They form blisters, and usually permit a considerable amount of fluid to accumulate under the skin. Third-degree burns, which penetrate still farther, are the most serious. However, they are not initially as painful as first- and second-degree burns, because the nerve endings have been made inactive.

Any doctor specializing in the treatment of burns will testify that serious burn injuries are among the most painful, among the most long-lasting and difficult to treat, and among the most costly for the patient.

People exposed to excessive heat may die if the heat is conducted into the lungs rapidly enough. This conduction can occur without visible signs of burning. Blood pressure will decline; circulation of the blood will fail; and the body temperature may rise sufficiently to damage the nerve centers of the brain.

Smoke: Smoke is most often defined as the airborne solid and liquid particulates and fire gases evolved when a material undergoes pyrolysis or combustion. Under the usual conditions of insufficient oxygen for complete combustion, wood, petroleum oil, paper, and other common combustibles give off particles of carbon that are visible as smoke.

Smoke, including the invisible poisonous gases it contains, is the major killer in fires, responsible for 50 to 75 percent of fire deaths. Smoke often creates hazardous conditions for life safety before temperatures in a fire building reach dangerous levels.

Smoke irritates the eyes and the lungs, and occasionally causes panic. Fire gases such as methane, formaldehyde, and acetic acid can be created under conditions of incomplete combustion, condensing on the smoke particles and being carried into the lungs with lethal results.

FIRE AND EXPLOSION CONTROL

The combustion process occurs in two modes: (1) the flaming mode, and (2) the flameless surface mode. The flaming mode, which includes explosions, is characterized by relatively high burning rates. This results in intense temperatures and high rates of heat release. On the other hand, an example of the flameless surface mode is the presence of glowing embers.

The flaming mode is represented graphically by the tetrahedron described earlier in this chapter. The four components of a tetrahedron as previously stated are heat, fuel, oxygen, and the chain reaction. By contrast, the flameless combustion mode can be represented graphically by the traditional fire triangle containing the three elements fuel, heat, and oxygen.

The flaming and flameless modes are not mutually exclusive; combustion may involve one or both modes. Often combustion occurs in the flaming mode and gradually makes a transition to the flameless mode. At one point in this process, both modes are in effect simultaneously.

However these modes occur, flaming modes require four considerations for extinguishment, and flameless modes require three. Four means of fire and explosion control will be described in the following subsections of this chapter. These four means are: (1) water (to reduce the heat), (2) removal of oxygen, (3) removal of fuel, and (4) flame inhibition by chemical means (to remove the chain reaction).

Use of Water

For the majority of common combustibles, such as wood, paper, and cloth, the simplest and most effective means of removing the heat of a fire is through the application of water. The water application can be varied and will depend on the type of fire. Water spray usually is the most effective application, but for a large fire a solid or straight stream of water will provide longer range and more powerful drenching action.

Applying water to the burning fuel cools the fuel to the point where the rate of release of combustible vapors and gases is reduced and ultimately stopped.

Heat developed by a fire tends to be carried away by radiation, conduction, and convection. This fact helps reduce the amount of heat and increases the water's effectiveness. Only a relatively small proportion of the heat evolved needs to be absorbed by the water in order to extinguish the fire.

Effective use of water spray or solid streams cannot be accomplished if the water cannot reach the burning fuel directly. For this reason, areas where fire fighters cannot readily reach the fire with water streams—such as in high-rise buildings and high-piled storage areas—must be provided with automatic sprinklers or other automatic fire protection systems.

When a water spray is used on a fire, the water may turn to steam. At one time it was thought that this steam was helpful in controlling a fire. However, steam diffuses rapidly, and while it helps dilute the oxygen concentration in the immediate fire area to some degree, the presence of steam is not a major factor in extinguishing a fire.

Some burning materials react violently with water. They can best be extinguished by smothering them with a suitable inert material.

Removal of Oxygen

The amount of dilution of oxygen necessary to stop the combustion varies greatly with the material that is burning. Ordinary hydrocarbon gases and vapors will not burn when the oxygen level is below 15 percent. The oxygen concentration in the atmosphere normally is about 21 percent. Acetylene will continue to burn unless the oxygen concentration is lowered, but will continue to glow on the surface even if the oxygen level is as low as 4 to 5 percent.

A fire in a closed space can extinguish itself by consuming the oxygen. However, incomplete combustion, which takes place when the oxygen is consumed, usually results in considerable generation of flammable gases. Fire fighters should use great caution and guard against violent flashbacks or explosions if such a space is improperly ventilated and a supply of oxygen suddenly rushes in.

A commonly used method of putting out a fire by removing or diluting the oxygen is by flooding the entire fire area with carbon dioxide or some other inert gas.

Fuel Removal

Fuel removal can be accomplished in a variety of ways. One of the most common examples is the practice of bulldozing a firebreak across the path of an advancing forest fire; the trees and brush are removed, and the fire runs out of fuel.

Fires in large piles of coal or wood pulp usually can be controlled only by moving the piles out of the fire zone. Fires in large oil storage tanks have been controlled by pumping the oil out of the burning tanks into empty ones. If a gas line is ruptured and the gas ignited, shutting off the supply of gas is usually the only way to stop the fire.

If it is not practicable to remove the fuel, extinguishment can sometimes be accomplished by shutting off the fuel vapors or by covering the burning or glowing fuel. Fire fighting foams and dry powder extinguishers are examples of materials used with good results to cover or coat a fire.

Forest fire fighters have been effective in using gelling agents in the water to coat burning wood and vegetable materials and also to retard water runoff.

Flame Inhibition

Extinguishment of fires by chemical inhibition of flame is the subject of much continuing research. Certain extinguishing agents—such as the gaseous and liquid halogenated hydrocarbons (halons), some of the alkali metal salts (dry chemicals), and ammonium salts—will extinguish flames with efficiency. Thus, flame extinguishment can occur without the accompanying action of the other traditional methods such as cooling by water, oxygen dilution, or fuel removal.

When the proper amounts of these extinguishing agents are injected into flames, the agents act as a negative catalyst, causing the flames to become "inhibited," and the fire is extinguished. This fourth method of extinguishment is used for the flaming mode only. The object of this type of extinguishment is to inhibit and halt the unrestrained combustion chain reaction.

Summary

Although the essential components of a fire are known, and extensive laboratory and field research has been conducted regarding fire and its causes and behavior, fire remains unpredictable.

Combustion, commonly referred to as fire, consists of four components: fuel, heat, oxygen, and chain reaction. These fires occur in the flaming mode. Some fires, however, consist only of fuel, heat, and oxygen and are considered to be in the flameless mode.

The heat of fire, its measurement, and its ability to transfer to other combustible materials are vital factors in fire control. Four primary sources of ignition that produce enough heat to generate a fire are chemical, electrical, mechanical, and nuclear. The basic difference between fires and explosions is the rate at which energy is released. Explosions can be caused by chemical, mechanical, atomic, or thermal reactions.

Contrary to popular opinion, the greatest life safety hazard from fire is neither flame nor heat, but the inhalation of heated, toxic, and oxygen-deficient gases and smoke. The gases formed in any fire depend on many variables, including the chemical composition of the material(s) burning, the amount of oxygen available for combustion, and the temperature of the heat generated. Fire fighters must be aware of the common fire gases and their lethal properties so they will maintain a healthy respect for any potential hazard based on the fire gases they encounter.

Fire and explosion control is based on an understanding of the basic components of fire—whether the fire is in the flaming or flameless mode—and of the appropriate means for reducing or eliminating the fire situation.

References

[1]Tuve, Richard L. 1976. *Principles of Fire Protection Chemistry*, National Fire Protection Association, Boston.

[2]*Fire Protection Handbook*, 16th ed. 1986. National Fire Protection Association, Quincy, MA.

[3]"The Explosibility of Agricultural Dusts," *RI 5753*. U.S. Department of Interior, Bureau of Mines, Washington, DC.

[4]"Explosibility of Dusts Used in the Plastics Industry," *RI 5971*. U.S. Department of Interior, Bureau of Mines, Washington, DC.

[5]"Explosibility of Metal Powders," *RI 6516*. U.S. Department of Interior, Bureau of Mines, Washington, DC.

Additional Reading

Clark, Frederic B. 1983. "Toxicity of Combustion Products: Current Knowledge," *Fire Journal*, Vol. 77, No. 5, September, National Fire Protection Association, Quincy, MA.

Drysdale, Dougal 1985. *An Introduction to Fire Dynamics*, John Wiley & Sons, Ltd.

4

Chapter

Fire Hazards of Materials

The world contains a great variety of highly flammable materials, both natural and synthetic, that can present a threat to both life and property. A wealth of knowledge exists concerning these substances, their physical and chemical properties, suitable methods for their handling and storage, and the proper steps to take when a fire involving them occurs.

INTRODUCTION

The two basic components of which everything in the physical world consists are matter and energy. Energy is the force that brings about both chemical and physical changes; it is the result of the interaction of matter under varying circumstances. Energy is "work," "performance," and is most readily perceived by our senses in terms of what it "does." Matter, the other basic component of our physical world, has two fundamental characteristics: (1) it has mass, and (2) it occupies space. Matter can exist in the form of a gas, a liquid, or a solid. For example, water normally is a liquid. When it is boiled, it becomes steam—a vapor or gas. And when it is frozen, it becomes a solid. It is "matter" in all its forms, with which this chapter is concerned.

Chapter 3 introduced the basic characteristics of fire and outlined fire's four components: fuel, heat, oxygen, and the chain reaction. "Energy" is represented by the heat necessary for ignition. "Matter" is represented by the fuel and the oxygen. The fuel may exist in one form or in any combination of the three forms of matter.

COMBUSTIBLE SOLIDS

Most people, when asked to name things that burn, generally name a solid form of matter, such as wood. Wood, or products of wood such as paper and cardboard, is in fact the most common form of combustible solid. Plastics and polymers, and textiles are other solids that require major consideration from a fire protection point of view.

Wood

As already noted in this text, fires in one- and two-family dwellings cause the highest percentage of death and injury from fire in the United States.

Wood, one of the most commonly used construction materials, generally is the predominant material used to build one- and two-family homes as well as many other types of occupancies. The wood used for framing, sheathing, flooring, and interior finish, as well as the wood products such as furniture which are installed once the house is built, provides a ready source of fuel in case of fire. When in contact with sufficient heat, all wood and wood-based products ignite sooner or later, the interval until ignition depending on the ignition source and the length of exposure.

Wood and wood-based products can be treated with fire-retardant chemicals. When so treated, the ease of ignition of these products is greatly reduced. Their combustibility also can be reduced when they are used in combination with other materials.

Physical Properties: The physical form of the wood is important from the standpoints of ease of ignition and rate of burning. A log or beam is hard to ignite; a pile of wood shavings will ignite easily and burn rapidly; and wood dust can explode violently when ignited by just a small spark. The shavings and dust expose the wood particles to more air, and provide less mass to conduct heat away from the wood particles. Wood is a poor conductor of heat, which will not readily penetrate wood when it is in log or beam form. Also, the char that develops on the surface of wood when attacked by fire provides an insulating effect. In his book *Principles of Fire Protection Chemistry*, Dr. Tuve explains the importance of this property of wood with regard to fire protection:[1]

> The low heat conductivity of wood is another important physical property of this combustible material, and is of considerable importance in some fire situations. If one considers the situation of a moderate heat or flame exposure of considerable thicknesses of wood, charring immediately takes place from the exposed surface inward. The char formation possesses an even lower heat conductivity than does wood, and the rate of heat penetration into the wood is progressively lowered by this mechanism. In the case of heavy timber construction or where large, laminated timber trusses are employed, the structural integrity of a building may be preserved for longer lengths of time under fire exposure than if the building had been constructed with steel framing of a type for similar stresses. The higher heat conductivity of the metal would be more quickly vulnerable to strength loss due to rapid penetration by heat attack.

Table 4.1 shows how the thermal conductivity of various materials can affect a fire situation.

Moisture Content: Wood is composed primarily of carbon, hydrogen, and oxygen. Live wood cells retain considerable moisture. When the wood is dead, air replaces most of the water in the cellular structure of the wood.

Table 4.1 Thermal Conductivity of Materials*

Material	Rate of Heat Conductivity (k) (in calories per sec per cm per °C) (*Note:* Multiply all values by 1/1,000)
Aluminum (metal)	500.0
Brick (common)	1.7
Charcoal	0.21
Concrete	4.1
Copper (metal)	910.0
Corkboard	0.1
Fiberboard ("celotex" type)	0.14
Glass	2.3
Iron (metal)	150.0
Marble	6.2
Mineral wool (blanket)	0.1
Paper	0.3
Plaster	1.7
Plastics (solid)	0.45
Vermiculite	0.14
Wood (oak)	0.41
Wood (white pine)	0.29
Wool (loose clothing)	0.8

*From *Principles of Fire Protection Chemistry,* by Richard L. Tuve.[1]

The fire behavior of wood and other combustible solids of the same size and shape will vary greatly with the moisture content. Obviously, wet wood is harder to ignite and will not burn as fast as dry wood. Thus, burning rate is influenced by moisture content.

The effects of moisture content can best be illustrated by the action of forest fires. When the forests are wet, the water content in the trees and forest vegetation absorbs much of the heat from the sun or other sources. The water evaporates as a result of absorbing the heat. Because water has a high specific heat, a great deal of heat is required to evaporate the moisture from wood. The large quantities of water vapor dilute the oxygen concentration in the air surrounding the combustible wood and forest duff (the partly decayed organic matter on the forest floor). This process affects both the development of combustible vapors and their ignition or continued burning. If the forest material is exposed to a long, hot dry spell, and if high winds develop to further dissipate the remaining moisture, the forest vegetation dries out and dies. At this point, forest materials can be ignited easily, and the ensuing flames can race through the forest area at incredible speed.

In a forest area, one of the major factors in determining whether "fire weather" is present or can be expected is the moisture content of the forest materials. Other important factors are atmospheric humidity, wind, the condition of the vegetation, and the season of the year.

Even when exposed to a relatively high temperature for a prolonged period of time, ignition generally is difficult when the moisture content of wood (and similar fuels) is above 15 percent. For example, the conditions of high humidity often present in the summer make wood less susceptible to ignition than do conditions in cold climates, where heated buildings cause indoor relative humidities to fall below the 15 percent level. Once ignition and resultant fire have begun, however, heat radiation and the rate of pyrolysis reduce the importance of the moisture factor; under some conditions, wood with a moisture content of 50 percent or more will burn.

Ignition of Wood: Wood can be ignited only if it is heated to the point where combustible gases are released from its surface. At normal temperatures and pressures, wood and similar combustible solids—unlike combustible liquids and gases—do not give off flammable vapors. Wood must be exposed to a heat source for a certain period of time before it begins to give off flammable vapors. Wood may not ignite if a blow torch is held momentarily to its surface, even though the temperature of the blow torch flame is much higher than the wood's ignition temperature. On the other hand, prolonged contact of wood with a steam pipe having a temperature much lower than the blow torch flame can well result in ignition of the wood.

The temperature at which wood will ignite varies widely, depending on the wood's species, form, size, and moisture content. Low-density softwoods, such as pine or spruce, will ignite at lower temperatures than high-density hardwoods, such as oak or maple. The rate of heating, the length of time that heat is applied, and the nature of the heat source attacking the wood also influence the ignition point. In addition, the amount of oxygen available will affect the ignition temperature.

Obviously, then, it is difficult to identify the specific ignition temperature of wood because of the large number of variables involved. Test results vary greatly. Generally, the average ignition temperature of wood is considered to be about 392°F (200°C). At this temperature, combustible vapors are produced in sufficient quantity to be ignited. As the burning progresses and the temperature increases, carbon monoxide begins to be emitted. Finally, the burning wood becomes charcoal and ash. The four stages of decomposition of wood and the corresponding temperatures that cause each stage are described in Table 4.2.[2]

Wood, however, can ignite at temperatures lower than 392°F (200°C). For example, if wood is in constant contact with a temperature source such as steam pipes over a long period of time, the wood undergoes chemical changes. These changes result in formation of charcoal, which can heat spontaneously. Research indicates that 212°F (100°C)—the boiling point of water—is the highest temperature to which wood can be exposed continually without the danger of ignition. Dr. Tuve explains this as follows:[1]

> Concerning the spontaneous ignition temperature of wood, the influence of the variables involved is even more indeterminate than

Table 4.2   The Four Stages of the Thermal Degradation of Wood[2]

Temperature	Reaction
392°F (200°C)	Production of water vapor, carbon dioxide, formic and acetic acids—all noncombustible gases.
392°F—536°F (200°C—280°C)	Less water vapor, some carbon monoxide—still primarily endothermic reaction.
536°F—932°F (280°C—500°C)	Exothermic reaction with flammable vapors and particulates. Some secondary reaction from charcoal formed.
Over 932°F (500°C)	Residue principally charcoal with notable catalytic action.

where an igniting flame for the evolved gases is present. Considering the fact that wood slowly changes its chemical composition under sustained heat attack lesser in amount than is necessary to cause ignition of the combustible gases evolved, the end point of such an exposure over long periods of time is the formation of extremely porous charcoal. Charcoal is a material that is capable of absorbing gases and vapors. If these gaseous materials are combustible and are capable of further slow breakdown with the evolution of heat within the charcoal, a temperature may be reached that is capable of spontaneous ignition of the charcoal and of any wood in contact with it.

Flashover: If fire occurs in a compartment or room, flashover conditions may be reached. Flashover takes place when the thermal radiation from the upper area of the room heats the combustible materials in the room to the point where simultaneous ignition of all combustibles in the room occurs. In seconds, temperatures can rise as much as fivefold, oxygen is greatly reduced, carbon monoxide is generated at lethal levels, and carbon dioxide increases rapidly. Experiments with room fires repeatedly have demonstrated this phenomenon. Fire fighters therefore need to be constantly aware of the dangers presented by flashover.

Flame Spread: The way flame spreads along the surface of wood or other combustible solids also affects the severity of a fire. Flame spread—which is heat transferred by convection—is the most frequent means of transferring fire in which combustible solids are involved. Combustible wallboard and other interior finish can spread flame so rapidly that life is endangered very quickly. Flame spread progresses across combustible tile ceiling or wall covering, making it different from "flashover," in which a whole room and its contents become involved simultaneously. The flame spread characteristics of wood and other common building materials have been measured and defined by the "Tunnel Test" apparatus designed by A.J. Steiner of Underwriters Laboratories Inc. (UL). Further details of this are covered in Chapter 6, "Fire Protection Through Building Design and Construction."

Fire Loading: Considering wood and similar combustible solids, it is possible to estimate the amount of fuel that could be involved in a fire. This is known as fire loading. Wood and similar materials will generate from 7000 to 8000 Btu/lb. (1 Btu = 1054 joules.) Using this information, the severity and duration of a building fire can be estimated. This subject is presented in greater detail in Chapter 6.

Smoke: In the early stages of a fire, wood can produce large quantities of smoke. The smoke develops when there is insufficient oxygen for complete combustion, allowing the vapors and gases from the burning wood to include droplets of tars. Research is constantly being conducted to understand more fully the hazards of smoke and to determine how control of the use of various building materials can limit the amount of smoke and toxic gases produced in fires. Smoke and toxic gases continue to be serious life safety hazards.

Storage: Wood and wood by-products generally are stored outdoors. At sawmills, paper mills, pulp mills, and lumberyards, wood is stored in quantities that may vary greatly in size during the year.

Logs can be stored in rank piles, i.e., in parallel form, or kept as they were dumped in a pile called a stack pile. The rank-pile method of log storage is preferable in terms of fire protection, because these piles usually do not exceed 15 ft (4.5 m) in height. Also, rank-piled logs can be arranged in aisles, allowing more adequate provision for fire fighting.

Storage of wood chips at paper and pulp mills requires fire protection considerations based on the physical properties of the wood (see subsection "Physical Properties" in this chapter). Chip fires can be surface fires or internal fires.

Lumber storage can pose serious problems in populated areas. Although generally prohibited in congested neighborhoods, some zoning laws permit lumber storage in or near these congested areas. The major fire hazard is, of course, exposure to nearby property.

Plastics and Polymers

Another common combustible solid is the large and varied group of materials called "plastics." Thousands of different plastic formulations (resins) are produced in a wide variety of shapes and sizes, such as solid shapes, films and sheets, foams, molded forms, fibers, pellets, and powders. There are about 30 major groups or classes of plastics and polymers. In addition, almost all finished plastic products contain additives, such as colorants, reinforcing agents, plasticizers, filler, stabilizers, and lubricants. These additives vary the chemical nature of the product still more, and may greatly alter the combustibility of the original resin. As a result, it is virtually impossible to assign a fire hazard or combustibility limit to any general plastic group. The only method of determining the fire hazard of a particular plastic is to test it under exact "end use" conditions.

Manufacture: The creation of a plastic article involves three basic steps: (1) manufacturing the resin, (2) molding, extruding, or casting the resin into the shape of the article itself (processing), and (3) bending, machining, cementing, decorating, and polishing the article (fabricating).

Manufacturing the basic plastic resin usually involves flammable ingredients or solvents. As previously stated, most finished plastic products also contain additives, such as colorants, plasticizers, stabilizers, reinforcing agents, etc., that influence the flammability characteristics of the plastic. Plants that make articles from plastic material by molding or machining do not have the same degree of fire hazard as plants making the basic plastics. Improper handling of the flammable organic solvents used in most plastic plants has resulted in many fires. Installation of vapor removal systems, use of explosionproof electrical equipment, and reduction or elimination of the possibility of static sparks are all steps that can and should be taken to minimize fire hazards.

Fire Behavior: The fire behavior of plastic materials is dependent on the chemical composition of the basic plastic, the kind of additives used, the material's size and shape, and its "end use." Almost all plastic materials are combustible; they will contribute fuel to any fire in which they are involved. For example, thermoplastics, such as polyethylene and plasticized polyvinyl-chloride, and thermosets, such as polyesters, present severe fire hazards. In a fire, thermoplastics will melt and then will behave and burn like combustible liquids. The ignition point will vary widely, as will the rate of flame propagation. While it is comparatively easy to predict what wood and wood products will do when exposed to fire, the same type of products made of plastic may display totally different and unpredictable burning behavior.

The smoke and gases given off by a burning plastic material likewise will depend on the particular plastic. Some plastics yield large quantities of dense smoke, while others burn more cleanly and produce less smoke.

The growing use of plastics for structural and finish materials in buildings has led to two developments: (1) incorporation of flame-retardant additives in the resin formulation, and (2) use of flame-retardant treatments for plastic products. Such additives and treatments may make the material more difficult to ignite, or reduce its rate of burning.

Storage: The chemical composition, the physical form (e.g., plastics in cellular form constitute the most severe hazard), and the manner and arrangement in which plastics are stored greatly affect the degree of fire hazard. Large quantities of smoke usually are generated when stored plastics are involved in a fire, a condition made more or less difficult by the amount of ventilation present or available in a given storage area. Automatic sprinkler systems with high discharge densities are necessary for adequate fire protection where plastics are stored.

Textiles

Textiles are a major component in everyday life. Clothing, bedding, carpets, upholstery, and many other items with which people come in frequent contact are textiles or textile-based products. Virtually all textile fibers are combustible. The chemical composition, the weight of the fabric, the type of weave, and the type of finishing treatment given to textiles are some of the many variables that affect the way a textile burns.

The combustibility of textiles, combined with the high degree of people's involvement with textiles in their daily lives, accounts for a number of textile-related fires and a corresponding number of deaths and injuries as a result. The NFPA estimates, from data collected in surveys of fire departments, that clothing ignition deaths represent 6.5 percent of all U.S. fire deaths.[3]

Any small ignition source, such as a match or a cigarette, can ignite cotton. The ensuing fire will develop and spread rapidly. The same is true of rayon. Cotton and rayon are a very large part of all clothing fabric materials in common use. The extent and severity of clothing burns is much greater than burns received on uncovered skin area, due to the chemical composition and finishing treatments of many textiles. Research has shown that clothing burn victims are four times more likely to die than burn victims not involved in clothing fires.

Natural Fiber Textiles: There are two basic types of natural fiber: (1) fibers derived from plants, and (2) fibers derived from animals.

Cotton and other plant fibers, such as flax and hemp, are composed mainly of cellulose, a chemical composition of carbon, hydrogen, and oxygen. Cotton and other plant fibers are combustible, but when involved in combustion do not melt. By-products of burning plant fibers are carbon dioxide, carbon monoxide, and water.

Fibers derived from animals are chemically different from plant fibers in that the basic component of animal fibers is protein, and a relatively high percentage of nitrogen is contained in the molecular composition. Animal fibers, such as wool, are not as likely to ignite and support combustion as are cotton and other plant fibers. For example, the ignition temperature of cotton is 752°F (400°C), while the ignition temperature of wool is 1,112°F (600°C).

Synthetic Textiles: The use of natural fibers in textiles is increasingly being challenged by synthetic fabrics. Synthetics are fabrics woven entirely or mostly from synthetic fibers. Rayon, for example, is a synthetic material produced by "reconstituting" pure cellulose made from cotton and wood fibers. Cellulose acetate is made by reacting pure cellulose with acetic acid.

Acetate, which previously was classed as a rayon, is now listed by the U.S. Federal Trade Commission as a separate category. Rayon and acetate and their analogs are the only major synthetic fibers that chemically resemble plant fibers.

Table 4.3 lists the various types of plastic resins used in the production of synthetic fibers. Also shown are some of the more common trade names associated with each type, and the fire hazard properties, including generalized statements about relative flammability.

Because the descriptions of burning synthetic fabrics are based on small-scale tests, they may be misleading. Some synthetic fabrics will give the appearance of being flame retardant when tested with a small flame source, such as a match. However, when the same fabrics are subjected to a larger flame or full-scale test, they may burst into flame and consume themselves while generating large quantities of black smoke.[3]

Finally, blends of two or more different fibers may exhibit burning behavior quite different from the component fibers. A good example is blends of cotton and polyester, very popular in permanent press fabrics, usually ignite more easily and burn more rapidly than either cotton or polyester alone.

Table 4.3   Synthetic Fibers

Plastic Resin Class	Trade Names	Fire Hazard Properties
Acetate	Chromspun, Celaperm, Arnel (Triacetate)	Burns and melts ahead of flame. Ignition temp., 475°C
Viscose	Avisco, Avril, Bemberg (Cuprammonium)	Burns about the same as cotton
Nylon	Antron, Caprolan	Supports combustion with difficulty. Melts and drips—melting point, 160-260°C. Ignition temp. 425°C and above
Polyester	Dacron, Fortrel, Vycron, Kodel, Terylene	Burns readily. Ignition temp., 450-485°C. Softens, 256-292°C, and drips
Acrylic	Acrilan, Orlon, Zefchrome, Zefran, Creslan	Burns and melts. Ignition temp., 560°C. Softens, 235-330°C
Olefin	Herculon	Burns slowly. Ignition temp., 570°C. Melts and drips
Modacrylic	Verel, Dynel	Burns very slowly. Melts
Saran	Rovana, Velon	Does not support combustion. Melts
Fluorocarbon	Teflon	Does not support combustion. Softens, above 327°C. Ignition temp. above 600°C
Spandex	Lycra, Vyrene	Burns and melts. Ignition temp., 415°C. Softens, 230-260°C
Rubber	Lastex	Burns
Phenolic	Kynol	Burns

Flame-Retardant Textiles: Because of the fire hazard of flammable clothing, much attention has been given to flame-retardant treatments for cotton and other textiles. So far, only sleepwear up to size 14 for children and highly flammable material, such as brushed rayon, have been subject to control. Most of the synthetic fibers used in manufacturing clothing are subject to melting when exposed to sufficient heat. This property can aggravate the severity of a burn. The textile industry resists producing flame-retardant materials because of the economic factors.

Many textile fires are caused by ignition of mattresses or overstuffed furniture through the careless use or disposal of burning cigarettes or other smoking materials. Some hospitals, nursing homes, and other health care facilities are now flameproofing the mattresses used by patients. The Consumer Product Safety Commission has promulgated regulations requiring the flameproofing of mattresses and mattress pads.

In most areas, the law requires that theater scenery, as well as curtains and draperies in places of public assembly, be treated with flame retardants. Disposable or nonwoven fabrics consist primarily of cellulose fibers and generally are treated with flame retardants.

Storage: As has been noted, both natural and synthetic textiles have varied combustibility and flammability characteristics. The method of packaging textiles for storage is the main factor in determining the textiles' level of fire hazard in the storage context. A solid wood case is preferable to combustible burlap wrappings. The general storage considerations of good aisles, stability of piles, and piling limitations are important factors in minimizing the fire hazards generated through storage.

FLAMMABLE AND COMBUSTIBLE LIQUIDS

Improper storage, handling, and use of flammable and combustible liquids has been the cause of many deaths, injuries, and disastrous fires. It is vital to understand the physical properties of such liquids, their classifications—particularly the fire and burning characteristics, and the procedures for fire prevention when dealing with storage and handling. This understanding is essential for the industries producing and using such liquids, and for fire fighters and fire inspectors.

In general, liquids can be considered a midpoint in the physical world of solids, liquids, and gases. Molecules in liquids move more freely than the molecules in solids. For example, most liquids can adapt themselves easily to the shape of the vessel containing them. However, the molecules in liquids do not have the tendency to separate themselves and expand indefinitely as do the molecules in gases.

Because their molecules are freer, changes in state are more common between liquids and gases. Liquids will become gases as the temperature increases or the pressure decreases. Conversely, a gas tends to become a liquid as the temperature decreases or the pressure increases. The critical temperature of a material is that temperature above which the material can exist only in a gaseous state.

Burning is of the vapor from the evaporation of a flammable or combustible liquid when exposed to air or under the influence of heat, rather than of the liquid itself. There is a flammable range below which the vapor mixture is too lean to burn or explode, or above which the vapor mixture is too rich to burn or explode. For gasoline—the most common and widely used flammable liquid—the flammable range is between 1.4 and 7.6 percent by volume in air. When the vapor-air mixture is near either the lower flammable limit (LFL) or upper flammable limit (UFL), the explosion is less intense than when the mixture is in the intermediate range. The violence of the explosion depends on the concentration of the vapor as well as the quantity of vapor-air mixture and the type of container in which it is stored. Thus, storing gasoline or other flammable liquid in the proper type of closed container and minimizing the exposure to air is of fundamental importance in controlling the fire hazard during storage or use. It must be noted, however, that a tank or other container, when exposed to heat from a fire, may rupture with dangerous results if properly designed vents are not provided and if the exposed tank or container is not cooled by hose streams.

Table 4.4 shows the wide variations in flammable range of some common liquids and gases. Those substances with a wide range of flammable limits are, of course, more hazardous than others. This wide range increases the possibility of confrontation with an ignition source that can occur in a great variety of circumstances.[1]

Fighting a fire involving flammable or combustible liquids requires, if possible, shutting off the fuel supply, excluding air from the area, cooling the liquid, or a combination of these measures. The principal fire and explosion prevention measures under such circumstances are: (1) excluding sources of ignition, (2) excluding air, (3) keeping the liquid in a closed container, (4) ventilating to prevent the accumulation of vapor, and (5) using an atmosphere of inert gas instead of air.

Classification by Properties

Liquids that burn are classified by NFPA 321, *Standard on Basic Classification of Flammable and Combustible Liquids*,[5] based on flash point, boiling point, and vapor pressure. Vapor pressure is the basic determinant in separating liquids from gases. Any fluid having a vapor pressure of 40 psia (2068.6 mm Hg) or less is considered a liquid. NFPA 321 establishes three categories of liquids: flammable (Class I liquids) and combustible (Class II and Class III liquids).

Table 4.4   Flash Points and Flammable Limits of Some Common Liquids and Gases*

Liquid (or gas at ordinary temps.)	Flash point °F	(°C)	Flammable limits (percent by volume)
Acetylene	(Gas)		2.5 to 81.0+
Benzene	12	(−11)	1.3 to 7.1
Ether (ethyl ether)	−49	(−45)	1.9 to 36.0
Fuel oil			
(Domestic, No. 2)	100 (min.)	(38)	None at ordinary temps.
(Heavy, No. 5)	130 (min.)	(54)	None at ordinary temps.
Gasoline (high test)	−36	(−38)	1.4 to 7.4
Hydrogen	(Gas)		4.0 to 75.0
Jet fuel (A & A-1)	110 to 150	(43 to 65)	None at ordinary temps.
Kerosine (fuel oil, No. 1)	100 (min.)	(38)	0.7 to 5.0
LPG (propane-butane)	(Gas)		1.9 to 9.5
Lacquer solvent (butyl acetate)	72	(22)	1.7 to 7.6
Methane (natural gas)	(Gas)		5.0 to 15.0
Methyl alcohol	52	(11)	6.7 to 36.0
Turpentine	95	(35)	0.8—(undetermined)
Varsol (standard solvent)	110	(43)	0.7 to 5.0
Vegetable oil (cooking, peanut)	540	(282)	(Ignition temp. = 833°F)

*From *Principles of Fire Protection Chemistry*, by Richard L. Tuve.[1]

Flammable liquids are defined as follows:[5]

Flammable liquids shall mean any liquids having a flash point below 100°F (37.8°C) and having a vapor pressure not exceeding 40 psia (2068.6 mm Hg) at 100°F (37.8°C).

Class I liquids shall include those having flash points below 100°F (37.8°C) and may be subdivided as follows:

Class IA shall include those having flash points below 73°F (22.8°C) and having a boiling point below 100°F (37.8°C).

Class IB shall include those having flash points below 73°F (22.8°C) and having a boiling point at or above 100°F (37.8°C).

Class IC shall include those having flash points at or above 73°F (22.8°C) and below 100°F (37.8°C).

Combustible liquids are defined as follows:[5]

Liquids with a flash point at or above 100°F (37.8°C) are referred to as combustible liquids and may be subdivided as follows:

Class II liquids shall include those having flash points at or above 100°F (37.8°C) and below 140°F (60°C).

Class IIIA liquids shall include those having flash points at or above 140°F (60°C) and below 200°F (93.4°C).

Class IIIB liquids shall include those having flash points at or above 200°F (93.4°C).

Some typical liquids, including their flash points and classifications, are:

Acetone	−4°F (−20°C)	Class IB
Diethyl ether	−49°F (−45°C)	Class IA
Gasoline	−36 to −45°F (−38 to −42°C)	Class IB
#2 fuel oil	126°F (52°C)	Class II
Peanut oil	540°F (284°C)	Class IIIB

UL has a useful system for grading the relative flammability hazards of liquids, which is based on the following scale:

Ether class	100
Gasoline class	90 to 100
Alcohol (ethyl) class	60 to 70
Kerosine	30 to 40
Paraffin oil class	10 to 20

Characteristics

Many terms are used to describe the characteristics of liquids. These characteristics also can be thought of in terms of their influence on fire behavior. The following terms describe some characteristics of liquids of particular interest to the fire protection professional.

Vapor Pressure: Molecules escape from the surface of all liquids. If the liquid is in a closed container, the molecules discharge into the space above the liquid, then condense back into the liquid. When the rate of discharge and the rate of return are in equilibrium, the resulting pressure exerted by the molecules in the space above the liquid is called vapor pressure.

Vapor pressure of a liquid is used to measure the vapor-air mixture above the liquid surface in a closed container. The percentage of vapor is directly proportional to the ratio of the vapor pressure of the liquid to the total pressure of the mixture. When heat is added, the molecules become more agitated, thus increasing the pressure. The vapor pressures of many liquids have been measured and are listed in various chemical handbooks. The vapor pressure is expressed in pounds per square inch absolute (psia). When the flash point and the vapor pressure of a liquid are known, it is possible to calculate the lower flammable limit of the vapor.

Flash Point: The flash point of a liquid is the lowest temperature at which the vapor pressure of the liquid will produce an ignitable mixture and resultant flame. The flame will not continue to burn if the source of ignition is removed, thus reducing the temperature. Dr. Tuve further explains:[1]

> Because it is an indicator of the degree of safety of a material, the flash point of a liquid is one of the most important fire characteristics of substances. At its flash point, a liquid continuously produces

flammable vapors at the right rate and amount (volume) to give a flammable and even explosive atmosphere if a source of ignition should be brought into the mixture.

Boiling Point: The boiling point of a liquid is the temperature at which the vapor pressure equals the total pressure on the surface. The normal boiling point is the temperature at which the liquid boils when under normal atmospheric pressure (14.7 psia). The boiling point increases as pressure increases and is dependent on the total surface pressure.

Specific Gravity: The specific gravity of a liquid is an important consideration in fighting a flammable or combustible liquid fire. The specific gravity of a liquid is the ratio of the weight of the liquid to the weight of an equal volume of water, the specific gravity of water being assigned a value of 1.0. Any liquid with a specific gravity of less than one will float on water (unless it is soluble in water). If a liquid has a specific gravity greater than one, the water will float on the liquid. Gasoline, with a specific gravity of 0.8, will float on water. Glycerine, with a specific gravity of 1.3, will sink to the bottom.

Evaporation Rate: As has been stated, molecules will escape from the surface of all liquids in the form of vapor. When a liquid is in an open container, molecules will escape. The evaporation rate is the rate at which the liquid is converted to vapor at any given temperature and pressure.

Viscosity: Some liquids, such as asphalt or wax, are on the borderline between liquids and solids. These liquids are considered viscous. The viscosity of a liquid can be measured in relation to the time required for the liquid to flow into a container or through an opening.

Latent Heat of Vaporization: Another common measure for liquids is called the latent heat of vaporization. This is measured by the heat required to convert one gram of liquid into vapor at the boiling point under one atmosphere pressure (14.7 psia). Latent heat usually is given in British thermal units per pound (Btu/lb) or in calories per gram (cal/g).

Ignition Temperature: The ignition temperature (also know as autoignition temperature) of a liquid is the minimum temperature to which the liquid must be heated in order to initiate self-sustained combustion independent of the heating element. The ignition temperature of kerosine is $410°F$ ($210°C$), and of motor oil is around $500°F$ ($260°C$).

Storage and Handling

Certain fundamentals in the proper storage and handling of flammable and combustible liquids must be observed to reduce the possibility of fire or explosion. Ventilation to prevent accumulations of flammable vapors is of primary importance. Liquids usually are exposed to air at some stage of a process, and there is always the possibility of breaks or leaks in a closed storage and handling system. It is important to eliminate possible sources of ignition in all areas where flammable liquids are stored, handled, or used.

Ventilation of an area where flammable liquids are manufactured or used can be accomplished by natural or mechanical means. Wherever possible, equipment such as compressors, stills, pumps, and the like should be located in a spacious, open area. Gasoline and most other flammable liquids produce heavier-than-air vapors that flow along the ground or floor and settle in depressions. Unless these vapors are removed at the ground level, they can travel long distances and be ignited and flash back from a point remote from the origin of the vapors. NFPA 30, *Flammable and Combustible Liquids Code*,[6] is the accepted standard for safe storage, handling, and use of such liquids.

Much attention has been given to the construction, installation, spacing, venting, and diking of aboveground and underground storage tanks, as well as container storage in buildings. NFPA 30 also covers loading and unloading practices, safe dispensing of liquids, and safe transportation of liquids.

GASES

As in the case of solids and liquids, a great variety and number of materials exist in the form of gas. A gas is made up of molecules in constant motion. The higher the temperature, the more rapid the motion. A solid has shape and volume; a liquid has volume but no shape of its own; a gas has neither shape nor volume of its own, but will take the shape and occupy the entire volume of whatever enclosure it occupies.

Since all substances can exist as gases if the temperature is high enough, a substance is a gas when it has an absolute pressure exceeding 40 lb/in.^{2} (184 kg/6.5 cm^2) at 70°F (21°C).

Classification by Properties

Gases can be classified broadly in various ways, such as by chemical properties, by physical properties, or by usage. Classification by chemical properties helps to define the hazards of gases to people and in fires.

Chemical Properties

Flammable Gases: Any gas that will burn in the normal concentrations of oxygen in air is a flammable gas. Like flammable liquid vapors, the burning rate of this gas in air is in a range of gas-air mixture (the flammable range).

The term "flash point," which is necessary information concerning flammable liquids, has no real bearing with regard to flammable gases. The flash point for flammable liquids is always below the normal boiling point. A flammable gas in its gaseous state and even in its liquid state usually exceeds its normal boiling point, and therefore already has exceeded its flash point. The ignition temperature of a gas is the temperature required to initiate combustion.

Nonflammable Gases: Nonflammable gases will not burn in air or in any concentration of oxygen. A number of nonflammable gases, however, will support combustion. Such gases often are referred to as "oxidizers" or "oxidizing gases." Common oxidizers are oxygen or oxygen in a mixture with other gases.

Nonflammable gases that will not support combustion usually are called "inert gases." Among the most common inert gases are nitrogen, argon, helium, and carbon dioxide.

Toxic Gases: Toxic gases are those gases that endanger life when inhaled. Gases such as arsine, chlorine, hydrogen sulfide, sulfur dioxide, ammonia, and carbon monoxide are poisonous and/or irritating when inhaled, and are serious considerations when fire fighting involves such gases.

Reactive Gases: Reactive gases are gases that will react with other materials or within themselves by a reaction other than burning. When exposed to heat and shock, some reactive gases rearrange themselves chemically. Such gases can produce hazardous quantities of heat or reaction products. For instance, fluorine is a highly reactive gas. At normal temperatures and pressures, it will react with most organic and inorganic substances, often fast enough to result in flaming. Other examples of reactive gases are acetylene and vinyl chloride.

Physical Properties

A second way of classifying gases is by their physical properties. Gases can be either liquefied or nonliquefied compressed gases. A nonliquefied compressed gas is one that at 70°F (21°C) cannot be liquefied by increasing the pressure of the gas in a container. Flammable limit ranges of nonliquefied gas in gas containers can vary greatly from a lower limit of 25 psig to an upper limit that may exceed 6000 psig. The common portable cylinders holding nonliquefied compressed gases do not contain very large quantities of gas.

When the gas in a container can exist with both a gas phase and liquid phase at 70°F (21°C), it is known as liquefied gas. The gas exists at the vapor pressure of the liquid as long as any liquid remains in the container.

When gas in a container is at a temperature below −238°F (−150°C) and it is still a liquid (not a solid), the gas is classified as a cryogenic liquid.

Compressed Gases: A nonliquefied compressed gas is one that is at normal temperature inside a gas container and exists solely in the gaseous state under pressure. Common nonliquefied compressed gases are hydrogen, argon, nitrogen, helium, oxygen, and ethylene.

Hydrogen has a high diffusion rate and is extremely flammable. It also has a wide flammability range, so that even very low energy sources, such as a friction spark, can produce ignition.

Oxygen, although a nonflammable gas, can support combustion and can accelerate flame and explosive conditions. Because of its ability to drastically alter the combustibility of other substances and materials, oxygen—although not flammable in itself—should be regarded as dangerous under some circumstances.

Ethylene is used as a fruit ripening agent in very low concentrations. Ethylene is highly flammable and reacts rapidly with oxidizing gases. When released into the atmosphere, ethylene mixes quickly with air and can produce an explosive mixture very rapidly.

Liquefied Gases: Liquefied gases are gases that can be liquefied at $70°F$ $(21°C)$ simply by increasing their pressure to the vapor pressure of that material at $70°F$ $(21°C)$. Liquefied gases are stored easily at ordinary temperatures at the vapor pressure of the gas. When stored, liquefied gas exists in both liquid and gaseous states. At storage pressure, both the liquid and gas in the liquefied gas container are in equilibrium and will remain so as long as any liquid remains in the container. Common liquefied gases are propane, butane, ammonia, and chlorine.

Acetylene, which is handled as a liquefied gas, is an extremely reactive flammable gas that must be stored in special containers to control its reactivity.

Cryogenic Liquids: Cryogenic liquids are made from gases liquefied by cooling to very low temperatures. The National Bureau of Standards defines cryogenic as a liquid having a boiling point below $-238°F$ $(-150°C)$. The common gases which are shipped as cryogenic liquids are oxygen (O_2), nitrogen (N_2), hydrogen (H_2), argon (Ar), helium (He), and liquefied natural gas (LNG). These gases must be maintained in their containers as low-temperature liquids at relatively low pressure. Cryogenic liquids must be stored at or below their boiling points in special containers that allow the gas, which is slowly boiling off from the liquid, to escape. This prevents a pressure buildup within the container, which would result in container failure. Liquefied gases and gases shipped as cryogenic liquids produce large volumes of gases when vaporized. For example, one volume of liquid nitrogen at its boiling temperature at 1 atm vaporizes to about 700 volumes of nitrogen when warmed to room temperature at 1 atm.

Classification by Usage

An understanding of gases as they are classified by usage is of great importance to everyone involved in fire protection, as the terms of these classifications are used in codes, standards, and general industrial and medical terminology. There are three broad classifications of gases based on usage:

(1) fuel gases, including natural gas or liquefied petroleum gas used for heating and cooking in the home; (2) industrial gases, customarily used for welding and cutting, refrigerating, heat treating, and chemical processing; and (3) medical gases, a specialized class of gases used for anesthesia and respiratory therapy.

The following are NFPA definitions of the three major classifications of gases, based on their use:[3]

Fuel Gases: Fuel gases are flammable gases customarily used for burning with air to produce heat, which in turn is used as a source of heat (comfort and process), power, or light. By far the principal and most widely used fuel gases are natural gas and the liquefied petroleum gases, butane and propane.

Industrial Gases: Industrial gases embrace the entire gamut of gases classified by chemical properties and are customarily used in industrial processes, for welding and cutting, heat treating, chemical processing, refrigerating, water treating, etc.

Medical Gases: By far the most specialized usage classification, the medical gases are used for health care purposes such as anesthesia and respiratory therapy. Cyclopropane, oxygen, and nitrous oxide are common medical gases. All medical gases and medical gas mixtures are labeled as either drugs or medical devices under Food and Drug Administration (FDA) regulations.

Gas Laws

Three basic physical laws predict the behavior of compressed gases under most conditions. These laws are based on three components: volume, temperature, and pressure of gases. Boyle's Law states that the volume occupied by a given mass of gas varies inversely with the absolute pressure if the temperature is constant. Charles's Law states that the volume of a given mass of gas is directly proportional to the absolute temperature if the pressure is kept constant. Gay-Lussac's Law states that when gas temperatures are raised and the volume in which they are confined stays the same, pressures are raised in proportion to the change in the absolute temperature of the gas.

Again, in very simple terms, Boyle's Law describes the behavior of gases when the temperature is constant; Charles's Law deals with the behavior of gases when the pressure is constant; and Gay-Lussac's Law describes the behavior of gas when the volume is constant. Although the following descriptions of these laws are general, the basic factors involved are important in terms of fire protection considerations and decisions.

Boyle's Law: Boyle's Law relates to gases under the pressure normally used in compressed gas containers. Boyle's Law states that the volume of a gas varies with the absolute pressure, if the temperature is kept constant. Very simply, if the absolute pressure on a given volume of gas is increased to twice the original level, the volume of gas will be compressed to one-half its original size. Conversely, if the pressure on a certain volume of gas is reduced by one-half, the volume of gas would increase to twice its original quantity.

Charles's Law: Whereas Robert Boyle's experiments dealt with the effect of pressure on the volume of gas if the temperatures were kept constant, Jacques Charles's experiments kept the pressure constant, and looked to the effect of temperature on the volume of gases.

Gases expand considerably when heated. Solids and liquids also will expand when exposed to heat, but not to the dramatic extent that gases expand.

Charles's Law is a result of experiments in which it was discovered that if the pressure were kept constant and the temperature became the variable, gases would expand in direct proportion to the temperature to which the gases were exposed. If the absolute temperature were doubled, the volume of gas would double.

Gay-Lussac's Law: Joseph Gay-Lussac experimented with the third aspect of the physical laws regarding gases. Keeping the volume of gases constant, he discovered that when gas temperatures are raised, gas pressures are raised in proportion to the change in temperature.

Gas Fires

The gas containers holding liquefied and cryogenic gases will expand with heat, as in the case of compressed gas. In addition, the liquid in the container will expand and the vapor pressure of the liquid will increase as the temperature increases. All of these physical changes combine to increase the pressure when the container is heated. To relieve the pressure in the container when it is exposed to fire, most compressed and liquefied gas containers are provided with relief valves or bursting discs or both. The danger is greater from failure of a liquefied gas container than from failure of a compressed gas container, because larger quantities of gas are released. Failure of a compressed gas container is more a flying missile hazard than an explosion danger.

BLEVEs

The acronym BLEVE stands for Boiling Liquid—Expanding Vapor Explosion. BLEVE describes a mechanism of failure of a container holding liquid at a temperature well above its boiling point at normal atmospheric pressure. The term was first used in 1957 to describe the sudden failure of a cast-iron reactor used to produce phenolic resin. The term came into widespread use in the 1970s after several transportation incidents involving failure of LP-Gas containers resulted in extensive fire.

A BLEVE occurs when a container containing a liquid above its boiling point is weakened so it no longer can hold the pressure for which it was designed. A common example is a container of LP-Gas—a liquid and gas in equilibrium. The boiling point of the liquid is $-51\,°F$ ($10\,°C$) at atmospheric pressure. If the container is exposed to fire above the liquid level, the container metal will be heated; as the LP-Gas vapor does not conduct heat well, the metal eventually will become softened. During this time, vaporizing LP-Gas will increase the pressure inside the container, and the relief valve may operate to keep the pressure from exceeding its preset operating pressure. Once the metal is softened, a crack will develop and propagate around the container. LP-Gas liquid is released to atmospheric pressure and flashes to vapor. A dramatic fireball usually is seen. Pieces of the container can travel as far as 3900 ft (1200 m). Death can result from the intense heat of the fireball and from flying tank fragments.

A BLEVE of this type can be prevented by cooling the tank surface at the point of flame impingement. Cooling can be with water from fixed sprays or hose or from monitor streams, or with insulation to slow the rate of heating of the tank metal. Many factors, including risk to emergency responders, must be taken into account before a decision is made to apply water to cool a tank at risk.

Other factors that can cause a BLEVE include corrosion or mechanical damage to a container.

Combustion Explosions

If a flammable gas or vapor from a liquefied flammable gas escapes or leaks from its container, piping, or other equipment, the gas mixes with the air. When the flammable limit is reached, the mixture is ignitable and will burn. A gas fire of this sort burns rapidly and produces heat rapidly. As virtually all materials expand when heated, materials in the vicinity of such a gas fire absorb heat and expand. Air will double its volume for every $459\,°F$ ($237.2\,°C$) it is heated, so if the heated air cannot disperse, the pressure will rise. If the room or structure is not strong enough to withstand the increasing pressure from the heated air, a combustion explosion will result. (See Figure 4.1.)

Most structures can withstand pressures of only one pound per square inch (psi) or less. Because a flammable gas-air mixture can develop pressures of from 60 to 110 psi (41.370 kPa to 758.540 kPa), a room or enclosed structure could blow apart if even 25 percent or less of the enclosed area were occupied by the flammable vapors.

Fig. 4.1   Aftermath of a BLEVE that occurred outside Mexico City on November 19, 1984. Almost 550 people died, and more than 2000 received burns. (Courtesy Skandia)

The basic safeguards against a combustion explosion are: (1) use of emergency flow-control devices, and (2) use of strong containers and equipment to reduce the possibility of leakage.

Odorizing

Natural gas and LP-Gas are odorized so that the average person can detect gas concentrations in air not exceeding one-fifth of the lower limit of flammability. The ability to smell odorant can be reduced by smoking, by having a cold or other health problems, and by the aging process.

Gas Standards

LP-Gas, acetylene, oxygen, and other common gases are contained in cylinders for transportation purposes and during use. Relevant standards are NFPA 50, 50A, 50B, 51, 54, 58, and Chapter 4 of NFPA 99.[7-13] The cylinders are built and used subject to regulations of the U.S. Department of Transportation (DOT) or the American Society of Mechanical Engineers (ASME) Boiler and Pressure Vessel Code.[14] Large volumes of gases are transported by pipelines, a method also under DOT regulation.

Comprehensive Reference

A reference with which all students of the fire hazard of materials should be familiar is the Table of Flammable Liquids, Gases, and Volatile Solids. Sponsored by the NFPA Committee on Flammable Liquids, the table is published in NFPA 325M, *Fire Hazard Properties of Flammable Liquids, Gases, and Volatile Solids.*[15] This table provides flash point, ignition temperature, flammable limits, specific gravity, vapor density, boiling point, water solubility, method of extinguishment, health, flammability, and reactivity ratings, and other data for more than 1,000 substances, listed alphabetically.

HAZARDOUS MATERIALS

Problems associated with hazardous materials ("hazmats") have escalated over the last decade. Increased attention is being given to the subject partly because of broadly applicable federal regulations. The U.S. Department of Transportation (DOT) considers a hazardous material to be any substance or material which has been determined to pose an unreasonable risk to health, safety, and property when transported in commerce. Long used as the basis for all operations relating to hazmats, DOT's definition may not be adequate in today's world. Concern cannot be limited solely to materials in transit; concern extends to hazardous wastes and hazardous substances, which must be included within the broader concept of hazardous materials.

Emergency personnel who deal with hazardous materials must be prepared to respond to fixed facilities as well as to transportation incidents; and the fixed facility may well be a waste disposal site. Note that here guidance is not limited to fire service responders, but includes all hazmat incident responders. Many incidents will not involve fire at all.

Official Definitions

Hazardous Substance: (1) A material and its mixtures or solutions that is identified by the letter "E" in column (1) of the Hazardous Materials Table, CFR 49, Section 172.101, when offered for transportation in one package, or in one transport vehicle if not packaged, and when the quantity of the materials therein equals or exceeds the reportable quantity. (2)(A) Any substance designated pursuant to Section 311(b)(2) of the Federal Water Pollution Control Act, (B) any element, compound, mixture, solution, or substance designated pursuant to Section 102 of this Act, (C) any hazardous waste having the characteristics identified under or listed pursuant to Section 3001 of the Solid Waste Disposal Act (but not including any waste the regulation of which under the Solid Waste Disposal Act has been suspended by the Act of Congress), (D) any toxic pollutant listed under Section 307(a) of

Table 4.5   Regulated Hazardous Materials—U.S. DOT

Classification	Examples
1. Explosives	
Class A	Dynamite
Class B	Propellent powders
Class C	Small-arms ammunition
Blasting agents	Nitro carbo nitrate
2. Compressed Gases	
Flammable	Propane
Nonflammable	Carbon dioxide
Liquefied	Nitrogen
Cryogenic	Hydrogen
3. Flammable Liquids	Gasoline
Combustible	Diesel fuel
Pyrophoric	Aluminum alkyl
4. Flammable solids	Magnesium
Water reactive	Calcium carbide
Spontaneously combustible	Phosphorous
5. Oxidizing Materials	
Organic peroxide	Benzoyl peroxide
6. Poisonous Materials	
Class A poison	Arsine
Class B poison	Arsenic
Irritant	Tear gas
Etiolozic agent	Anthrax
7. Radioactive Materials	
Radioactive I	Uranium
Radioactive II	Uranium hexafluoride
Radioactive III	
8. Corrosive materials	
Acids	Hydrofluoric acid
Bases	Caustic soda
9. Other radioactive materials (ORMs)	
ORM A	Carbon tetrachloride
ORM B	Quicklime
ORM C	Bleaching powder
ORM D	Consumer commodities
ORM E	Hazardous wastes

the Federal Water Pollution Control Act, (E) any hazardous air pollutant listed under Section 112 of the Clean Air Act, and (F) any imminently hazardous chemical substance or mixture with respect to which the Administrator has taken action pursuant to Section 7 of the Toxic Substances Control Act. The term does not include petroleum, including crude oil or any fraction thereof, which is not otherwise specifically listed or designated as a hazardous substance under subparagraphs (A) through (F) of this paragraph, and the term does not include natural gas, natural gas liquids, liquefied natural gas, or synthetic gas usable for fuel (or mixtures of natural gas and such synthetic gas).

Hazardous Waste: Any material that is subject to the hazardous waste manifest requirements of the Environmental Protection Agency specified in the CFR, Title 40, Part 262 or would be subject to these requirements in the absence of an interim authorization to a State under Title 40, CFR, Part 123, Subpart F.

When the terms "hazardous materials" and "hazmats" are used in this text, they will include hazardous substances and hazardous wastes.

Community Right-to-Know

Two federal agencies play a role in what has come to be known as "community right-to-know" reporting requirements. In addition, many states, and even some municipalities, have enacted similar regulations.

On the federal level, the U.S. Department of Labor's Occupational Safety and Health Act (OSHA) rules require Material Safety Data Sheets (MSDS) along with labels and warnings on containers. The regulations assign responsibility for providing and maintaining the MSDS, outline how employees can gain access to them, and provide for how they are to be updated. Relative to labels and warnings, the OSHA regulations assign responsibility for provision of labels and describe the labeling system to be used. Other aspects of the OSHA regulations deal with training requirements and a list of hazardous chemicals. The OSHA regulations can be found in 29 CFR 1910.1200.

Regulations of the U.S. Environmental Protection Agency (EPA) deal with both emergency planning and community right-to-know. The Superfund Amendments and Reauthorization Act of 1986 (SARA), in a part known as Title III, has provisions for emergency planning, emergency modification, right-to-know, and release reporting.

Under emergency planning, SARA rules require each state to designate an emergency response commission, which in turn establishes local emergency planning committees. The local committee must develop an emergency response plan, based on an evaluation of available resources for preparing for and responding to a potential hazardous material accident. The plan should include the identification of facilities, transportation routes, response and notification procedures, evacuation plans, training programs, and designation of a community coordinator.

Two reporting requirements appear under the right-to-know regulation. One specifies that facilities required to have MSDS available under OSHA must submit the MSDS or a list of MSDS hazardous materials to the state commission, local planning committee, and local fire department. The second requirement involves submission of a hazardous material inventory to the same three entities.

Subsequent to the enactment of SARA, a rule entitled Hazardous Waste Operations and Emergency Response was promulgated by OSHA. Among its many requirements are four that have a direct effect on fire departments:

1. The fire chief is required to develop emergency response plans for any response relating to hazardous substances.
2. "Off-site" emergency response training requires that regular response personnel receive 24 hours of training annually in procedures for handling hazardous substances.
3. Procedures for handling "off-site" emergency incidents apply when the senior officer responding declares the incident to be an emergency response involving hazardous substances, based on that officer's best judgment.
4. The requirements for hazardous materials teams apply only to designated members of such teams, and not to every emergency responder who may work at an emergency incident involving hazardous substances.

Officially designated hazardous materials response teams, be they from fire departments, governmental agencies, industry, or waste sites, have special requirements imposed by the OSHA rule. These include special training, medical surveillance, work practices, personal protective clothing, monitoring, decontamination, and emergency response.

Response

It is generally recognized that there are three levels of responders to hazardous materials incidents:

First Responders: The first trained personnel to arrive on the scene of a hazmat incident. These usually are officials from local emergency services, such as fire fighters and police officers.

Second Responders: Personnel required to assist or relieve first responders at a hazmat incident due to their specialized knowledge, equipment, or experience. These people can include state environmental protection or health officials, commercial response and cleanup companies, and appropriate industry representatives.

Third Responders: Personnel required to help the first or second responders handle special situations or to conduct the cleanup, removal, and associated activities. These people can include federal environmental protection and health officials, experts from other federal agencies, commercial response and cleanup companies, and appropriate industry representatives.

In addition to the three levels of responders, hazmats can be under three levels of incidents. Releases of hazardous substances create incidents with a wide variation of consequences, ranging from incidents with little if any impact to those affecting thousands of people and with the potential for severe environmental effects. To effectively utilize response resources and to provide a uniform alerting and notification system, standard designations denoting the severity level of an incident are recommended. The distinction between levels may not be clearly delineated; also, a lower-level incident may escalate to a more severe situation.

Level I Incident: A quantity of hazmats that easily can be controlled or contained. No evacuation involved other than of the immediate area. May or may not pose a threat to life or property.

Level II Incident: Involves larger quantities of hazmats that can be confined and reduced to a relatively small area. Environmental problems usually are present; evacuation of a small area may be necessary. Poses some threat to life and/or property.

Level III Incident: A quantity of hazmats capable of catastrophic, immediate, or long-term damage to life, property, and environment. Not readily controllable, involves a large area, and usually requires large-scale evacuation. Magnitude of the incident may exceed the capabilities of the local emergency response agencies.

Incident levels indicate both the magnitude of the situation and the level of responder required. It should be recognized that the procedures used to mitigate the incident, the safety practices followed, and the training of responders are related to the incident level and what is expected of those who respond. Personnel responding to incidents of levels II and III (second- and third-level responders) require more complete training and more rigid safety procedures than do first-level responders.

Hazmat spills are generally classified as minor, major, or catastrophic. Severity of an incident depends upon the hazardous materials' physical and chemical properties. It is affected by the type and a variety of the hazmats involved, as well as by the acute or chronic exposure conditions.

Minor spills (Level I) can be described as incidents confined to the immediate area of the occurrence, readily controllable, not posing a life-threatening situation.

Major spills (Levels II and III) are incidents which cannot be readily controlled, which occur over large areas, and which can pose a threat to life or the environment.

Levels of Responders: Different levels of incidents require different levels of responders. Each level of response may require responders with different capabilities and technical expertise. As the incident becomes more severe (or has this potential), the resources to control the situation become more critical. The incident may be beyond the control of local agencies and may require additional aid from a variety of other governmental agencies and private industry.

Level I Responders: Local government agencies (generally the fire service) should be capable of responding. They may or may not all be trained hazardous materials team members. Other local organizations such as police, emergency medical services, and the health department may be needed.

Level II Responders: These are personnel from a trained hazmat team and/or from outside the local government agencies who are trained in handling hazardous materials incidents. They may be from other municipalities, counties, state or federal government, or the private sector.

Level III Responders: These are specialists from state and federal government agencies, private industry, and other organizations.

Response Planning: Plans for responding to hazardous material incidents should be prepared in advance. To respond safely and effectively, appropriate equipment should be available, personnel should be trained, and necessary resources should be readily available. Advance plans should be reviewed, tested, and kept current.

Personal Protective Equipment

When response activities are conducted where atmospheric contamination is known or suspected to exist, personal protective equipment must be worn. Such equipment is designed to prevent/reduce skin and eye contact with the chemical substance, as well as inhalation or ingestion of it.

U.S. EPA has established several levels of protection, along with some guidance on when each level should be used. Personal equipment to protect the body against contact with known or anticipated chemical hazards has been divided into four categories:

1. *Level A protection* should be worn when the highest level of respiratory, skin, eye, and mucous membrane protection is needed. Personal protection equipment should include:

    Positive-pressure (pressure-demand), self-contained breathing apparatus (OSHA/NIOSH-approved).
    Fully encapsulating chemical-resistant suit.
    Gloves, inner, chemical resistant.
    Gloves, outer, chemical resistant.
    Boots, chemical resistant, steel toe and shank (depending on suit boot construction, worn over or under suit boot).

Underwear, cotton, long-john type.*
Hard hat (under suit).*
Coveralls (under suit).*
Two-way radio communications (intrinsically safe).

*Optional.

2. *Level B protection* should be selected when the highest level of respiratory protection is needed, but a lesser level of skin and eye protection. Level B protection is the minimum level recommended on initial site entries. It should prevail until the hazards have been further identified and defined by monitoring, sampling, and other reliable methods of analysis, and until personal equipment corresponding with those findings can be utilized. Personal protective equipment should include:

Positive-pressure (pressure-demand), self-contained breathing apparatus (OSHA/NIOSH approved).
Chemical-resistant clothing (overalls and long-sleeved jacket, coveralls, hooded two-piece chemical splash suit, disposable chemical-resistant coveralls).
Coveralls (under splash suit).*
Gloves, outer, chemical resistant.
Gloves, inner, chemical resistant.
Boots, outer, chemical resistant, steel toe and shank.
Boots, outer, chemical resistant.*
Two-way radio communications (intrinsically safe).
Hard hat.*

*Optional.

3. *Level C protection* should be selected when the type of airborne substance is known, the concentration has been measured, criteria for using air-purifying respirators have been met, and when skin and eye exposure is unlikely. The air must be monitored periodically. Personal protective equipment should include:

Full-face, air-purifying respirator (OSHA/NIOSH approved).
Chemical-resistant clothing (one-piece coverall, hooded two-piece chemical splash suit, chemical-resistant hood and apron, disposable chemical-resistant coveralls).
Gloves, outer, chemical resistant.
Gloves, inner, chemical resistant.*
Boots, chemical resistant, steel toe and shank.
Boots, outer, chemical resistant.*
Cloth coveralls (inside chemical protective clothing).*
Two-way radio communications (intrinsically safe).
Hard hat.*

Escape mask.*

*Optional.

4. *Level D protection* is primarily a work uniform. It should not be worn on any site where respiratory or skin hazards exist.

It is most important to note that fire fighters' protective clothing is not designed nor intended for exposure to hazardous materials; they offer only limited protection in a hazmats incident.

IDENTIFICATION OF HAZARDOUS MATERIALS

DOT Placards

The Hazardous Materials Regulations of the Department of Transportation (DOT) require placarding of trucks, trailers, and railway cars carrying dangerous materials. Hazardous materials must be identified on shipping papers regardless of the quantity being shipped.

Present DOT Requirements for Hazardous Materials

Over-the-Road Equipment: Equipment must be marked on front, rear, and both sides with the hazardous material's name at least 4 in. (102 mm) tall. Explosives, extremely toxic materials, and high-strength radioactives in any amount must be marked, along with highly toxic, flammable, and oxidizing materials, corrosive liquids, and both flammable and nonflammable compressed gases, in quantities of 1,000 lb (454 kg) or more. A bill of lading indicating mixed hazards totaling over 1,000 lb (454 kg) carries a ''DANGEROUS'' marking, and is used as well for explosives, extremely toxic, and high-strength radioactive materials when they are present. Other than for radioactive materials, a multiple hazard material is marked only with the name of its most severe hazard. A mixed shipment does not identify hazards.

Railway Equipment: Equipment must be placarded on both sides and ends. Placards for explosives and poison gases are rectangular, the smallest being about 10 by 14 in. (254 by 356 mm). For other hazards, 10¾ in. (273 mm) diamond-shaped placards are used. Radioactives show the name on the placard, while all remaining hazards are identified only by a placard with the word DANGEROUS in large red letters. Ladings which have been fumigated in the car and could bear dangerous residues carry a FUMIGATED placard on or near the door. Placards may be attached directly to the car or placed in special holders. Tank cars which have been emptied, but contain residues, must carry DANGEROUS—EMPTY placards. These are the same size as originals, but the right half is solid black.

Air Shipments: These generally are small quantities, since many hazardous materials cannot be shipped by air—particularly in passenger aircraft. Placarding of equipment is not required.

Transportation Emergencies—Identification Problems

Certain deficiencies in current DOT placarding and labeling practices contribute to difficulties in identifying the hazardous nature of some commodities found in transit. For example, placards are not required on the following:

1. Class C explosives. These are principally manufactured articles containing explosives, such as fireworks, squibs, and small-arms ammunition.
2. Moderate- and low-level radioactives.
3. Tear gases.
4. Corrosive solids, including those which could dissolve in water to produce corrosive liquids.
5. Over-the-road equipment containing less than 1000 lb (454 kg) of highly toxic, flammable, or oxidizing materials; corrosive liquids; or flammable and nonflammable compressed gases, either singly or as a mixed shipment.

DOT-regulated color coding of existing placards and package labels is inconsistent for the various modes of transportation, so the use of colors for quick identification is of doubtful value.

DOT Placard and Labeling System

DOT regulations require that each tank-truck vehicle carrying hazardous materials display a designated four-digit number. The identification numbering system is based on the system adopted for worldwide use by the United Nations Committee of Experts on the Transportation of Dangerous Goods.

The numbers are assigned by governmental authorities under the aegis of the Economic and Social Council of the United Nations, and each number has the same meaning throughout worldwide commerce. Each four-digit number identifies a specific hazardous material, and each has no other meaning or use. For example, 1294 will always signify Toluene (Toluol). (See Tables 4.6 and 4.7.)

As an adjunct to the identification system, a manual has been prepared to associate the identification number with a brief, concise, instruction intended to assist emergency personnel during the first minutes of a hazmat accident. The manual, called the *Fire Officer's Guide to Emergency Action*,[16] is carried by all emergency response vehicles.

To use the *Fire Officer's Guide to Emergency Action*,[16] first identify the material by finding either the four-digit ID number (on a placard or orange

Table 4.6   Sample Page from the ID Number Index. The Four-Digit ID Number 1294 Indicates that the Cargo is Toluene and that the Correct Guide Number Is 27.

ID No.	Guide No.	Name of Material	ID No.	Guide No.	Name of Material
1246	27	METHYLPROPENYL KETONE, inhibited	1266	26	PERFUMERY PRODUCTS, with flammable solvent
1247	27	METHYL METHACRYLATE, monomer, inhibited	1267	27	PETROLEUM CRUDE OIL
1248	26	METHYL PROPIONATE	1268	27	NAPHTHA DISTILLATE
1249	26	METHYL PROPYL KETONE	1268	27	PETROLEUM DISTILLATE, n.o.s.
1250	29	METHYL TRICHLOROSILANE	1268	27	ROAD OIL
1251	28	METHYL VINYL KETONE	1270	27	OIL, petroleum, n.o.s.
1255	27	NAPHTHA, PETROLEUM	1270	27	PETROLEUM OIL
1256	27	NAPHTHA, SOLVENT	1271	26	PETROLEUM ETHER
1257	27	CASINGHEAD GASOLINE	1271	26	PETROLEUM SPIRIT
1257	27	NATURAL GASOLINE	1272	26	PINE OIL
1259	28	NICKEL CARBONYL	1274	26	PROPANOL
1261	26	NITROMETHANE	1274	26	PROPYL ALCOHOL
1262	27	ISOOCTANE	1275	26	PROPIONALDEHYDE
1262	27	OCTANE	1276	26	PROPYL ACETATE
1263	26	COMPOUND, PAINT, etc., removing, reducing, or thinning liquid	1277	68	MONOPROPYLAMINE
			1277	68	PROPYLAMINE
			1278	26	PROPYL CHLORIDE
1263	26	ENAMEL	1279	27	DICHLOROPROPANE
1263	26	LACQUER	1279	27	PROPYLENE DICHLORIDE
1263	26	LACQUER BASE, liquid	1280	26	PROPYLENE OXIDE, inhibited
1263	26	PAINT, etc., flammable liquid	1281	26	PROPYL FORMATE
1263	26	PAINT RELATED MATERIAL, flammable liquid	1282	26	PYRIDINE
			1286	26	RESIN OIL
1263	26	POLISH, liquid	1286	26	ROSIN OIL
1263	26	SHELLAC	1287	26	RUBBER SOLUTION
1263	26	STAIN	1288	27	SHALE OIL
1263	26	THINNER	1289	26	SODIUM METHYLATE, solutions in alcohol
1263	26	VARNISH			
1263	26	WOOD FILLER, liquid	1292	29	ETHYL SILICATE
1264	26	PARALDEHYDE	1292	29	TETRAETHYL SILICATE
1265	27	AMYL HYDRIDE	1293	26	TINCTURE, medicinal
1265	27	ISOPENTANE	1294	27	TOLUENE
1265	27	PENTANE	1295	30	TRICHLOROSILANE
			1296	68	TRIETHYLAMINE

SEE "HOW TO USE THIS GUIDEBOOK" ON THE FIRST PAGE, IF YOU HAVE NOT YET BECOME FAMILIAR WITH THE DETAILS OF USING THESE INDEXES TO THE GUIDES.

panel posted on the vehicle or after the letters UN or NA on the shipping papers) or the name of the material on the shipping papers, placard, label, or package. Next, look up either the four-digit number or the material name in the *Guide*. The four-digit number will identify the material by name and refer to a two-digit guide number. (See Table 4.6.) Under the material name are references to the same two-digit guide number, with the four-digit ID number as a cross-reference. (See Table 4.7.) The two-digit guide number

Table 4.7   Sample Page from the Materials Name Index. Once Again the Guide Number For Toluene is 27.

Name of Material	Guide No.	ID No.	Name of Material	Guide No.	ID No.
TETRAHYDROPHTHALIC ANHYDRIDE	60	2698	THIOUREA	53	2877
TETRAHYDROPYRIDINE	26	2410	THIRAM	55	2771
TETRAHYDROTHIOPHENE	26	2412	THORIUM METAL, pyrophoric	65	2975
TETRALIN HYDROPEROXIDE, technical pure	48	2136	THORIUM METAL, pyrophoric	65	9170
			THORIUM NITRATE, solid	64	2976
TETRAMETHYL AMMONIUM HYDROXIDE	60	1835	THORIUM NITRATE, solid	64	9171
1,1,3,3-TETRAMETHYLBUTYL HYDROPEROXIDE, technical pure	48	2160	TIN CHLORIDE, fuming	39	1827
			TINCTURE, medicinal	26	1293
1,1,3,3-TETRAMETHYLBUTYL-PEROXY-2-ETHYL HEXA-NOATE, technical pure	52	2161	TIN TETRACHLORIDE	39	1827
			TITANIUM, metal, powder, dry	37	2546
TETRAMETHYL LEAD	56	1649	TITANIUM, metal, powder, wet with not less than 20% water	32	1352
TETRAMETHYLMETHYLENE-DIAMINE	58	9069	TITANIUM HYDRIDE	32	1871
TETRAMETHYL SILANE	29	2749	TITANIUM SPONGE, granules or powder	32	2878
TETRAPROPYL-ortho-TITANATE	27	2413	TITANIUM SULFATE SOLUTION	60	1760
TETRANITROMETHANE	47	1510	TITANIUM TETRACHLORIDE *	39	1838
TEXTILE TREATING COMPOUND	60	1760	TITANIUM TRICHLORIDE, pyrophoric	37	2441
TEXTILE WASTE, wet, n.o.s.	32	1857	TITANIUM TRICHLORIDE MIXTURE	60	2869
THALLIUM CHLORATE	42	2573	TITANIUM TRICHLORIDE MIXTURE, pyrophoric	37	2441
THALLIUM COMPOUND, n.o.s.	53	1707	TOE PUFFS, nitrocellulose base	32	1353
THALLIUM NITRATE	42	2727			
THALLIUM SALT, n.o.s.	53	1707	TOLUENE	27	1294
THALLIUM SULFATE, solid	53	1707	TOLUENE DI-ISOCYANATE (T.D.I.)	57	2078
THIAPENTANAL	55	2785	TOLUENE SULFONIC ACID, liquid	60	2584
THINNER	26	1263	TOLUENE SULFONIC ACID, liquid	60	2586
THIOACETIC ACID	26	2436	TOLUENE SULFONIC ACID, solid	60	2583
THIOGLYCOL	53	2966	TOLUENE SULFONIC ACID, solid	60	2585
THIOGLYCOLIC ACID	60	1940	TOLUIDINES (o-, m-, and p-)	55	1708
THIOLACTIC ACID	59	2936			
THIONYL CHLORIDE	39	1836			
THIOPHENE	27	2414			
THIOPHOSGENE	55	2474			
THIOPHOSPHORYL CHLORIDE	60	1837			

* Look for information next to this **NAME** in the TABLE OF EVACUATION DISTANCES in the back of this book. Use this in addition to the Guide Page if there is NO FIRE.

refers to the appropriate detailed instruction sheet in the *Guide*. (See Table 4.8.)

Tank trucks carrying hazardous materials must have the hazardous identification (HI) number on all four sides. The number can be applied in two ways: (1) as an identification number label in addition to the existing placard, with the orange-colored label having a black border and black numbers 4 in. (102 mm) high; or (2) the number can be displayed in the existing placard required by DOT.

Table 4.8   Guide Number 27 Refers to a Set of Detailed Instructions

Guide 27

POTENTIAL HAZARDS

FIRE OR EXPLOSION

Flammable/combustible material; may be ignited by heat, sparks or flames.
Vapors may travel to a source of ignition and flash back.
Container may explode in heat of fire.
Vapor explosion hazard indoors, outdoors or in sewers.
Runoff to sewer may create fire or explosion hazard.

HEALTH HAZARDS

May be poisonous if inhaled or absorbed through skin.
Vapors may cause dizziness or suffocation.
Contact may irritate or burn skin and eyes.
Fire may produce irritating or poisonous gases.
Runoff from fire control or dilution water may cause pollution.

EMERGENCY ACTION

Keep unnecessary people away; isolate hazard area and deny entry.
Stay upwind; keep out of low areas.
Wear self-contained (positive pressure if available) breathing apparatus and
    full protective clothing.
Isolate for 1/2 mile in all directions if tank car or truck is involved in fire.
FOR EMERGENCY ASSISTANCE CALL CHEMTREC **(800) 424-9300.**
If water pollution occurs, notify appropriate authorities.

FIRE

Small Fires: Dry chemical, CO_2, water spray or foam.
Large Fires: Water spray, fog or foam.
Move container from fire area if you can do it without risk.
Cool containers that are exposed to flames with water from the side until well
    after fire is out.
For massive fire in cargo area, use unmanned hose holder or monitor
    nozzles; if this is impossible, withdraw from area and let fire burn.
Withdraw immediately in case of rising sound from venting safety device or
    any discoloration of tank due to fire.

SPILL OR LEAK

Shut off ignition sources; no flares, smoking or flames in hazard area.
Stop leak if you can do it without risk.
Use water spray to reduce vapors.
Small Spills: Take up with sand or other noncombustible absorbent material
                and place into containers for later disposal.
Large Spills: Dike far ahead of spill for later disposal.

FIRST AID

Move victim to fresh air; call emergency medical care.
If not breathing, give artificial respiration.
If breathing is difficult, give oxygen.
In case of contact with material, immediately flush eyes with running water for
    at least 15 minutes. Wash skin with soap and water.
Remove and isolate contaminated clothing and shoes at the site.

Note that a tank vehicle transporting gasoline or fuel oil has a third option, because special placarding rules have been established for these products. In order to eliminate the need for changes, a single placard may be used with the HI number associated with the lowest flash point of any distillate fuel ever carried in the tank.

In addition, the words "gasoline" or "fuel oil" on the diamond-shaped placard in place of the HI number also are authorized. If a tank contains only gasoline or fuel oil and has the words "gasoline" or "fuel oil" on each side and rear in 2-in. (50-mm) letters, or uses placards having the words "gasoline" or "fuel oil" in place of the HI number, the HI number will not be required.

Package and Container Labels

A word of warning: Absence of warning labeling on a package or container does not guarantee that the contents are not hazardous. Also, a warning may be printed on a label separate from the label bearing the product name or the manufacturer's signature, and may be on a different face of the package.

Two separate, but related, information systems require labels on shipping packages and on product containers. These systems are DOT labeling and warning labels on immediate containers.

DOT Labels

Current DOT hazmats regulations require diamond-shaped labels on shipping containers holding an amount of hazardous material which could cause a hazardous condition in transportation. These color-coded labels bear a hazard symbol, or pictograph, and the name of the hazard. They carry no warning text. Present regulations generally require only a single label on a multi-hazard exposure, that label showing the most dangerous hazard present. Where a material is extremely toxic, explosive, or highly radioactive, the proper label must be used in addition to any other hazard label. Also, the package must be marked with the name of the contents as it appears on a commodity list required by DOT regulations. Because this may be a chemical name or a class name such as "corrosive liquid N.O.S.," it is of secondary usefulness. On compressed-gas cylinders, the label may be attached to a wire-tied tag.

The presence of a DOT label on a package or container indicates a dangerous hazard. Because of exemptions, based mostly on the size of inner containers placed in outer shipping containers, a package *without* a DOT label *may contain* significant amounts of fairly dangerous materials, including flammable liquids.

See Figures 4.2(a), (b), and (c) for illustrations of DOT labels.

Warning Labels on Immediate Containers

Containers for hazardous materials in interstate commerce usually bear necessary warning labels, but products of small local distributors may not. Warehouses, particularly those of contract packagers, may contain unlabeled stocks of filled containers intended for labeling to customers' orders. Where the immediate container is also the shipping container, such as a carboy, a

HIGHWAY SHIPMENTS

OXYGEN placard required for pressurized liquid oxygen.

OXYGEN placard may also be used to identify liquefied oxygen contained in a manner that does not meet the definition of a compressed gas in Sec. 173.300.

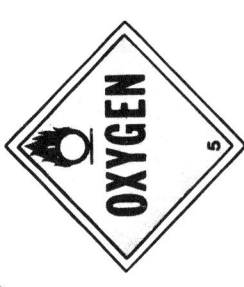

CARGO TANKS AND PORTABLE TANKS

Above placard may be used in place of FLAMMABLE placard when gasoline is being transported (Sec. 172.542(c)).

Above placard may be used in place of COMBUSTIBLE placard when FUEL OIL that is not classed as a Flammable liquid is being transported (Sec. 172.544(c)).

RAIL SHIPMENTS

CHLORINE placard required only for a packaging having a rated capacity of more than 110 gallons. Use the NON-FLAMMABLE GAS placard for packagings having a rated capacity of 110 gallons or less.

EMPTY placard. Each "empty" tank car must be placarded with an EMPTY placard that corresponds to the placard that was required for the material the tank car last contained unless the tank car last contained a Combustible liquid. This placard is required for the following hazardous materials.

Non-Flammable Gas	Flammable Solid
Oxygen	Flammable Solid W
Flammable Gas	Oxidizer
Chlorine	Organic Peroxide
Poison Gas	Poison
Flammable	Corrosive

FREIGHT CONTAINERS

FREIGHT CONTAINERS—640 CUBIC FEET OR MORE
(Placard each end and each side)

AIR OR WATER—Placard any quantity

HIGHWAY OR RAIL

1. Placard any quantity of hazardous material classes listed in TABLE 1 (Sec. 172.512(a)).

2. Placard 1,000 pounds or more (aggregate gross weight) of hazardous material classes in TABLE 2 (Sec. 172.512(a)).

NOTE: For placarding options for freight containers of less than 640 cubic feet, see Sec. 172.512(b).

CARGO TANKS AND PORTABLE TANKS

1. Cargo tanks containing any quantity of hazardous material must be placarded.

2. Portable tanks having a rated capacity of 1,000 gallons or more must be placarded.

3. Portable tanks having a rated capacity of less than 1,000 gallons need be placarded on only two opposite sides (Sec. 172.514(a)).

4. Cargo tanks and portable tanks must remain placarded when emptied unless reloaded with a material not subject to CFR, Title 49, Parts 100-199 or sufficiently cleaned and purged to remove any potential hazard. (Sec. 172.514(b)).

5. For Combustible liquids, a FLAMMABLE placard may be used on cargo tanks or portable tanks when transported by highway or water.

USE OF THIS CHART DOES NOT RELIEVE PERSONS INVOLVED FROM COMPLYING WITH THE DOT HAZARDOUS MATERIALS REGULATIONS AS PUBLISHED IN TITLE 49, CODE OF FEDERAL REGULATIONS, PARTS 100-199.

Fig. 4.2(a) DOT Hazardous Materials Warning Placards. The "explosive" labels are black on orange. The "poison gas" labels are black on white. The "radioactive" label is yellow, black, and white. The "flammable solid" label is red, white, blue, and black (Department of Transportation Chart 5).

PLACARDING ANY QUANTITY—

TABLE 1

MOTOR VEHICLES, FREIGHT CONTAINERS AND RAIL CARS

Placard motor vehicles, freight containers, and rail cars containing "any quantity" of hazardous materials listed in TABLE 1.

HAZARDOUS MATERIAL CLASSED OR DESCRIBED AS	PLACARDS
Class A explosives	EXPLOSIVES A.
Class B explosives	EXPLOSIVES B.
Poison A	POISON GAS.
Flammable solid (DANGEROUS WHEN WET label only)	FLAMMABLE SOLID `W_
Radioactive material	RADIOACTIVE.
Radioactive material:	
Uranium hexafluoride, fissile (containing more than 0.7 pct U^{235})	RADIOACTIVE AND CORROSIVE.
Uranium hexafluoride, low specific activity (containing 0.7 pct. or less U^{235})	RADIOACTIVE AND CORROSIVE.

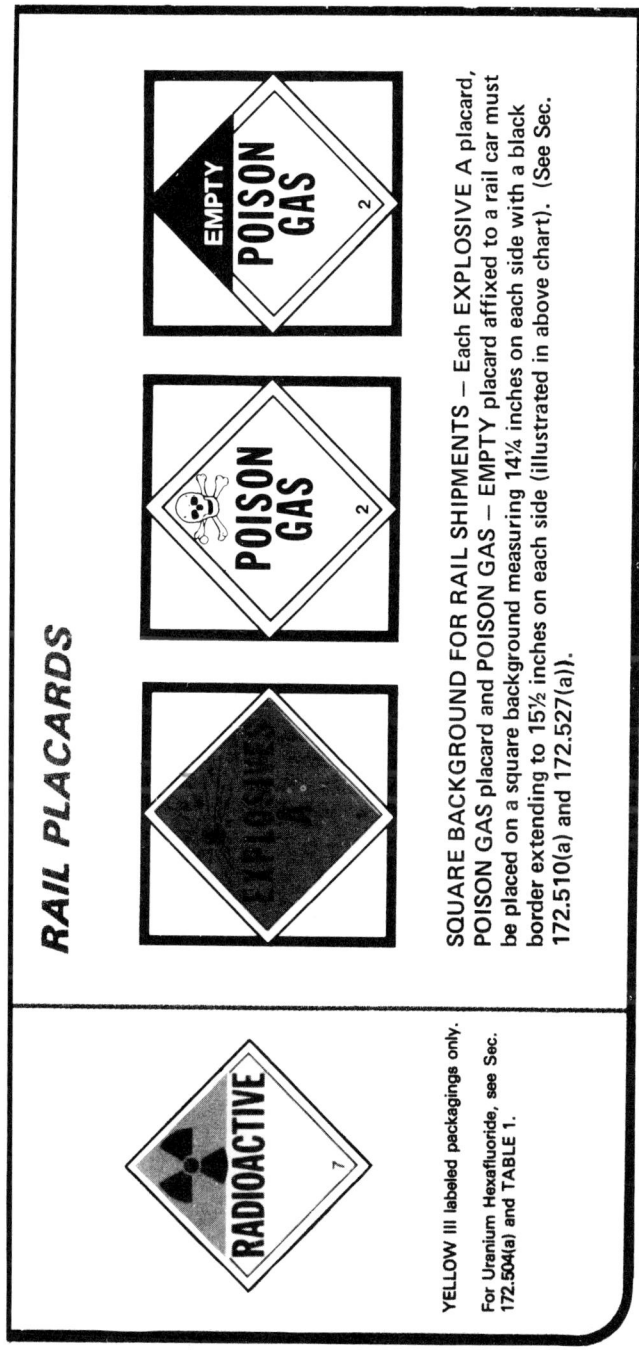

Fig. 4.2(b) The "flammable," "flammable gas," and "combustible" labels are white on red. The "poison" and "corrosive" labels are black and white. The "organic peroxide" and "oxidizer" lables are black on yellow. The "nonflammable gas" label is white on green. The "flammable solid" and "dangerous" labels are red, white, and black. (Department of Transportation Chart 5).

TABLE 2

OTHER PLACARDING REQUIREMENTS

MOTOR VEHICLES, RAIL CARS AND FREIGHT CONTAINERS

1. Placard motor vehicles and freight containers containing 1,000 pounds or more gross weight of hazardous materials classes listed in TABLE 2.

2. Placard any quantity of hazardous materials classes listed in TABLES 1 and 2 when offered for transportation by air or water.

3. Placard rail cars containing any quantity of hazardous materials classes listed in TABLE 2 except when less than 1,000 pounds gross weight of hazardous materials is transported in TOFC (Trailer on flat car) or COFC (Container on flat car) service.

HAZARDOUS MATERIAL CLASSED OR DESCRIBED AS	PLACARDS
Class C explosives	FLAMMABLE.
Nonflammable gas	NONFLAMMABLE GAS.
Nonflammable gas (Chlorine)	CHLORINE.
Nonflammable gas (Fluorine)	POISON.
Nonflammable gas (Oxygen, pressurized liquid)	OXYGEN.
Flammable gas	FLAMMABLE GAS.
Combustible liquid	COMBUSTIBLE.
Flammable liquid	FLAMMABLE.
Flammable solid	FLAMMABLE SOLID.
Oxidizer	OXIDIZER.
Organic peroxide	ORGANIC PEROXIDE.
Poison B.	POISON.
Corrosive material	CORROSIVE.
Irritating material	DANGEROUS.

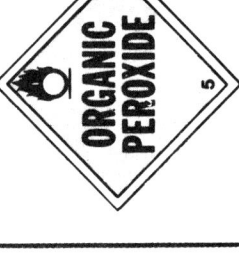

Also used for Class C Explosives labeled with an Explosive C label.

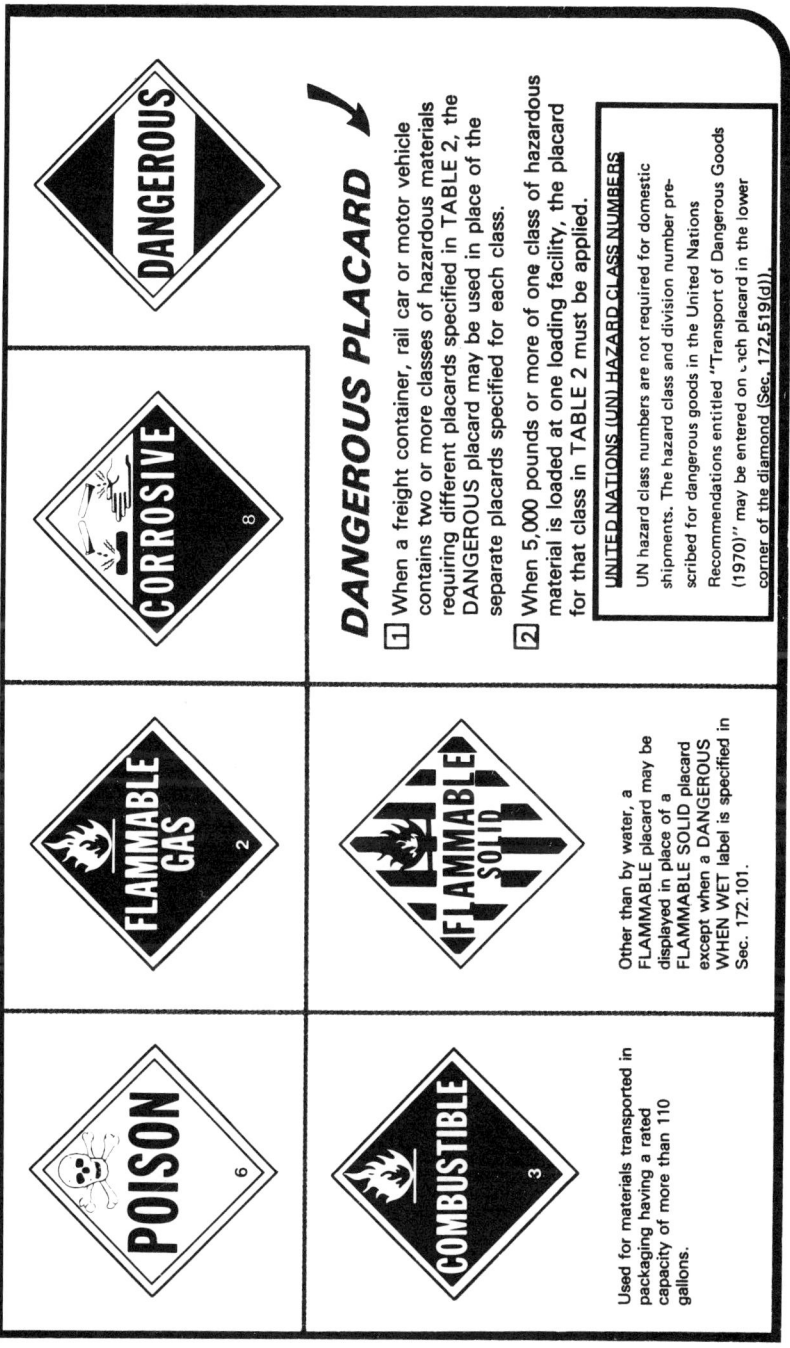

DANGEROUS

CORROSIVE
8

FLAMMABLE GAS
2

POISON
6

FLAMMABLE SOLID

COMBUSTIBLE
3

DANGEROUS PLACARD

☐1 When a freight container, rail car or motor vehicle contains two or more classes of hazardous materials requiring different placards specified in TABLE 2, the DANGEROUS placard may be used in place of the separate placards specified for each class.

☐2 When 5,000 pounds or more of one class of hazardous material is loaded at one loading facility, the placard for that class in TABLE 2 must be applied.

UNITED NATIONS (UN) HAZARD CLASS NUMBERS

UN hazard class numbers are not required for domestic shipments. The hazard class and division number pre-scribed for dangerous goods in the United Nations Recommendations entitled "Transport of Dangerous Goods (1970)" may be entered on ₁ach placard in the lower corner of the diamond (Sec. 172.519[d]).

Other than by water, a FLAMMABLE placard may be displayed in place of a FLAMMABLE SOLID placard except when a DANGEROUS WHEN WET label is specified in Sec. 172.101.

Used for materials transported in packaging having a rated capacity of more than 110 gallons.

Fig. 4.2(c)   The "oxygen" label is black on yellow. The "chlorine" label is black on white. The "gasoline" and "fuel oil" labels are white on red. The "empty" sticker is black. (Department of Transportation Chart 5).

drum, or a compressed-gas cylinder, DOT and warning labels may both be present and may be printed on a single label sheet.

Immediate container labeling practice is to use the signal word "DANGER" for corrosive liquids, extremely corrosive solids, poisons, flammable gases, and extremely flammable liquids. The signal word "WARNING" is used on flammable liquids, less-corrosive solids, toxic materials, and similar levels of hazard. The signal word "CAUTION" is used for the least-severe hazards, including combustible liquids and solids and nonflammable compressed gases. The rest of the warning label is self-explanatory, except to note that warning labels cover only hazards arising from normal use and handling. Hazards under fire conditions are not included in the warning.

Consumer Products: Substances or articles intended or suitable for household use are labeled in accordance with the Federal Hazardous Substances Act which vests authority in the Consumer Product Safety Commission (CPSC).

Shipping Papers

Transport regulations require that a truck driver, a train conductor, or an aircraft pilot have a shipping paper for every shipment of hazardous material in the transport unit. This shipping paper lists names and quantities of all hazardous materials, including those exempt packages which bear no labels, and should include hazard and emergency handling information.

Hazard Emergency Teams

Emergency assistance in identifying hazards of materials and for guidance on handling emergencies involving them is available on short notice from a variety of sources. The assistance ranges from immediate advice via telephone to the dispatch of emergency teams to assist in field operations. Among these services are:

CAER: The Community Awareness and Emergency Response Program was established by the Chemical Manufacturers Association (CMA) to integrate industry's emergency response plan with the community's. For information, write to CMA, 2501 M St., N.W., Washington, DC 20037.

CHEMTREC: Telephone (800) 424-9300. CHEMTREC stands for Chemical Transportation Emergency Center, and is operated by the Chemical Manufacturers Association, 2501 M St., N.W., Washington, DC 20037. It handles only transportation emergencies. When called, a CHEMTREC operator provides immediate advice from CHEMTREC files, then notifies the shipper. CHEMTREC maintains close liaison with the Department of Transportation.

NACA Pesticide Safety Team Network: A service sponsored by the National Agricultural Chemicals Association, 1155 15th St., N.W., Washington, DC 20005, which covers only pesticides. A network of more than 40 teams of specially trained personnel go to the scene of emergencies involving pesticides. Telephone answering service is supplied by CHEMTREC.

TEAP: The Canadian Chemical Producers' Association operates a Transportation Emergency Assistance Program through regional teams prepared to give telephone and field response.

CHLOREP: The Chlorine Institute, 342 Madison Ave., New York, NY 10017, operates the Chlorine Emergency Plan under which the nearest chlorine producer responds to a problem.

U.S. Coast Guard "React Teams": These teams operate on the east and west coast, giving skilled help in hazardous material spills on water.

Environmental Protection Agency: The EPA sponsors teams which go to hazardous material emergencies.

Many manufacturing companies have organized response capabilities for their own products. In many cases, an emergency telephone number appears on the bill of lading.

The NFPA 704 System of Hazard Identification

NFPA 704, *Standard System for the Identification of the Fire Hazards of Materials*,[17] is a symbol system intended for use on fixed installations, such as chemical processing equipment, storage and warehousing rooms, and laboratory entrances. The system is designed to tell a fire fighter what must be done to protect himself/herself from injury while fighting a fire in the area.

The NFPA 704 Diamond

The information system based on the "704 Diamond" (see Figure 4.3) is the vehicle for visually presenting information on health, flammability, and reactivity hazards. The diamond also can convey special information associated with the hazards.

The NFPA 704 diamond symbol provides immediacy, but at some sacrifice of adequacy; also, there is a tendency to read more into it than it says. The five degrees of hazard, in the order of their descendency, have these general meanings to fire fighters:

4—Too dangerous to approach with standard fire fighting equipment and procedures. Withdraw and obtain expert advice on how to handle.

3—Fire can be fought using methods intended for extremely hazardous situations, such as using unmanned monitors or personal protective equipment which prevents any bodily contact.

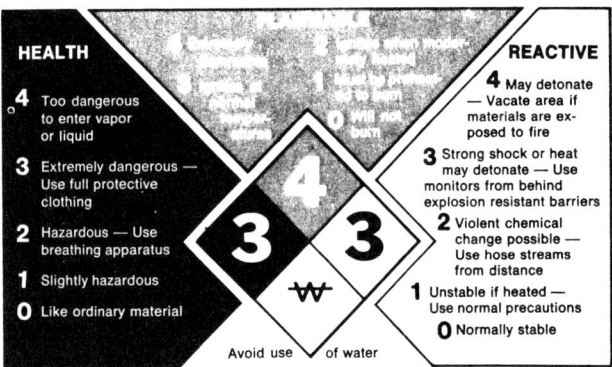

Fig. 4.3   The NFPA 704 diamond.

2—Can be fought with standard procedures, but hazards are present which require certain equipment or procedures to handle safely.

1—Nuisance hazards present which require some care, but standard fire fighting procedures can be used.

0—No special hazards, therefore no special measures.

The numbers from 0 through 4 are placed in the three upper squares of the diamond to show the degree of hazard present for each of the three hazards. Zero indicates the lowest degree of hazard, and four, the highest. The fourth square, at the bottom of the diamond, is used for special information. Three symbols for this bottom space are suggest by NFPA 704. (See Figure 4.4.) They are:

1. The letter W with a bar through it to indicate that a material may have a hazardous reaction with water. This does not mean "Do Not Use Water," since some forms of water—fog or fine spray—may be used in many cases. What it does say is: "Water may cause a hazard, so use it very cautiously until you have proper information."
2. The "radioactive pinwheel" for radioactive materials.
3. The letters OX to indicate an oxidizer.

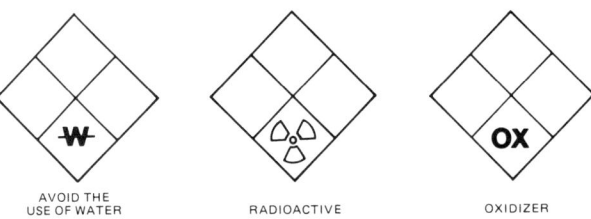

Fig. 4.4   Special information is presented in the bottom square of the NFPA 704 diamond. This square has a white color code.

NFPA 704 describes in detail the hazards and hazard levels which the various numbers indicate for the three hazards: health, flammability, and reactivity. Adapted from the appendix of the standard, the following material summarizes the hazard information and recommends protective actions.

Health Hazards

In general, the health hazard in fire fighting is that of a single exposure which may vary from a few seconds up to an hour. The physical exertion demanded in fire fighting or other emergencies may be expected to intensify the effects of any exposure. In assigning degrees of danger, local conditions must be considered. The following explanation is based upon use of the protective equipment normally worn by fire fighters. (See Figure 4.5.)

4—Materials which are too dangerous to health for fire fighters to be exposed to them. A few whiffs of the vapor could cause death, or the vapor or liquid could be fatal on penetration of the fire fighter's normal protective clothing. Protective clothing and breathing apparatus available to the average fire department will not provide adequate protection against inhalation or skin contact with these materials.

3—Materials which are extremely hazardous to health, but fire areas may be entered with extreme care. Full protective clothing, self-contained breathing apparatus, rubber gloves, boots, and bands around legs, arms, and waist should be provided. No skin surface should be exposed.

2—Materials which are hazardous to health, but fire areas may be entered freely with self-contained breathing apparatus.

1—Materials which are only slightly hazardous to health.

0—Materials which on exposure under fire conditions would offer no health hazard beyond that of ordinary combustible materials.

Flammability Hazards

Susceptibility to burning is the basis for assigning degrees within this category. The method of attacking the fire is influenced by this susceptibility factor. (See Figure 4.6.)

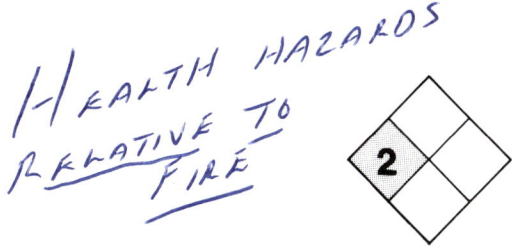

Fig. 4.5 Hazards to health are presented in the left-hand square of the NFPA 704 diamond. This square has a blue color code.

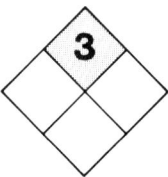

Fig. 4.6   Flammability hazards are presented in the top square of the NFPA 704 diamond. This square has a red color code.

4—Very flammable gases or very volatile flammable liquids. If possible, shut off flow and keep cooling water streams on exposed tanks or containers. Withdrawal may be necessary.

3—Materials that can be ignited under almost all normal temperature conditions. Water may be ineffective because of the low flash point of the materials.

2—Materials that must be moderately heated before ignition will occur. Water spray may be used to extinguish the fire, because the material can be cooled below its flash point.

1—Materials that must be preheated before ignition can occur. Water may cause frothing if it gets below the surface of the liquid and turns to steam. If this is the case, water fog gently applied to the surface will cause a frothing which will extinguish the fire.

0—Materials that will not burn.

Reactivity (Stability) Hazards

The assignment of relative degree of hazard in the reactivity category is based upon the susceptibility of materials to release energy either by themselves or in combination with other materials. Fire exposure is one of the factors considered along with conditions of shock and pressure. (See Figure 4.7.)

4—Materials which are readily capable of detonation at normal temperatures and pressures. If they are involved in a massive fire, vacate the area.

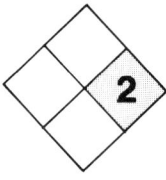

Fig. 4.7   Reactivity (stability) hazards are presented in the right-hand square of the NFPA 704 diamond. This square has a yellow color code.

3—Materials which when heated and under confinement are capable of detonation and which may react violently with water. Fire fighting should be conducted from behind explosion-resistant locations.

2—Materials which will undergo a violent chemical change at elevated temperatures and pressures but do not detonate. Use portable monitors, hoseholders, or straight hose streams from a distance to cool the tanks and the material in them. Use caution.

1—Materials which normally are stable but which may become unstable in combination with other materials or at elevated temperatures and pressures. Use normal precautions as in approaching any fire.

0—Materials which normally are stable and, therefore, do not produce any reactivity hazard to fire fighters.

Special Information

When ₩ appears at the bottom in the 4th space (see Figure 4.4), the following apply:

4— ₩ is not used with reactivity hazard 4.

3—In addition to hazards above, these materials can react explosively with water. Explosion protection is essential if water in any form is used.

2—In addition to hazards above, these materials may react violently with water or form potentially explosive mixtures with water.

1—In addition to hazards above, these materials may react vigorously but not violently with water.

0— ₩ is not used with reactivity hazard 0.

Methods of Presentation

Considerable leeway is allowed in the presentation of the NFPA 704 diamond numbers. The only basic requirement is that numbers be spaced as though they were in the diamond outline. Several methods which have been used are shown in Figure 4.8. Chapter 5 of NFPA 704 gives recommended layout and sizes for the symbol, and a distance-legibility table, as well as several examples of the symbol's use.

Assigning Degrees of Hazard

Numbers (degrees of hazard) for use in the diamond are assigned on the basis of the worst hazard expected in the area, whether from hazards of the original material or of its combustion or breakdown products. The effects of local conditions must be considered. For instance, a drum of carbon tetrachloride sitting in a well-ventilated storage shed presents a different hazard from a drum sitting in an unventilated basement.

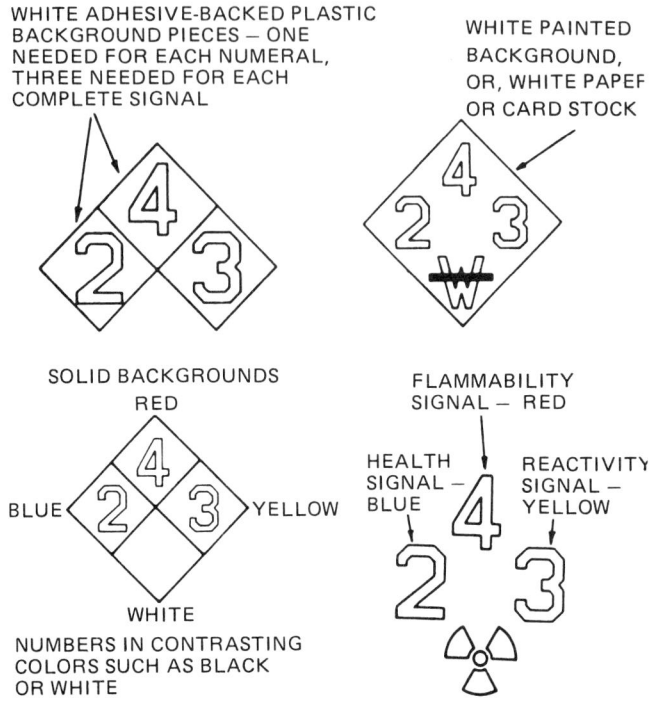

Fig. 4.8   Methods of presenting the NFPA 704 system hazard information.

Summary

The three basic types of materials that provide fuel for all fires are (1) combustible solids, (2) flammable and combustible liquids, and (3) flammable gases. The number and variety of these materials are in the thousands.

Wood is still the most "basic" combustible solid fuel for fires. However, even the combustible properties of wood are not a simple subject. The physical properties of wood, the moisture content, and the smoke factor involved in burning wood are but a few of the necessary considerations in fire protection planning and fire control decisions. Plastics and textiles, which also are combustible solids, present special fire hazards because of their chemical nature and properties.

The characteristics of flammable and combustible liquids are many and varied. Temperature and pressure have a marked effect on the fire hazards that might be presented by flammable and combustible liquids. In addition, the special characteristics of these liquids—such as ignition temperature and evaporation rate—affect all fire protection considerations and decisions.

Gases can be classified by their physical properties—whether they are flammable, nonflammable, toxic, and/or reactive—as well as by their physical state—whether they are liquefied, nonliquefied, or cryogenic liquids. These properties or states of gases have direct bearing on fire protection considerations and decisions. In addition, gases can be classified by usage—fuel, industrial, or medical. Knowledge of the physical laws regarding the behavior of gases as well as gases' chemical reactivity and behavior is a vital decision-making factor for all fire fighting personnel.

Hazardous materials such as corrosive chemicals and vapors and radioactive materials require a special knowledge and consideration because of the high life safety factors involved. Inorganic acids, halogens, and nuclear materials all require special handling, storage, and fire fighting procedures.

Fire fighting and control with regard to fire hazards and materials is dependent on knowledge about the materials involved. In addition, more specialized knowledge and information may be required regarding materials stored or used in buildings or plants in a particular fire protection district. Preplanning is a necessary factor where hazardous materials are known to be used in relatively large quantities. In addition, members of the fire service should be aware of the kinds of hazardous materials likely to be transported through a given area. With this information, preplanning and any necessary special firesafety and life safety procedures may be planned well in advance of any emergency.

References

[1]Tuve, Richard L. 1976. *Principles of Fire Protection Chemistry*, National Fire Protection Association, Boston.

[2]Beall, F.C., and Eichner, H.W. 1970. "Thermal Degradation of Wood Components: A Review of the Literature," *Report No. 130*, Forest Products Laboratory, Madison, WI.

[3]*Fire Protection Handbook*, 16th ed. 1986. National Fire Protection Association, Quincy, MA.

[4]"America Burning," 1973. The National Commission on Fire Prevention and Control, Washington, DC.

[5]NFPA 321-1987. *Standard on Basic Classification of Flammable and Combustible Liquids*, National Fire Protection Association, Quincy, MA.

[6]NFPA 30-1987. *Flammable and Combustible Liquids Code*, National Fire Protection Association, Quincy, MA.

[7]NFPA 50-1985. *Standard for Bulk Oxygen Systems at Consumer Sites*, National Fire Protection Association, Quincy, MA.

[8]NFPA 50A-1984. *Standard for Gaseous Hydrogen Systems at Consumer Sites*, National Fire Protection Association, Quincy, MA.

[9]NFPA 50B-1985. *Standard for Liquefied Hydrogen Systems at Consumer Sites*, National Fire Protection Association, Quincy, MA.

[10]NFPA 51-1987. *Standard for the Design and Installation of Oxygen-Fuel Gas Systems for Welding, Cutting and Allied Processes*, National Fire Protection Association, Quincy, MA.

[11]NFPA 54-1984. *National Fuel Gas Code*, National Fire Protection Association, Quincy, MA.

[12]NFPA 58-1986. *Standard for the Storage and Handling of Liquefied Petroleum Gases*, National Fire Protection Association, Quincy, MA.

[13]NFPA 99-1987. *Standard for Health Care Facilities*, National Fire Protection Association, Quincy, MA.

[14]ASME *Boiler and Pressure Vessel Code* 1985. American Society of Mechanical Engineers, New York.

[15]NFPA 325M-1984. *Fire Hazard Properties of Flammable Liquids, Gases, and Volatile Solids*, National Fire Protection Association, Quincy, MA.

[16]Bahme, Charles W. 1976. *Fire Officer's Guide to Emergency Action*, 3rd ed., National Fire Protection Association, Boston.

[17]NFPA 704-1985. *Standard System for the Identification of the Fire Hazards of Materials*, National Fire Protection Association, Quincy, MA.

Additional Reading

Purington, Robert G., and Patterson, Wade 1977. *Handling Radiation Emergencies,* National Fire Protection Association, Boston.

NFPA 801-1986. *Recommended Fire Protection Practice for Facilities Handling Radioactive Materials,* National Fire Protection Association, Quincy, MA.

Chapter **5**

Investigating the Fire Loss Problem

Very seldom can the cause of fire be determined without some degree of investigation. Current codes, standards, and inspection and suppression procedures were developed mainly from the investigation of fires and from the careful analysis of the important information that these investigations yielded.

THE NEED FOR INVESTIGATIONS

Fire always has been a major concern of society. Although originally fire often was thought of as a supernatural act, its destructive qualities and the tragedies it caused soon made people want more advanced knowledge on how to protect themselves from fires. In order to better understand fire protection needs, investigations were made to determine the causes of past fires and to discover ways to eliminate similar occurrences in the future.

Although it is not known exactly when the first fire investigations took place, it can be speculated that they may have resulted from the curiosity of early human beings. Following unset fires that occurred in their dwellings, early people often returned to the fire scene to sift through the rubble in search of salvageable possessions. As they searched, they probably thought about how and why the fires started. The earliest recorded instances of what can be considered organized fire investigation date from the days of the Roman Empire when the *Quarstionarius*—the equivalent of today's state fire marshal—was assigned to determine the causes of all destructive fires.

As early civilizations expanded and the use of various fire-prone materials for building construction became more common, some of the reasons for the simultaneously increasing number of fires became more obvious. Overcrowded building conditions, thatched roofs, and wood shingles soon were recognized as being directly related to fire occurrences. Still later, medical research revealed that specific by-products of fire, such as toxic gases, were the actual cause of many deaths from fire. Such information eventually led to in-depth testing of the burning characteristics of manufactured materials.

The Purpose of Fire Investigations

Because local fire departments are present at most fires, they have the best opportunity to compile information on fire incidents. Fire investigations

113

are an important function in determining that information. Fire investigations by the public fire service serve the following three basic purposes:

1. *To determine what happened, so that preventive measures can be taken in the future*: Too often, fire investigations are conducted strictly to affix blame. A fire occurrence generally is a failure of either a code enforcement program or a public education program, except where criminal activity is involved. Fire investigations can lead to improved pre-fire activities that exact better control over the fire ignition sequence and over fire development and growth.

2. *To ascertain whether any criminal activity was involved*: The rate of incendiarism is on the increase. Only through proper fire investigation can incendiarism be detected and the proper evidence secured for conviction of arsonists. Criminal activity also can include negligent disregard for codes and regulations, which can result in deaths or injuries even if a fire is of accidental cause.

3. *To provide accurate information for the fire report*: (By definition, the fire report is the legal record of a fire incident.) The fire service must maintain an accurate report of each fire occurrence, the circumstances surrounding it, and the damage and/or casualties resulting from it. Accurate information for the fire report can come only from an investigation of the fire incident.

Scope of the Investigation

The three most significant areas in fire investigation are: (1) fire ignition sequence, (2) fire development, and (3) fire casualties.

Fire Ignition Sequence: It is important to determine the location within the property where the fire began, and the ignition sequence that caused the fire to start. Identifying this ignition sequence involves three factors: (1) a heat source, (2) a combustible material, and (3) an event, human action, or natural act that combines the heat source with the combustible material to start the fire. Each of these three factors must be identified separately in order to fully explain the ignition sequence.

Information provided by fire-scene witnesses can be helpful in determining the fire ignition sequence. However, when the fire ignition sequence cannot be determined so easily, the point of origin needs to be ascertained, as accurately as possible. This process is started by first reconstructing the layout of the room or area of origin. Information provided by surviving occupants, or by individuals who are familiar with the interior of the property, can be of value in determining the position of furniture and the types of combustible materials that were in the area.

When the area of origin has been identified and then reconstructed, the investigator can begin to identify potential points of fire origin within the area of origin. An examination of the contents of the room or area of origin can help determine the level of fire origin and the direction of burning or fire travel. The point of origin is sometimes fairly obvious, but many times it is not the area of most severe burning. The investigator must be able to identify a heat source and a combustible material at or near any potential point of origin.

Each heat source needs to be evaluated carefully to determine (1) if it could have provided enough heat energy to ignite the fuels present, (2) if it was in a proper physical relationship with the fuel to ignite it, and (3) in situations where the heat was from a piece of equipment, if the equipment was on or operating in such a way to have provided the heat. Each heat source/combustible material relationship in the general area of the point of origin must be carefully examined and either eliminated as responsible for the fire or verified as the possible ignition point. This process of elimination eventually should lead the investigator to a point of origin, a heat source, a combustible material, and—it is hoped—a reason the heat source and the combustible material were present in a configuration which allowed the fire to start. When the fire ignition sequence has been established, investigators can begin to determine the fire's development.

Fire Development: Once a fire starts, its growth or development is based on a number of factors, each of which is important in understanding why a fire grew as it did or remained as small as it did. The contents of a building often provide fuel to the fire, thus allowing the fire to grow and spread. These materials should be identified, and their role in the overall fire development should be evaluated. Additionally, contents of a room may effect fire fighting conditions or may create fire casualties because of gases or smoke they created.

Another factor directly involved in fire development is the compartmentation, or other subdivision, of a structure. Because these physical barriers play an important part in limiting the development of fire and smoke conditions, their performance should be evaluated after the fire. If fire walls, doors, or other smoke- or fire-limiting devices were present and did not perform as intended, the reasons for unsatisfactory performance should be explored.

Furthermore, the presence of conditions which allowed unusual fire spread should be investigated and documented. These can include highly combustible interior wall and ceiling finishes and voids or concealed spaces that can allow horizontal or vertical fire spread. Such conditions provide serious threats to life safety in buildings. Their role in a fire should be understood.

The length of time the fire burned prior to its detection and the time between the fire's detection and resultant transmission of an alarm to the fire department are important factors in the ultimate extent of damage from the fire. The effect that automatic fire detection and extinguishing equipment had on the extent of damage or on the survivors also should be analyzed. If there were delays in the transmission of alarms from automatic detection equipment, the reasons for these delays should be evaluated as part of the overall fire development sequence. Likewise, any delay or failure of fire extinguishing equipment needs to be thoroughly investigated to understand why it occurred.

Fire department tactics at the fire scene can affect the outcome of a fire. To help determine whether or not a fire department had the available resources to properly tackle a fire or if it properly utilized available resources, which ultimately helps determine whether any part of the fire's development or spread resulted from inadequate tactical procedures, a thorough evaluation of fire-scene tactics should be made after each major fire.

Fire Casualties: Each injury and fatality associated with a fire incident should be thoroughly investigated to determine the reasons it occurred and any information that could lead to prevention of future casualties. All casualties should be traced for a period of time to determine their outcome. To evaluate the reasons for the casualties, an understanding of the fire development sequence may be necessary. Finally, the fire department's public education program should be studied to determine its effect upon the reasons for the safe survival of some occupants, or for the lack of knowledge of those who became casualties.

Responsibility for Fire Investigation

In many American states, the primary responsibility for investigation of incendiary fires is assigned to the office of the state fire marshal, and in Canada, to the provincial fire commissioner or fire marshal. In states that have no such office, or where the fire marshal's office does not have the responsibility for arson investigation, such investigation is conducted by the local police and fire departments.

The preliminary investigation, however, is the responsibility of the local fire department. This work often will be the basis for establishing whether the fire is suspicious or not and thus whether the appropriate state or provincial authorities should be notified.

Ideally, a fire department should have a team of specialists available for assistance when investigation circumstances are too complex to handle, or when the situation is too time-consuming for the fire officer. In any event, the fire officer present during the extinguishment of the fire should remain involved to lend support to the investigative team in reconstructing the scene and identifying the time sequence of certain events.

Loss-of-life fires, large fires, and arson fires are likely to be investigated by agencies in addition to the local fire department. For example, besides the state fire marshal who often is called in by the fire department in these cases, insurance companies and insurance bureaus often investigate fires of special interest to them. Also, manufacturers of products involved in the fire may send investigators to report on the performance of their products.

NFPA for many years has sent specialists to the scenes of significant fires to work with local officials in documenting the incidents. This effort has produced many well-documented accounts of major fires—accounts that have become part of fire protection literature.

Conducting the Investigation

Every fire should be investigated. The depth of investigation will vary with the size and toll of the fire, the scope of the investigation, and the length of time it takes to fully understand all the facts about the incident. In some fires, the fire department officer-in-charge can readily determine what happened and can adequately give the facts on the fire report. In other cases, however, the size of the fire, the degree of burning, and/or the complexity of the situation indicate that trained fire investigators should be called to assist with or conduct the investigation.

The following is a general procedure which can be followed for an investigation:

1. Review exterior of structure, documenting fire damage, fire fighting damage, and other outside physical evidence.
2. Review interior of structure, documenting fire damage, fire fighting damage, areas of most severe burning, and any unusual conditions observed.
3. Reconstruct as much as possible to identify the pre-fire location and position of contents as well as the types and amounts of fuel.
4. Study burn patterns and burning times to identify an area of origin.
5. Identify all potential heat sources in the area of origin.
6. Talk with fire fighters to get their observations on conditions on arrival, smoke color, property security, etc.
7. Interview witnesses or occupants and corroborate their statements with physical evidence.
8. Conduct tests of materials or laboratory analysis of equipment to determine burning characteristics, composition of residues, or reasons for mechanical failures.
9. Collect other pertinent information, such as service records, inventory records, time information from fire dispatcher, etc.
10. Analyze all information. Determine the point of origin and most probable sequence of events and failure mode.

For a small simple fire, many of the above steps can be omitted. The evidence will be very obvious or occupants of the area will have observed what happened. However, many fire reports carry an erroneous fire cause, because the steps were not followed and someone jumped to a conclusion. Proper reconstruction of the scene would have revealed information leading to a different, more accurate fire cause determination.

It is important that fire investigation efforts be properly supported. Fire scenes must be photographed to record locations of evidence, extent of fire development, relationships of materials, and details in the area of origin and at the point of origin. Some fire departments are using videotape to supplement still photographs.

Adequate laboratory facilities should be available. Samples from the fire scene may need to be checked for presence of accelerants or may just need to be identified and have their burning characteristics documented. Most state fire marshal offices or state police departments have laboratories for chemical analysis. In addition, numerous commercial laboratories are available to conduct the tests.

Some fires may be beyond the skills of even experienced fire investigators. In such cases, experts should be called for assistance. Lists of consultants or experts can be obtained from the National Fire Protection Association, International Association of Arson Investigators, National Association of Fire Investigators, and most state fire marshal offices. Faculty at a local university science, physics, or engineering department may also be of assistance in certain aspects of the investigation.

RECORDING THE FIRE PROBLEM

All the information gathered during the fire investigation should be properly recorded in a fire report. The fire report—the legal record of a fire incident—includes information on the time of the incident, response to the incident, action taken, details of the fire, and damage or casualties resulting from the incident. Some fire incident reports may be brief, while others may be extensive and include photographs, physical evidence, and laboratory test results. Whatever its length, scope, or form, the fire report should be in the words of the fire officer present at the incident, and should be complete and clear so that another fire service member not present at the incident can understand what happened. This section of this chapter will explore the purposes of fire reports and fire reporting systems, their application to fire investigations, and how these reporting methods are used.

Purpose of the Fire Report

The three basic purposes of a fire report at the local level are:

1. It is the legal record of the fact that the fire occurred, and provides official notification to those who may be legally required to know of the incident (e.g., the state or local fire marshal). It reports facts about the particular property affected, why the fire occurred, how building components and fire protection devices performed, casualties or damage that resulted, and fire department action.

2. It provides information to senior officers and fire department managers so they are kept informed about what is happening within their area of responsibility. This allows them to evaluate the performance of their units at the incident and to talk knowledgeably about the incident.
3. The report provides data on the fire problem to fire service management so they can track trends, gauge the effectiveness of current fire prevention and fire suppression measures and practices, evaluate the impact of new methods, and indicate areas that may require further attention.

The first two purposes can be served by any report that is an accurate description of the incident. The third purpose, however, requires that information be recorded in a consistent format that will permit meaningful aggregation of the data from reports on many incidents.

It also is important for a single report to serve the data needs of all potential users. The data required at the state and national levels must originate with what is collected locally. At the same time, it is important for locally collected data to have a visible, significant use at the local fire service level. If the data are collected only for the benefit of those outside the local area, the motivation and commitment to quality and completeness may diminish, with a resulting reduction in the usefulness of the data.

Uniformity in Fire Reporting

To maintain uniformity in fire reporting, the NFPA Technical Committee on Fire Reporting has developed NFPA 901, *Uniform Coding for Fire Protection*.[1] This standard establishes basic definitions and terminology for use in fire reporting and a means of classifying data so they can be aggregated either manually or automatically.

NFPA 901[1] provides the common language used by nearly all large-scale (e.g., state, national) data bases in the United States and around the world. It is recognized that not every fire department will want to collect every data element; likewise, there may be additional data elements that a fire department wishes to collect. A fire department that uses NFPA 901 will find itself in a position to most efficiently contribute data to larger data bases and, in turn, to use data from these larger systems in its management of the local fire problem. It also will find that the quality, uniformity, and usefulness of information within the fire department will improve.

NFPA 901[1] contains data elements that provide a classification of practially every type of property (whether fixed or mobile); a description of a specific structure prior to an incident; a description of the ignition sequence, including the area of origin of the fire; conditions found upon arrival; what action was taken; and why the fire grew as large or stayed as small as it did. There also are data elements for describing injuries or fatalities to both civilians and fire fighters, extent of damage, loss of property, and the investment in personnel provided by the community to control the incident. Useful analysis can be adapted to the particular data elements—no matter how few or how many—adopted by the fire department.

FIRE REPORTING SYSTEMS

There is a major difference between a fire report and a fire reporting system. A fire report is only part of a fire reporting system. A complete fire reporting system has to address these stages of operation: fact finding, fact processing, and fact use. Each of these stages is discussed below.

Fact Finding

The traditional functions of a fire report can be satisfied with a minimal written narrative of the basic facts of the incident. To serve as input to a fire reporting system, however, an incident report must be clearly structured and must use uniform definitions and terminology. The collection of such data requires not only the form or forms upon which to record the information desired, but also a set of instructions and related training regarding the forms so the information is provided in a uniform manner. A procedure must be established for forwarding the information for centralized use and storage.

Equally important is making sure that the people responsible for reporting the data are trained and capable of investigating a fire to determine its scenario and cause.

Fact Processing

Once data have been recorded, they must be processed into a form useful for legal, statistical, planning, and management purposes. The first step in information processing is to check the incident reports for accuracy, clarity, consistency, and completeness. A procedure of quality-control screening and followup corrections is needed for this purpose. For best results, visual screening of the reports is needed even when computerized edit checks are employed. The second step is to combine information about one incident from several reports into a composite record. The third step is to create a fire fact file of all records of all reported incidents. This file then becomes the basic source of information about similar incidents. Use of the file will determine the facts that must be recorded on an individual incident report.

In the last several years, fire departments have begun using computers to assist with fact processing. Computer programs quickly check the consistency and completeness of data about an incident. These same computers build and maintain a data base which becomes one part of the fire fact file.

Fact Use

Once an incident file has been created, whether as a paper file or a computerized data base, it has many potential uses. At a minimum, the file should meet all the informational needs of the local fire service. Legal and statistical information required to manage the department, to spot trends in fire incidence, and to provide documentation for program evaluation and fire research will be readily accessible.

Small fire departments might have special problems in creating fire incident fact files due to their low fire incidence level. This means they can analyze their fire problems only by working with several other small departments within a geographic region. Therefore, it is especially important that small fire departments use uniform terminology and uniform coding in collecting information so data from different departments can be merged.

Benefits of a Fire Reporting System

At the local level, a fire department can derive many benefits from a good fire incident reporting system, particularly if it is based on NFPA 901.[1] Some of the following uses involve no more than totaling data from the system. Others require more extensive analysis. Many of these benefits also apply to users at the state and national levels.

Describing a Community's Fire Problem: It is possible to pinpoint where fires are occurring, what factors are most responsible for ignitions, and what casualties and damage are occurring as a result of fires. With the problem placed in proper perspective, the most serious and vulnerable aspects of the fire problem can be tackled first.

Supporting Budget Requests: In this era of increasing concern about taxes, municipal officials are quick to cut budgets and slow to add new programs. Frequently, fire department managers do not have the statistics to support their requests for additional funds. Good statistics will put the fire problem in perspective with other municipal concerns, and help community officials realize the consequences of budget cuts or the value of new programs for the fire department.

Supporting Code Refinements: A good data base permits fire departments to identify and describe fire incidents that would have developed differently or might not have occurred at all if certain code changes had been in place. Loss statistics from other areas with more stringent codes can be used for comparison. Estimation of the likely impact of a code change can involve complex analysis, however, and no incident data base can address all the subtleties of code impact.

Evaluating Code Enforcement Programs: It is not sufficient to have codes on the books if they are not properly enforced. In evaluating loss experience, it is possible to see whether certain losses are occurring because existing codes are not being properly enforced. The reason for the improper enforcement then can be analyzed and corrected.

Evaluating Public Firesafety Education Programs. Not all problems can be solved by establishing and enforcing codes. Certain aspects of the fire problem can be controlled best by public firesafety education programs—educating people about the dangers of fire, how to reduce the occurrence, and how to react when hazardous situations arise. Such programs require considerable time and money. It is important to know the exact problem that needs to be addressed. Appropriate evaluation criteria must be in place to measure whether an educational program is helping to solve that aspect of the fire problem.

Planning Future Fire Protection Needs: Many communities and fire departments are becoming active in planning and are developing master plans. It is essential that the fire service be involved in such planning. A good data base will allow a fire department to compute fire rates relative to population and building inventory. This, along with other characteristics of the community fire problem and planning, will support better fire protection in the future, based on changing demography and planned community growth. It also will provide input to decisions on the type and level of fire protection a community will provide, so requirements can be established for developers who propose construction of properties that exceed fire department capabilities.

Improving Allocation of Resources: It is not always possible for a fire department to grow at the rate necessary to protect a changing community. In fact, it may not be necessary for a department to expand at all. Proper analysis may show where redeployment of existing resources can provide the same level of protection or even improve it.

Scheduling Nonemergency Activities: Training sessions, in-service inspections, and other activities are important aspects of a fire department's function. A fire department which tracks times and severity of fires can schedule nonemergency activities when they are least likely to be interrupted by emergency calls, or when the normal delay from such activities will have the least impact on emergencies.

Regulating Product Safety: Particularly at the national and state levels, a fire reporting system can be useful in measuring the size and severity of problems associated with various types of consumer products. By identifying the most common ways these products become involved in fire, more accurate precautions and information can be disseminated to consumers.

Support for Major Fire Engineering Analysis: The NFPA Systems Concepts Committee and other users of sophisticated engineering models depend upon the output of fire reporting systems on a continuing basis. Each time a method of fire defense works well and fire loss and danger are confined to a small area, success will increase confidence in that particular method of fire defense. Conversely, each time a method of fire defense fails, as indicated by injuries, deaths, or an expensive loss, this failure needs to be recorded so that method of fire defense can be reevaluated.

STATUS OF UNIFORM FIRE REPORTING

Considerable work has been and continues to be performed in developing fire reporting systems based on NFPA 901.[1] The NFPA Fire Reporting Committee also has established a basic system, published as NFPA 902M, *Fire Reporting Field Incident Manual*.[2] Figure 5.1 is an example of a fire incident report form used in NFPA 902M. The NFPA 902M system provides the necessary instructions for completion of an incident report, supplementary reports on civilian or fire fighter casualties suffered at incidents, and EMS reports which can be used by those fire departments providing Emergency Medical Services. Work sheets are provided for summarizing incident data. This system can be used manually or adapted for electronic data processing.

The U.S. Fire Administration (USFA), within the Federal Emergency Management Agency (FEMA), has developed the National Fire Incident Reporting System (NFIRS). This is an automated system based on the work of the NFPA Fire Reporting Committee, as published in NFPA 901.[1] The system now has been installed in approximately 40 states and the District of Columbia, and includes a number of larger fire departments.

NFIRS is also an incident-based system. Data is reported on an incident-by-incident basis. Computer programs edit the data, maintain a data base, and generate a variety of summary reports. The data collected is similar to that collected by the basic system published in NFPA 902M.[2]

At the state level, NFIRS provides for the collection of written reports on incidents to which local communities responded. Communities with automated data systems based on NFPA 901[1] can participate by providing their data on computer disks to the state. Summary reports are then generated based on both individual community and statewide fire experience.

Nationally, NFIRS provides for the merging of data bases from individual states to form a U.S. data base. FEMA/USFA analyzes this data base and publishes the analysis. The data base also is available to businesses or individuals who wish to perform their own analyses.

ANALYZING FIRE LOSSES

Analysis in the U.S. of patterns and trends in real-fire experiences is a powerful method by which to focus on the major problems in firesafety, check hypotheses, examine impacts of new technology and new purchase or use behavior, and answer what-if questions about the likely effects of new strategies. In the last decade, new national fire incident data bases have made possible more valid, detailed analysis of real-fire experience than was possible earlier. In the next decade, powerful new computer-based risk assessment tools will be available to provide swift manipulation of these data bases, product performance data bases, and scientifically based models of fire

BASIC INCIDENT REPORT   902F

Fill In this Report In Your Own Words _____ **Fire Department** ☐ Revised Report

A | FD ID | Incident No. | Index No. | Mo. | Day | Year | Alarm Time | Time on Scene | Time Last Unit Clear |

B | Location/Address | City/Town | Zip Code | Property No. |

C | Occupant Name (Last, First, MI) | Telephone No. | Room or Apt. |

D | Owner Name (Last, First, MI) | Address | Telephone No. |

E | Method of Alarm to Fire Department | Type of Incident |

F | Type of Action Taken | District | Shift | No. Alarms | Mutual Aid ☐ Rec'd. ☐ Given ☐ N/A |

G | General Property Use | Specific Property Use | County | Census Tract |

H | No. Injuries* Fire Service | Other Emerg. | Civilian | No. Fatalities* Fire Service | Other Emerg. | Civilian |

I | No. Fire Service Personnel Responded | No. Engines Responded | No. Aerial Apparatus Responded | No. Other Vehicles Responded |

J | Condition of Fire upon Arrival of First Unit | Time from Alarm to Agent Application | Area of Fire Origin |

K | Equipment Involved in Ignition | Year | Make | Model | Serial No. |

L | Form of Heat of Ignition | Material First Ignited Form/Use | Type |

M | Ignition Factor | Method of Extinguishment |

N | Property Damage Classification | No. Buildings Damaged | Termination Stage |

O | Construction Type | No. of Stories | Level of Origin |

P | Structure Status | No. of Occupants at Time of Incident |

Q | Material Generating Most Flame Form/Use | Type | Factor Contributing to Flame Travel |

R | Material Generating Most Smoke Form/Use | Type | Avenue of Smoke Travel |

S | Detector Type | Detector Power Supply |

T | Detector Performance | Reason for Detector Failure |

U | Sprinkler System Performance | No. of Sprinkler Heads Opened | Reason for Sprinkler System Failure |

V | Extent of Flame Damage | Extent of Smoke Damage | Extent of Extinguishing Agent Damage |

W | Mobile Property Type | Year | Make | Model | Serial No. | License No. |

X | No. of Private Acres Burned | No. of Federal Acres Burned | No. of Other Public Acres Burned |

Y | Fuel Model |

Z | Member Making Report | Date | Officer in Charge (Name, Position, Assignment) | Date |

AA | Remarks: |

☐ Remarks continued on reverse side.

Side markings: COMPLETE ON ALL INCIDENTS / ON ALL FIRES TI 10-19 / COMPLETE IF FIRE TYPE OF INCIDENT (TI) 10-19 / COMPLETE IF FIRE FOR STRUCTURE FIRE TI 11-13 / TI 12-14 / TI 15 / COMPLETE ON ALL INCIDENTS

* A Form 902G must be completed for each Fire Casualty.
This form is for use with NFPA 902M, *Field Incident Manual.* Users should also refer to NFPA 901, *Uniform Coding for Fire Protection,* for information on fire reporting systems and classifications for information entered on this form.

Fig. 5.1 A Basic Incident Report Form (see NFPA 902M).[2]

growth, smoke spread, and toxic impact. This section will focus on what is available now.

In analyzing fire patterns, two questions usually emerge. How did the fire begin (ignition cause)? How and why did the fire become or not become severe?

Analyzing Ignition Cause

The major national fire data bases collect information on five aspects of ignition cause: (1) the form of heat of ignition; (2) the type of equipment, if any, involved in ignition; (3) the ignition factor, i.e., the human error or oversight or the mechanical or electrical problem that brought the heat source and ignited material together; (4) the form of material first ignited, i.e., the form of use, such as upholstered furniture or trash; and (5) the type of material first ignited (e.g., paper, wood). Also, information on the area of fire origin is available. It is possible to identify the major parts of the fire problem in any particular property or other class of interest by calculating percentages of fires (or deaths or injuries or dollars of property damage) having each of the ignition cause properties.

Use of percentages to identify the major parts of the fire problem is not as easy as it sounds. Judgment is involved in selecting the categories for which percentages will be calculated. For example, suppose an analysis showed the largest subgroup of a particular occupancy group's fire problem involved electrical short circuits. This finding does not fit well into most strategies for attacking the fire problem. If someone is going to examine every piece of equipment that could be subject to short circuits, that person might as well also check for other electrical faults in that equipment. The available strategy may dictate a broader category, covering fires involving all electrical equipment faults. Or maybe it is easier to design strategies around particular types of electrical equipment—for instance, fixed wiring, heating systems, ranges, and ovens—and address all fires involving such equipment, whether the equipment or the user was at fault.

Take another example. Which of the following findings is more useful: (1) 20 percent of all home fires begin in the kitchen; or (2) 15 percent of all home fires involve cooking equipment? The latter finding identifies a smaller part of the problem, and uses a completely different categorization (equipment involved in ignition versus area of origin), thus locating the problem much more specifically. In a kitchen, there could be dozens of fire hazards to check out; home cooking equipment has only a few hazards.

Finally, suppose a categorization includes equipment involvement, material ignited, and the human elements that brought them together. Such a categorization will produce detailed scenarios, but the "leading" category might account for only 2 to 4 percent of all fires. It would be necessary to develop strategies for several dozen scenarios in order to make any noticeable impact upon the total fire problem. If that is so, it may make more sense to select scenarios that fit together logically and can be addressed by variation of the same strategy—such as use of less specific categories—than to select several dozen distinct scenarios and tackle each one separately. To put it another way, the scenario technique of analyzing several fire characteristics simultaneously is most useful as a second stage analysis, performed after one characteristic has been used to establish basic clustering.

Therefore, in selecting categories for calculating percentages, it is important to match the structure of the categories to the structure of the strategies available to attack the fire problem. Remember that the size of the problem being attacked is no more or less important than the anticipated leverage of the problem. For example, a 7 percent reduction in a problem that causes 500 deaths a year is worth just as much as a 70 percent reduction in a problem that causes 50 deaths a year. So be sure when selecting the biggest parts of the problem that they collectively account for a sizable share of the total, but also remember that some parts of the problem may be much more preventable than others.

It is important not only to choose categories that match available strategies, it also is important to understand that the so-called "leading causes" depend in part on how the possible cause categories are grouped. Each of the fire cause dimensions has nearly 100 possible values, which means billions of cause categories could be constructed. This is unworkable, so some aggregation is necessary, and the question becomes how to do the aggregation.

One approach to aggregation is to divide fire causes into incendiary, suspicious, all known causes that are accidental in nature, and unknown. For many property types, such a partition sufficiently identifies the major fire causes to justify concentrating on arson problems. If not, a different approach is needed.

A second approach has been to use one of the NFPA 901[1] cause dimensions to sort the fires. Use of the first digit alone can create 10 categories, or use of both digits can create nearly 100 categories for each dimension. This approach often proves frustrating, however, because the structure of the coding elements does not always coincide with the issues to be analyzed. For example, suppose the "ignition factor" dimension is used. Incendiary and suspicious fires will be easy to identify. Fires caused by children playing, however, involve two code values (36 and 48) that do not have the same first digit and therefore do not fall into a natural common grouping. An analysis using "ignition factors" also will produce a large number of entries under "abandoned, discarded material." A novice analyst might not recognize that these are nearly all smoking-related fires and can best be characterized as such.

If "ignition factor" has problems when used by itself, what about the other cause dimensions? Problems arise here because incendiary and suspicious fires are not always characteristically unique. For example, under "form of heat of ignition," arson fires set with matches (the most common type) are indistinguishable from accidental fires involving matches.

A third approach has been to use scenarios based on two or three of the NFPA 901[1] cause dimensions. This avoids the tendency for major problems to be mixed together (e.g., arson and accidental match fires) or disguised (e.g., smoking-related fires listed as discarded or abandoned material). The use of even two dimensions of NFPA 901, however, or the use of three dimensions with first digit only, creates thousands of cause categories, and again there is the likelihood that even the largest parts of the total will not represent a very large share of the total.

The fourth approach has been to use a small number of major, easily recognizable categories based on NFPA 901[1] data elements, but not necessarily grouped along the lines laid out in the NFPA 901 coding structure. The most widely used scheme based on this approach is a set of approximately a dozen categories developed at the USFA. This scheme uses a hierarchical process to sort fires into categories. For example, one will start with "ignition factor" and remove all the incendiary, suspicious, and children-playing fires. Then the analyst may switch to "form of heat of ignition" and use it to separate fires involving smoking materials.

This last approach provides what is probably the most flexible overview. Here one can determine that the leading ignition causes in home fires are heating, cooking, smoking, and arson or suspected arson, even though three different data elements are required to identify these four groups. (See Table 5.1.)

Another matter to consider when analyzing fire data by cause is how to handle incidents for which the cause was unknown or unreported. A sizable share of fires are reported without cause information—particularly those severe fires that involve deaths or significant loss. For example, roughly one-third of all civilian fire fatalities in residential properties with known cause involve smoking materials. Yet, just under one-fourth of all civilian fire fatalities in residential properties are coded as involving smoking materials. A calculation of percentages based only on cases where cause was reported implicitly assumes (in the absence of contrary evidence) that the cause profile of the fires reported without known cause would look the same (if those causes were known) as the cause profile of the fires reported with known causes. (The two types of percentages are sometimes referred to as causes based on "allocating unknowns over known causes" and causes based on "unallocated unknowns.")

Why is it desirable to allocate unknowns? Suppose two neighboring states have relative cause profiles that look the same, except that fire deaths due to unknown causes appear as 10 percent in one state and 40 percent in the other. State A will seem to be doing better than State B with respect to every known cause, but if unknowns were allocated, then both states would have the same problems to the same degree. If it is necessary or desirable to compare two groups, differences in rates of unknowns need to be taken into consideration.

Another reason to allocate unknowns is that many knowledgeable individuals believe that assessments of probable cause are subject to persistent biases. In fact, no credible evidence, consistent among a variety of fire departments, exists to support the notion that fire officers regularly label fires as suspicious, smoking-related, electrical in origin (all very popular suspicions in certain quarters), or anything else when they really are not sure of the cause. (See Table 5.2.)

A more serious argument goes as follows: Probable cause is most difficult

Table 5.1   How "Leading Cause" of Fires Can Depend on How Data Is Sorted*

Emphasis on Accidental vs. Intentional

Incendiary	9%
Suspicious	5%
Accidental	77%
Unknown	9%

Emphasis on Equipment

No equipment involved	44%
Cooking equipment involved	10%
Appliances involved	8%
Electrical system involved	8%
Other known equipment type	14%
Unknown equipment	16%

Emphasis on Behavior

Abandoned material	26%
Mechanical failure or malfunction	20%
Other misuse of heat source	15%
Incendiary or suspicious	14%
Operational deficiency	8%
Other known	8%
Unknown ignition factor	9%

Emphasis on Heat Source

Smoking materials involved	32%
Electrically powered equipment involved	24%
Matches involved	11%
Fueled equipment involved	7%
Other known	13%
Unknown form of heat	11%

Emphasis on Item Ignited

Rubbish or trash	21%
Bedding, mattress, pillow, or linen	21%
Electrical wires	11%
Cooking materials	6%
Supplies or stock	6%
Other known	28%
Unknown form of item	6%

Hierarchical Sorting Approach

Smoking related	32%
Incendiary or suspicious	14%
Cooking equipment involved	10%
Appliances involved	8%
Electrical system involved	8%
Matches, lighters, other open flame	6%
Other equipment	13%
Other known or unknown	10%

*Figures in some groupings may not sum to 100 percent because of rounding errors.
Source: 1980–82 NFIRS data on hospital fires.

Table 5.2   The Problem of Comparisons with Unknown Cause Fires Unallocated

Cause Category	Percentage of Civilian Fire Fatalaties in Residences	
	State A	State B
Smoking	36%	24%
Heating	18%	12%
Incendiary/Suspicious	18%	12%
Other Known	18%	12%
Unknown	10%	40%

to assess in the largest fires because more of the evidence of fire origin is destroyed, and the known-cause fires indicate that cause profiles are different for larger versus smaller fires. For example, incendiary/suspicious and electrical distribution system fires tend to have greater property loss per fire, and incendiary/suspicious and smoking-related fires are more likely than other fires to result in deaths. Therefore, if unknowns are to be allocated, the allocation should take into account the cause profiles for fires of comparable size. This is a fire data analysis issue around which no consensus has yet formed. While the ideal solution would be to find ways to reduce the rate of unknowns, there also may be better ways to arrive at cause estimates in the face of the unknowns that do exist.

Having spent time on how to analyze ignition cause, it may be of interest to know what some of the leading causes and patterns are. (The figures presented here include proportional allocation of deaths in fires with cause unreported, except where indicated.)

- Smoking materials account for by far the largest share of fire deaths. As of 1984, the number was 1,600 civilian fire deaths per year, or nearly one-third of the total. Most (roughly four-fifths) of these deaths began with ignition of upholstered furniture, mattresses, or bedding.
- Arson and suspected arson are by far the leading cause of property damage due to fire. Even without allocating unknowns, arson and suspected arson fires account for roughly one of every four dollars lost to fire each year and claim hundreds of lives.
- Heating equipment has provided the most notable growth portion of the fire problem in the early 1980s. This results chiefly from the sharp increase in the use of portable and space heating equipment, which has a considerably higher risk of fire than central heating equipment. Heating ranks second to smoking as the leading cause of home fire deaths (roughly 700 in 1984). Heating and arson or suspected arson are close as the second leading cause of fire deaths overall.
- Cooking equipment is involved in by far the leading share of fire incidents not reported to fire departments. It is a major, though not leading, cause of fire deaths (roughly 300 in 1984).

- Electrical distribution systems are a major cause of fire damage and deaths overall, although ranking behind smoking, heating, and arson or suspected arson.
- Fires resulting from vehicular crashes account for one of the largest shares of overall fire deaths (just over 600 in 1984).

In all these cause categories, the principal contributing factor is usually human, either directly or indirectly. Heating fires are primarily caused by poor installation, inadequate maintenance, and insufficient distance between the heating device and combustibles. Cooking fires are primarily caused by unattended cooking. Electrical distribution fires contain a very large share—probably a majority—of non-code modifications or alterations, or other unsafe practices. Fire safety education and motivation—or their absence—are crucial to any analysis of why fires happen.

Causes of Severe Fire Development

Computerized national fire data bases are not nearly so helpful on the subject of why and how fires become severe. Although the possibilities are, if anything, more numerous and more complex, there are fewer associated data elements, and those data elements are significantly under-used by reporting personnel. However, it may be useful to note that either a statistical analysis of what is reported or a casual review of published reports on investigations of major fires will show that most of the possible factors occur with some regularity.

One class of factors has to do with provisions for compartmentation. This might concern basic design (e.g., failure to enclose stairwells, or failure to subdivide attic spaces or large open work areas), or it might be related to deviations from the basic design (e.g., pipe openings, doors blocked open). Such factors could lead to horizontal or vertical spread of flame or smoke.

Another class of factors has to do with interior finishes and coverings. Ceilings, walls, and floors all can present problems, such as causing fire acceleration.

A third class of factors involves contents. A building may be quite firesafe for an ordinary use, but quite inadequate if the types or quantities of contents pose special dangers. Changes in owners or tenants, leading to changes in use, are particularly likely to create such problems. Also, the procedures for handling or storing contents can pose dangers of fire spread that would not be present if safe practices were observed.

A fourth class of factors has to do with the presence or absence of active protective systems, notably automatic detection and suppression systems. The national fire data bases provide some information on the considerable effectiveness of these systems, although the data bases are not detailed enough to address fine points of system design or applications.

A fifth class of factors concerns provisions for egress (e.g., number and capacity of exits and routes leading to them, smoke control systems). It is important to recognize that the degree of severity is not only a matter of slowing, confining, or stopping the fire, but is also a matter of removing the occupants from danger.

The sixth and final class of factors involves prevention of building-to-building fire spread, which can result in a group fire or conflagration. These factors include separation considerations (e.g., buildings from each other, buildings from wildlands), exterior finish problems (e.g., untreated wood shingles), and considerations of the fire protection resources of the community (e.g., water supply).

AVAILABLE MAJOR DATA BASES ON FIRE

Three major data bases are available to analyze patterns in U.S. fire experience: (1) the annual NFPA survey of fire departments, (2) the FEMA/USFA National Fire Incident Reporting System (NFIRS), (3) and the NFPA Fire Incident Data Organization (FIDO). Together, these three data bases can provide valid, detailed information on national and regional fire problems, overall or by specific property type and cause. The characteristics of the three data bases and the best ways of using each are the subject of this section.

Annual NFPA Survey of Fire Departments

The NFPA survey is based on a stratified random sample of roughly 3,000 U.S. fire departments (or just over one of every 10 fire departments in the country). The survey collects the following information: (1) the total number of fire incidents, civilian deaths, and civilian injuries, and estimated property damage (in dollars) for each of the major property use classes defined by NFPA 901,[1] the standard for fire incident reporting; (2) similar tallies specifically for incendiary and suspicious fires, separated only into structure versus vehicle; (3) the number of on-duty fire fighter injuries, by type of duty and nature of injury or illness; (4) information on the type of community protected (county, township, city, etc.) and the size of the population protected, which is used in the statistical formula for projecting national estimates from sample results; and (5) leads on multiple-death fires, large-loss fires, and fire fighter fatalities, which if cited are then captured under FIDO.

The totals in (1) and the special incendiary and suspicious fire results in (2) are analyzed and reported in NFPA's annual study, "Fire Loss in the United States," which traditionally appears in the September issue of *Fire Journal*. The fire fighter injury information in (3) is analyzed and reported in NFPA's annual report "U.S. Fire Fighter Injuries," traditionally published in the November or December issue of *Fire Command*.

The NFPA survey begins with the NFPA Fire Service Inventory, a computerized file of nearly 30,000 U.S. fire departments which is the most complete and thoroughly validated such listing in existence. The survey is stratified by size of population protected to reduce the uncertainty of the final estimate. Small, rural communities protect fewer people per department and are less likely to respond to the survey, so a larger number must be surveyed to obtain an adequate sample of those departments. (NFPA also makes followup calls to a sample of the smaller fire departments that do not respond, to confirm that those which did respond are truly representative of fire departments their size.) On the other hand, large city departments are so few in number and protect such a large proportion of the population that it makes sense to survey all of them. Most respond, resulting in excellent precision for their part of the final estimate.

These methods have been used in the NFPA survey since 1977 and represent a state-of-the-art approach to sample surveying. Because of the attention paid to representativeness and the use of appropriate weighting formulas for projecting national estimates, the NFPA survey provides a valid basis for measuring national trends in fire incidents, civilian deaths and injuries, and direct property loss, as well as for determining patterns and trends by community size and major region.

FEMA/USFA's National Fire Incident Reporting System (NFIRS)

NFIRS provides annual computerized data bases of fire incidents, with data classified according to a standard format based on NFPA 901.[1] Roughly three-fourths of all states have NFIRS coordinators, who receive fire incident data from participating fire departments and combine the data into a state data base. These data are then transmitted to FEMA/USFA. Participation by the states and by local fire departments is voluntary. NFIRS captures roughly one-third of all U.S. fires each year. More than one-third of all U.S. fire departments are listed as participants in NFIRS, although not all of these departments provide data every year.

One of the strengths of NFIRS is that it provides the most detailed incident information of any national data base not limited to large fires. NFIRS is the only data base capable of addressing national patterns of fires of all sizes by specific property use and specific fire cause. (The NFPA survey separates out fewer than 20 of the hundreds of property use categories defined by NFPA 901[1] and provides no cause-related information except for incendiary and suspicious fires.) NFIRS also captures information on the construction type of each involved building, avenues and extent of flame spread and smoke spread, and performance of detectors and sprinklers.

One weakness of NFIRS is that its voluntary character produces annual samples of shifting composition. Despite the fact that NFIRS draws on three times as many fire departments as the NFPA survey, the NFPA survey is more suitable as a basis for projecting national estimates, because its sample is truly random and is systematically stratified to be representative.

Analyses based on NFIRS have been widely calculated only since 1982, the year of publication of the second edition of FEMA/USFA's "Fire in the United States"[3]—the first study based primarily on NFIRS. Because consensus on how best to address the weaknesses of NFIRS does not yet exist, the next few years may see further revisions in the formulas used to calculate statistics from this system. In the meantime, most analysts use NFIRS to calculate percentages, such as the percentage of residential fires that occur in apartments, or the percentage of apartment fire deaths that involve discarded cigarettes. Some analysts combine NFIRS-based percentages with NFPA survey-based totals to produce estimates of numbers of fires, deaths, injuries, and dollar loss for subparts of the fire problem. This is the simplest approach now available to compensate in the area where NFIRS is weak.

NFPA's Fire Incident Data Organization (FIDO) System

The FIDO system is a computerized data base that provides the most detailed incident information available, short of a full-scale fire investigation. The fires covered are those deemed to be of high technical interest. The system that identifies fires for inclusion in FIDO is believed to provide virtually complete coverage of incidents reported to fire departments involving three or more civilian deaths, one or more fire fighter deaths, or large dollar loss (redefined periodically to reflect the effects of inflation, and defined since 1980 as $1,000,000 or more in direct property damage).

FIDO also captures a selection of smaller incidents as technical interests dictate. These are useful primarily because of the type of property involved (e.g., high-rise buildings), the presence of hazardous materials, or the performance of detectors or sprinklers.

The FIDO system covers fires from 1971 to date, contains information on more than 53,000 fires, and adds 3,000 to 4,000 fires per year. NFPA learns of fires that may be candidates for FIDO through a newspaper clipping service, insurers' reports, state fire marshals, NFIRS, respondents to the NFPA annual survey, and other sources. Once notified of a candidate fire, NFPA seeks standardized incident information from the responsible fire department and solicits copies of other reports prepared by concerned parties, such as the fire department's own incident report and results of any investigations.

The strength of FIDO is its depth of detail on individual incidents. Information captured by FIDO, but not by NFIRS, includes types and performance of all built-in systems for detection, suppression, and smoke and flame control; details on factors contributing to flame and smoke spread; estimates of time between major events in fire development (e.g., ignition to detection, detection to alarm, etc.); reasons for any unusual delay at various points; indirect loss and detailed breakdowns of direct loss; and escapes, rescues, and number of occupants. (Building height—necessary to analysis of high-rise building fires—will be included in Version IV of NFIRS but to date has been available only in FIDO.) Additional uncoded information often is available in the hard-copy FIDO files, which are indexed for use in research and analysis.

One weakness of FIDO is that it mostly covers larger incidents. Many questions can best be answered by comparing the characteristics of large and small fires that involved similar types of properties and similar causes of ignition. FIDO does not permit such comparisons.

FIDO supports three annual NFPA reports: "U.S. Fire Fighter Deaths," typically published in the May or June issue of *Fire Command*; "Multiple-Death Fires in the United States," usually published in the July issue of *Fire Journal*; and "Large-Loss Fires in the United States," typically published in the November issue of *Fire Journal*. FIDO also supports the anecdotal summaries published in the "Bimonthly Fire Record" of *Fire Journal* and in the annual NFPA study cited earlier, "U.S. Fire Fighter Injuries."

Summary

The increased amount of combustibles present in our daily lives has contributed considerably to the national and international fire problem—a problem of such magnitude that only recently has the general public come to realize its seriousness. Accordingly, the scope of fire department responsibility in the area of fire investigation has broadened immensely. With the use of more sophisticated analytical methods, investigations into the fire problem have become more accurate, and fire reports and fire reporting systems have been better able to pinpoint particular hazards, although the value of uniform reporting procedures has only recently begun to be appreciated. From the results of fire investigation data, fire departments are better able to understand the basis of local fire problems and are, therefore, better able to cope with the varied types of fire hazards.

With society's increased knowledge of the seriousness of the fire problem has come an awareness of the need to discover new methods for combating the fire problem. Such new methods will, undoubtedly, result from still further investigation and analysis.

Analysis of fire investigation sequence, fire development, fire casualties, and fire loss can yield information the fire protection weaknesses of a community, as well as on how to prevent the cause of fires. Investigation of the fire problem is also important in helping to detect and combat the growing problem of arson.

References

[1]NFPA 901-1986. *Uniform Coding for Fire Protection*, National Fire Protection Association, Quincy, MA.

[2]NFPA 902M-1986. *Fire Reporting Field Incident Manual*, National Fire Protection Association, Quincy, MA.

[3]"Fire in the United States," 2nd ed., Federal Emergency Management Agency/U.S. Fire Administration, Washington, DC.

Additional Reading

Fire Protection Handbook, 16th ed. 1986. National Fire Protection Association, Quincy, MA.

NFPA 903M-1986. *Fire Reporting Property Survey Manual*, National Fire Protection Association, Quincy, MA.

NFPA 904M-1986. *Incident Follow-up Report Manual*, National Fire Protection Association, Quincy, MA.

NFPA 907M-1983. *Manual on the Investigation of Fires of Electrical Origin*, National Fire Protection Association, Quincy, MA.

NFPA 1031-1982. *Standard for Professional Qualifications for Fire Inspector, Fire Investigator and Fire Prevention Education Officer*, National Fire Protection Association, Quincy, MA.

Carlson, Gene, ed. 1982. "Fire Cause Determination," Fire Protection Publications, Stillwater, OK.

Carroll, John R. 1983. *Physical and Technical Aspects of Fire and Arson Investigation*, C.C. Thomas, Springfield, IL.

DeHaan, J.D. 1983. *Kirk's Fire Investigation*, 2nd ed., John Wiley & Sons, New York.

Dennett, M.F. 1980. *Fire Investigation: A Practical Guide for Fire Students and Officers, Insurance Investigators, Loss Adjusters, and Police Officers*, Pergamon Press, Elmsford, NY.

Fire Investigation Handbook 1980. National Bureau of Standards, U.S. Department of Commerce, Washington, DC.

Kennedy, John and Patrick 1985. *Fire and Explosions: Determining Cause and Origin*, Investigations Institute, Chicago.

KODAK, 1976. *Using Photography to Preserve Evidence*, Eastman Kodak, Rochester, NY.

KODAK, 1977. *Fire and Arson Photography*, Eastman Kodak, Rochester, NY.

Lyons, Paul R. 1978. *Techniques of Fire Photography*, National Fire Protection Association, Boston.

National Fire Incident Reporting System Handbook 1984. National Fire Information Council, United States Fire Administration, Federal Emergency Management Agency, Washington, DC.

Phillips, Calvin, and McFadden, David 1982. *Investigating the Fireground*, R.J. Brady, Bowie, MD.

Roblee, C., and McKechnie, A. 1981. *Investigation of Fires*, Prentice-Hall, Englewood Cliffs, NJ.

6

Fire Protection Through Building Design and Construction

Because fire presents one of the greatest threats to property, building design and construction must take into account a wide range of firesafety features. Not only must the interiors and contents of buildings be protected from the dangers of fire; the building site itself must be planned to ensure accessibility of both fire departments and water supplies.

FUNDAMENTALS OF FIRESAFETY DESIGN

Building design and construction practices have changed significantly during the past century. A century ago, design techniques and materials, such as structural steel, reinforced concrete, and slab construction, were unknown and unimagined.

A hundred years ago, major fires were common occurrences in cities. Because of combustible construction and poor city planning, whole cities often were destroyed by fire. As a result of those disasters, increased attention was given to firesafety in building design. The following excerpt from "America Burning" briefly summarizes the changes in building design:[1]

> Around the turn of the century, in the wake of many conflagrations, so-called "fireproof" buildings began to be constructed. They had thick walls and floors to keep fire from spreading. Like older buildings, they still had windows that could be opened to allow heat and smoke to escape. They had fire escapes or internal fire stairs, and seldom were they too tall for the topmost occupants to escape.

> Fires, some of them disastrous, occurred in these buildings, nonetheless. Then, after World War II, a new generation of buildings began to appear: the modern high-rise building. Lighter construction systems and many new materials were used, especially for interiors. Windows were permanently sealed so that central air conditioning would operate efficiently. Walls and floors were left with openings for air conditioning ducts and utility cables. Each of these features compromised the firesafety of these buildings.

Objectives of Firesafety Design

"America Burning,"[1] the report of The National Commission on Fire Prevention and Control, points out a major weakness in building design and construction practice; namely, architects, builders, and building owners all too often think of expense, utility, and appearance, and do not consciously think of fire protection. Many architects and their clients are content to meet the minimum fire protection standards of the local building code. These same persons often assume, incorrectly, that the code provides maximum measures rather than minimum requirements for fire protection.

Before a building designer can make effective decisions relating to firesafety design, the specific needs of the client regarding the function of the building and the general and unique conditions that are to be incorporated into the building must be clearly identified. Decisions regarding the firesafety design and construction of the building should be made in the following areas: (1) life safety, (2) property protection, and (3) continuity of operations. These objectives describe the degree to which the building should protect its occupants, property contents, continuity of operations, and neighbors. The objectives should be quantified wherever possible, rather than stated in broad or general terms.[2]

Life Safety: Design considerations for life safety must address two major questions: (1) Who are the occupants of the building? (2) What will the occupants be doing most of the time? The identification of specific function patterns and constraints is vital in designing specific fire protection features that recognize occupant conditions and activities. The following are additional life safety considerations as described in *Operation Skyline:*[3]

> The occupied building provides a great potential for fire because of the presence of large numbers of people, any one of whom could perform a careless or malicious act resulting in fire. Appliances and mechanical or electrical equipment are a potential hazard through misuse, faulty construction, or substandard installation. Accumulations of combustibles, either waiting for disposal or in storage, frequently provide a ready means by which otherwise controllable fires could spread.

Property Protection: There is an important question to be asked about the design of buildings with regard to protection of property: Is there any specific high-value content or one-of-a-kind feature that will need special design protection? The requirements with regard to protection of property within a building are often fairly easy to identify. Materials of high value that are particularly susceptible to fire and/or smoke damage can usually be identified in advance of building design. For example, vital records that cannot be replaced easily or quickly can be identified in advance as needing special fire protection design considerations.

Continuity of Operations: Continuity of operations, the third major area of building design decision-making, must take into consideration those specific functions conducted in a building that are vital to continuing operation and that cannot be transferred to another location. In this regard the owner must identify for the designer the amount of "downtime," or the amount of time an operation can be suspended without completely suspending total operations. Indirect loss or loss of business income must be considered as a part of this downtime. The degree of protection required in firesafe building design varies with the number and scope of vital operations that are nontransferable.

Fire Hazards in Buildings

When the designer and owner either consciously or unconsciously overlook or ignore the possibility of fire in the building to be constructed, the building and its occupants are endangered. The broad approach to the firesafe design of a building requires a clear understanding of the building's function, the number and kinds of people who will be using it, and the kinds of things they will be doing. In addition, appropriate construction and protection features must be provided for the protection of the contents and, particularly for mercantile and industrial buildings, to ensure the continuity of operations if a fire should occur. Too many fires which harm people and property have occurred, and will continue to occur, because no one has given proper consideration to the threat of potential fire. (See Figure 6.1.)

Building Elements and Contents: If a building has combustible interior finish and combustible furnishings, including drapes and curtains, flames and toxic gases may spread so rapidly that occupants may not be able to escape. Poor design and construction practices, such as failing to enclose shafts and other vertical openings, make the work of fire fighters more difficult. Collapse of structural members causes a significant number of deaths and injuries to fire fighters.[2]

> The collapse of structural building elements can be a serious life safety hazard. Although statistically it has not resulted in many deaths or injuries to building occupants, structural collapse is a particular hazard to fire fighters. A number of deaths and serious injuries to fire fighters occur each year because of structural failure. While some of these failures result from inherent structural weaknesses, many are the result of renovations to existing buildings that materially, though not obviously, affect the structural integrity of the support elements. A building should not contain surprises of this type for fire fighters.

Fig. 6.1   A fire destroyed this unsprinklered plastic products manufacturing plant, exemplifying the large loss that can occur in unsprinklered buildings. (Courtesy: Providence Journal.)

Elements of Building Firesafety

The firesafety of a building will depend first on what is done to prevent a fire from starting in the building, and second on what is done through design, construction, and management to minimize the spread of fire if and when it happens. Good housekeeping is perhaps the major factor in both fire prevention and control. Keeping the fuel load down not only lessens the amount of material that can be ignited but provides less material that can be consumed if a fire breaks out.

Once a fire has started, its spread will depend on the design of the building, the materials used in construction, building furnishings and contents, methods of ventilation, and fire suppression systems, if any. Table 6.1 describes the building design and construction features that influence firesafety. These elements are within the decision-making authority of various members of the design team, based on the assumption that their firesafety objectives are clearly defined by management, the owners, and other responsible parties, both from the public and private sectors. The design and construction elements should be organized in a manner that can give a quick overview of the major aspects that must be considered for firesafety. They should also show features that include both active and passive design and construction considerations.

The persons responsible for fire prevention may not be the same persons responsible for the building design. Table 6.2 describes the elements that comprise firesafety from a prevention consideration. Decisions concerning these elements are predominantly under the control of the building owner or occupant, or both. Table 6.2 includes the elements of emergency prepared- ness in case of fire that are the responsibility of the owner and occupant. A further note from *Operation Skyline*:[3]

> ... some owners, operators, and managers [boast] about certain "fireproof" or "fire-resistive" materials in their buildings. Those materials provide a definite advantage, but they should not be confused with "firesafe." They simply indicate that a building with "fireproof" or fire-resistive materials can withstand a burnout of its contents without subsequent structural collapse. Firesafe, on the other hand, indicates that if a fire starts it can be confined and extinguished without jeopardizing life and property elsewhere in the structure.

BUILDING AND SITE PLANNING FOR FIRESAFETY

Two major categories of decisions should be made early in the design process of a building in order to provide effective firesafe design. Early considerations should be given to both the interior building functions and exterior site planning. Building fire defenses, both active and passive, should be designed in such a way that the building itself assists in the manual suppression of fire.

Firesafety Planning for Buildings

Interior layout, circulation patterns, finish material, and building ser- vices are all important firesafety considerations in building design. Building design also has a significant influence on the efficiency of fire department operations. As a result, manual fire suppression activities should be consid- ered during all architectural design phases. Table 6.3 lists pertinent data regarding some recent high-rise fires. It can be assumed that design decisions or poor construction practices undoubtedly affected the statistics in this table. The dollar loss values given in this table reflect the direct loss due to property damage. It does not represent loss associated with business interruption or liability awards. These expenses usually outweigh direct loss expenses by a factor of 20:1.

Table 6.1   Elements of Building Firesafety

Building Design and Construction Features Influencing Firesafety
1. Fire Propagation
   a. Fuel load and distribution
   b. Finish materials and their location
   c. Construction details influencing fire and products of combustion movement
   d. Architectural design features

2. Smoke and Fire Gas Movement
   a. Generation
   b. Movement
      —Natural air movement
      —Mechanical air movement
   c. Control
      —Ventilation
      —Heating, ventilating, air conditioning
      —Barriers
      —Pressurization
   d. Occupant Protection
      —Egress
      —Temporary refuge spaces
      —Life support systems

3. Detection, Alarm, and Communication
   a. Activation
   b. Signal
   c. Communication systems
      —To and from occupants
      —To and from fire department
      —Type (automatic or manual)
      —Signal (audio or visual)

4. People Movement
   a. Occupant
      —Horizontal
      —Vertical
      —Control
      —Life support

   b. Fire Fighters
      —Horizontal
      —Vertical
      —Control

5. Suppression Systems
   a. Automatic
   b. Manual (self-help; standpipes)
   c. Special

6. Fire Fighting Operations
   a. Access
   b. Rescue operations
   c. Venting
   d. Extinguishment
      —Equipment
      —Spatial design features
   e. Protection from structural collapse

7. Structural Integrity
   a. Building structural system (fire endurance)
   b. Compartmentation
   c. Stability

8. Site Design
   a. Exposure protection
   b. Fire fighting operations
   c. Personnel safety
   d. Miscellaneous (water supply, traffic, access, etc.)

Fire Emergency Considerations
1. Life Safety
   a. Toxic gases
   b. Smoke
   c. Surface flame spread

2. Structural
   a. Fire propagation
   b. Structural stability

3. Continuity of Operations
   a. Structural integrity

Fire Fighting Accessibility to Building's Interior: One of the more important considerations in building design is access to the fire area. This includes access to the building itself as well as access to the interior of the building.

In larger and more complex buildings, serious fires over the years have brought improvements in building design to facilitate fire department operations. The larger the building, the more important access for fire fighting becomes. In some buildings where fire fighters cannot function

Table 6.2   Fire Prevention and Emergency Preparedness

1. **Ignitors**
   a. Equipment and devices
   b. Human accident
   c. Vandalism and arson

2. **Ignitable Materials**
   a. Fuel type and quantity
   b. Fuel distribution
   c. Housekeeping

3. **Emergency Preparedness**
   a. Awareness and understanding
   b. Plans for action
      —Evacuation or temporary refuge
      —Self-help extinguishment
   c. Equipment
   d. Maintenance—operating manuals available

effectively, the best solution is provision of a complete automatic sprinkler system, supplemented by a standpipe system for fire department use. The following excerpt from the NFPA *Fire Protection Handbook*[2] explains the benefits of sprinklers:

> Spaces in which adequate fire fighting access and operations are restricted because of architectural, engineering, or functional barriers should be provided with automatic extinguishing systems. A complete automatic sprinkler system with a fire department connection is probably the best solution to this problem. Other methods that could be incorporated in appropriate design situations include access panels in interior walls and floors, fixed nozzles in floors with fire department connections, automating smoke venting and pressurization of certain building areas, although none of these items can take the place of or be considered equivalent to a complete automatic sprinkler system.

Ventilation: Ventilation is of vital importance in removing smoke, gases, and heat so that fire fighters can reach the seat of a blaze. It is difficult, if not impossible, to ventilate a building unless appropriate skylights, roof hatches, and similar devices are provided when the building is constructed. The following is a description of the importance of the ventilation factor in building design:[2]

> Ventilation of building spaces performs the following important functions:
> 1. Protection of life by removing or diverting toxic gases and smoke from locations where building occupants must find temporary refuge.

2.   Improvement of the environment in the vicinity of the fire by removal of smoke and heat. This enables fire fighters to advance close to the fire to extinguish it with a minimum of time, water, and damage.

3.   Control of the spread or direction of fire by setting up air currents that cause the fire to move in a desired direction. In this way occupants or valuable property can be more readily protected.

4.   Provision of a release for unburned, combustible gases before they acquire a flammable mixture, thus avoiding a backdraft or smoke explosion.

Connections for Sprinklers and Standpipes: Connections for sprinklers and standpipes must be carefully located and clearly marked. The larger and taller the building becomes, the greater the volume and pressure of water that will be needed for a potential fire. Water damage may be very costly unless adequate measures, such as floor drains and scuppers, have been incorporated into the building design. Confinement of a fire in a high-rise building can only be accomplished by careful design and planning for the whole building. As buildings increase in size and complexity, more dependence on suppression systems is necessary. Such systems will be described in detail in subsequent chapters.

Firesafety Planning for Sites

Proper building design for fire protection should include a number of factors outside the building itself. The site on which the building is located will influence the design. Among the more significant features are traffic and transportation conditions, fire department accessibility, and water supply. Inadequate water mains and poor spacing of hydrants have contributed to the loss of many buildings.

Traffic and Transportation: Fire department response time is a vital factor in building design considerations. Traffic access routes, traffic congestion at certain times of the day, traffic congestion from highway entrances and exits, and limited access highways have significant effects on fire department response distances and response time, and must be taken into account by building designers in selecting appropriate fire defenses for a building.

Fire Department Access to the Site: Building designers must ask an important question. Is the building easily accessible to fire apparatus? Ideal accessibility occurs where a building can be approached from all sides by fire department apparatus. However, such ideal accessibility is not always possible. Congested areas, topography, or buildings and structures located appreciable distances away from the street make difficult or prevent effective use of fire apparatus. When apparatus cannot come close enough to the building to be used effectively, equipment such as aerial ladders, elevating platforms, and

Table 6.3   Partial Listing of Recent High-rise Fires

Date of Fire	City	Occupancy	Stories	Time of Fire	Place of Origin	Fatalities	Injuries	Dollar Loss
Jan. 1970	Chicago	Hotel	25	6:45 am	Elevator (lobby)	2	36	$70-100,000
Aug. 1970	New York	Offices	50	5:45 pm	Office (33rd floor)	2	30	$10 million
Dec. 1970	Tucson	Hotel	11	12:20 am	Fourth floor	28	71	$1.5 million
Dec. 1970	New York	Offices	47	9:50 am	Showroom (5th floor)	3	20	$2.5 million
Mar. 1971	Los Angeles	Offices	21	2:11 am	Restaurant (21st floor)	0	0	$378,000
July 1971	New Orleans	Hotel	17	2:00 am	Room (12th floor)	6	2	$150,000
Nov. 1972	Atlanta	Home for elderly	11	2:00 am	Apartment (7th floor)	10	30	$250,000
Nov. 1972	New Orleans	Offices	16	1:28 pm	Restaurant (15th floor)	6	0	$887,000
Nov. 1972	Chicago	Apt. & other	100	4:40 am	Restaurant (96th floor)	0	1	$40,000

Date	Location	Occupancy		Time	Origin			Damage
Dec. 1972	Ventnor, NJ	Apartments	19	12:45 pm	Apartment (4th floor)	1	3	$325,000
Dec. 1972	Dallas	Apartments	16	7:30 am	Eighth floor	0	0	$340,000
June 1973	Tucson	Offices	11	3:30 pm	Storage (4th floor)	0	0	$565,000
Dec. 1973	Omaha	Home for elderly	13	5:10 pm	Apartment (13th floor)	0	4	$30,000
Sept. 1974	Virginia Beach	Hotel	11	11:35 am	Room (9th floor)	1	21	$145,000
Nov. 1980	Las Vegas	Hotel	21	7:00 am	The Deli	85	600	$50 million
Mar. 1982	Houston	Hotel	13	2:00 am	Guest Room (4th floor)	12	3	$1.1 million
Dec. 1984	Waukegan, IL	Hotel	9	2:30 am	Clothing Store (1st floor)	8	9	Not available
Jan. 1986	Boston	Offices	52	5:13 pm	(14th floor)	0	12	$1.5 million
Dec. 1986	San Juan, Puerto Rico	Hotel	20	3:00 pm	South Ballroom	97	146	Not available

water tower apparatus can be rendered useless. The importance of the complications resulting from the accessibility factor is further emphasized in the following statement from the NFPA *Fire Protection Handbook*:[2]

> The matter of access to buildings has become far more complicated in recent years. The building designer must consider this important aspect during the planning stage. Inadequate attention to site details can place the building in an unnecessarily vulnerable position. If its fire defenses are compromised by preventing adequate fire department access, or more complete internal protection, the building itself must make up the difference.

Water Supply to the Site: A building designer must ask another important question. Are the water mains adequate and are the hydrants properly located? The more congested the area where the building is to be located, the more important it is to plan in advance what the fire department may face in its attack if a fire occurs on the property. An adequate water supply delivered with the necessary pressure is required to control a fire properly and adequately. The number, location, and spacing of hydrants and the size of the water mains are vital considerations when the building designer plans fire defenses for a building. The designer must also account for the water supply demands for fixed suppression systems, such as sprinklers and standpipes. These demands may reduce the supply available to the fire apparatus.

EXPOSURE PROTECTION

Still another consideration in the design of the building is the possibility of damage from a fire in an adjoining building. The building may be exposed to heat radiated horizontally by flames from the windows of the burning neighboring building. If the exposed building is taller than the burning building, flames coming from the roof of the burning building can impinge on and damage the exposed building.

The damage from an exposing fire can be severe. It is dependent upon the amount of heat produced and the length of exposure, the fuel load in the exposed building, and the construction and protection of the walls and roof of the exposed building. Other factors are the distance of separation, wind direction, and accessibility of fire fighters.

Fire severity is a description of the total energy of a fire, and involves both the temperatures developed within the exposing fire and the duration of the burning. NFPA 80A, *Recommended Practice for Protection of Buildings from Exterior Fire Exposures*,[4] describes estimated minimum separation distances under light, moderate, or severe exposures. The severity of the exposure is calculated on the width and height and the percentage of openings in the exposing wall areas and the estimated fire loadings of the buildings involved. Building designers should be aware that the separation distance hazards between the exposing buildings can be reduced by blank walls, closing wall openings, use of automatic deluge water curtains which discharge water directly on one of the vertical surfaces of the exposed building, and use of wired glass instead of ordinary glass. (See Figure 6.2.)

INTERIOR FINISH

The way a building fire develops and spreads, and the amount of damage that ensues, is greatly influenced by the characteristics of the interior finish in a building. The types of interior finish used in buildings are numerous, varied, and serve many functions. Primarily they are used for aesthetic and acoustical purposes. However, insulation and protection against wear and abrasion are also considered major functions by building designers. The

Fig. 6.2   The effectiveness of exposure protection from fire is evidenced in a reinforced concrete fire wall, which protected the dwelling from a fire in an adjacent lumberyard.[2]

following statements from the National Commission on Fire Prevention and Control's report titled "America Burning" point out the need for greater concern and attention to the potential fire hazards of interior finishes:[1]

> The modern urban environment imparts to people a false sense of security about fire. Crime may stalk the city streets, but certainly not fire, in most people's views. In part, this sense of security rests on the fact there have been no major conflagrations in American cities in more than half a century. In part, the newness of so many buildings conveys the feeling that they are invulnerable to attack by fire. Those who think only of a building's basic structure (not its contents) are satisfied, mistakenly, that the materials—concrete, steel, glass, aluminum—are indestructible by fire. Further, Americans tend to take for granted that those who design their products, in this case buildings, always do so with adequate attention to their safety. That assumption, too, is incorrect.

Types of Interior Finish

Interior finish is usually defined as those materials that make up the exposed interior surface of wall, ceiling, and floor constructions. The common interior finish materials are wood, plywood, plaster, wallboards, acoustical tile, insulating and decorative finishes, plastics, and various wall and ceiling coverings.

Interior floor coverings are also within this category. While floor coverings are subjected to tests different than those for interior wall finish, NFPA *101, Life Safety Code*,[5] does stipulate interior floor finishes for certain occupancies. Hard-surface floors as well as carpet and rugs are subjected to fire tests in order to evaluate their behavior under fire conditions.

Variations are usually permissible for some of the fixed or decorative items contained in buildings. The following excerpt from the *Fire Protection Handbook* explains:[2]

> Many codes, such as NFPA *101, Life Safety Code*,[5] exclude trim and incidental finish from the requirements for wall and ceiling finish; less rigid requirements are set for trim of a type acceptable to the code in question if they comprise less than ten percent of the aggregrate wall and ceiling area.

Cellular plastics sprayed in walls for insulation have become popular. Fire retardants can be incorporated in many of these plastics so they can meet building code requirements. However, some plastics containing polyurethane or polystyrene have been involved in serious, rapidly spreading fires. Most building codes have limitations and restrictions on the use of such materials. Untreated, combustible wallboards have been a major factor in a number of fires causing large loss of life in hotels, hospitals, and nursing homes over the years. Fire-retardant treatments are now required by most codes for this type of material. Without such fire-retardant treatments, combustible wallboards not only enable a fire to spread so fast that people may become trapped, but also contribute fuel to the fire and create hazardous concentrations of smoke and toxic gases.

BUILDING MATERIALS

Wood

The physical size of wood and its moisture content are important factors that determine whether this material will provide reasonable structural integrity. Wood is the most prevalent material used in the construction of dwellings. If a wood frame house is subject to a serious fire, either from burning combustibles inside the house or from an exposure fire, it will not withstand much heat and will have little structural integrity.

Heavy timber construction can resist fire very well. The timbers will char, and the resulting coating of charcoal provides an insulation for the unburned wood. Heavy timber maintains its structural integrity during a fire for a relatively long time, thus providing an opportunity for extinguishment. Much of the original strength of the members is retained and reconstruction is possible.

Steel

The most common building material for larger buildings is structural steel. While steel is noncombustible and contributes no fuel to a fire, it loses its strength when subjected to the high temperatures that are easily reached in a fire. The stress in a steel beam determines its load-carrying capacity. The normal critical temperature of steel is 1,100°F (593°C). At this temperature the yield stress of steel is about 60 percent of its value at room temperature. Buildings constructed of unprotected steel will collapse relatively quickly when exposed to a contents fire or an exposure fire. The lighter the steel members, the quicker they will fail.

Another property of steel that influences its behavior in fires is expansion when the steel is heated. Walls can collapse from the movement caused by expansion of steel framing.

Encasement of the structural steel member has become a very common and effective way of insulating steel to increase its fire resistance. The NFPA *Fire Protection Handbook* describes some of these methods:[2]

> Because unprotected structural steel loses its strength at high temperatures, it must be protected from exposure to the heat produced by building fires. This protection, often referred to as

"fireproofing," insulates the steel from the heat. The more common methods of insulating steel are encasement of the member, application of a surface treatment, or installation of a suspended ceiling as part of a floor-ceiling assembly capable of providing fire resistance. In recent years, additional methods, such as sheet steel membrane shields around members and box columns filled with liquid, have been introduced.

Structural steel members can also be protected by sheet steel membrane shields. The sheet steel holds inexpensive insulation materials in place, thus providing a greater fire resistance. In addition, polished sheet steel has been used in tests to protect sprandrel girders. The shield reflects radiated heat and protects the load-carrying spandrel.

Concrete

The resistance of reinforced concrete to fire attack will depend on the type of aggregate used to make the concrete, fire loading, and moisture content. In general, lightweight concrete performs better at elevated temperatures than normal-weight concrete.

Usually, reinforced concrete buildings resist fire very well; however, the heat of a fire will cause spalling (chipping and peeling away), some loss of strength of the concrete, and other deleterious effects.

Prestressed concrete is stronger than reinforced concrete and provides better fire resistance. However, prestressed concrete has a greater tendency to spall, with the result that the prestressing steel may become exposed. The type of steel used for prestressing is more sensitive to elevated temperatures than the type of steel that is usually used in reinforced concrete construction. In addition, the steel used for this type of reinforced concrete construction does not regain its strength upon cooling.

Glass

Glass is a commonly used building material. Modern high-rise buildings commonly contain large amounts of glass. Glass is utilized in three primary ways in building construction: (1) for glazing, (2) for fiberglass insulation, and (3) for fiberglass-reinforced plastic building products.

Glass used for windows and doors has little resistance to fire. Labeled glass for fire windows provides a slightly higher resistance to fire, but no glazing should be relied upon to remain intact in a fire.

Fiberglass insulation is widely used in modern building construction. Fiberglass is popular because it does not burn and is an excellent insulator. However, fiberglass is often coated with a resin binder that is combustible and can spread flames.

Fiberglass-reinforced plastic building products, such as translucent window panels, are common. The fiberglass acts as reinforcement for a thermosetting resin. The resin, combustible even with fire retardants incorporated in the composition, frequently comprises about 50 percent or more of the material. Thus, while the fiberglass itself is noncombustible, the products are quite combustible.

Gypsum

Gypsum, as reflected in products such as plaster and plasterboard, has excellent fire-resistive qualities. Gypsum is widely used because it has a high proportion of chemically combined water. Evaporation of this water requires a great deal of heat energy. The result is a material which possesses a high endurance to fire, is inexpensive, and is easy to install.

Masonry

Masonry (such as brick, tile, and sometimes concrete) provides good resistance to heat, and usually retains its integrity. Because of the prevalence of brick construction in European dwellings, as compared to American wood-frame construction, the dwelling fire record in Europe is much more favorable than the dwelling fire record in America.

Plastics

Plastic products are increasing in use by the building industry. Lower cost and aesthetic considerations make the use of plastic building materials desirable. However, all plastics are combustible. Presently, there is no known treatment that is able to make plastics noncombustible. There are, however, some chemical treatments that can result in higher ignition temperatures and inhibit the flame spread characteristics of these products.

TEST PROCEDURES

Because of the importance of assessing the fire resistance of buildings and building materials, a great deal of work has been done to develop standard test procedures that assess the way building materials and structural assemblies perform under fire conditions. It is important to keep in mind that these tests must be repeatable in order to obtain uniform results from different testing laboratories. It is possible to estimate the damage that fire can cause to a building by studying: (1) the amount and kind of combustible materials in the building, and (2) the way they are distributed throughout the building. These two factors not only indicate the rate of combustion and the duration of the fire, but also the difficulty that might be encountered when automatic suppression systems are activated or when manual suppression is attempted.

The effects of fire on the components of a building (such as the columns, floors, walls, partitions, and ceiling or roof assemblies) are tested against both time and temperature. Results of the tests are recorded in hours or minutes, and indicate the duration of fire resistance.

Test procedures require the loading of the elements being tested to simulate actual conditions under normal building use. In some cases, in order to qualify for a certain rating, the construction, after the prescribed fire exposure, is subjected to hose streams.

Various criteria are established and used to determine the acceptance of the material or construction being tested. Such criteria include failure to support the load, temperature increase on the unexposed surface, passage of heat or flame sufficient to ignite cotton waste, excess temperature on steel members, and structural failure under hose streams.

The standard fire test is designed to define the ability of the test structure to perform its intended function during fire exposure, as well as its subsequent load capacity. The fire test does not measure the suitability of the test structure for future use.

Full-scale room tests with various wall and ceiling finishes demonstrate that flashover can occur as noted previously. This transition may also be described as a change from localized flaming to volumetric flaming of the space.

Flame spread ratings of interior finish materials have been established by the use of the 25-ft (7.6-m) tunnel developed by A.J. Steiner at Underwriters Laboratories Inc. These ratings are used in NFPA *101, Life Safety Code,*[5] and in other codes to indicate the areas in which finishes of varying flame spread characteristics may be used. The three classifications used in the *Life Safety Code* are:[5]

HOSPITALS

Class	Flame Spread Range	Smoke Developed
A	0–25	0–450
B	26–75	0–450
C	76–200	0–40

The greater the flame spread range, the greater the hazard. For example, in a new hospital, Class A materials would be required for most areas. (See Figure 6.3.)

The flame spread rating is measured in a relative scale with cement asbestos board having a flame spread rating of 0 and red oak flooring having

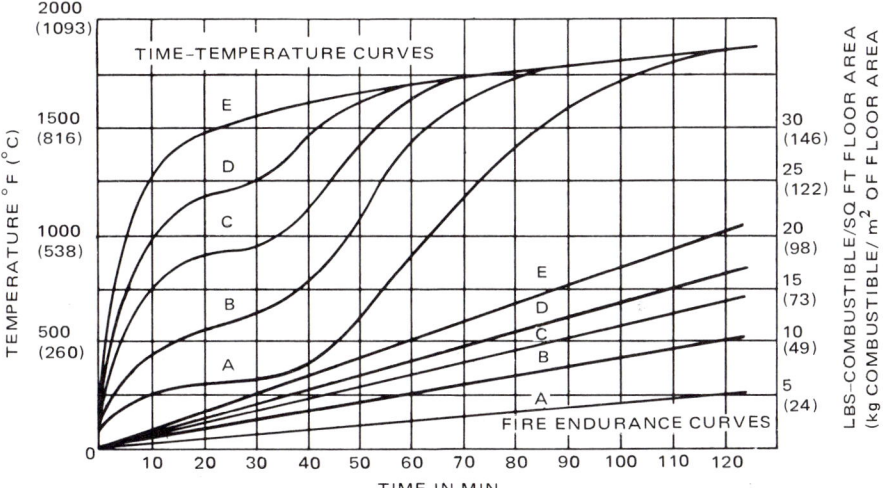

Fig. 6.3 Possible classification of building contents for fire severity and duration. The straight lines indicate the length of fire endurance based upon amounts of combustibles involved. The curved lines indicate the severity expected for the various occupancies. (See Table 6.5.) There is no direct relationship between the straight and curved lines, but, for example, 10 lb of combustible per sq ft (48.8 kg/m²) will produce a 90-minute fire in a "C" occupancy, and a fire severity following the time-temperature curve "C" might be expected.

a rating of 100. Some of the highly combustible wallboards involved in fatal fires have received ratings as high as 1500.

The principal United States organizations that test assemblies of building construction materials for flame spread and fire resistance are Underwriters Laboratories Inc. and Factory Mutual Research Corporation. Results of their tests, as well as tests of other laboratories and of building manufacturers, are published and made available to interested parties. One of the best sources of information showing the wide variety of building assemblies and giving the fire-resistance ratings of beams, columns, floors, walls, and partitions is the Underwriters Laboratories Inc. *Fire Resistance Directory*[6] which is published annually. The *Building Materials Directory*,[7] also published by Underwriters Laboratories Inc., lists flame spread ratings of assorted building construction materials.

Because of the various circumstances involving fires and the wide range of materials that might be involved in a fire, testing procedures cannot provide exact statistics applicable to all fire situations. The problem is summarized in the NFPA *Fire Protection Handbook* as follows:[2]

> The nature of materials and the fire environment vary so widely that the development of a fire test becomes a highly complex matter. Three factors must be taken into account in this development process.

1. The start and growth of fire in a building is affected by the ignition source, space geometry, ventilation, and the nature, amount, and location of other processes and materials in the building.

2. The changing conditions during a fire, such as oxygen concentration, rate of heat release, protection systems, etc.

3. Variations in form, composition, density, and application of the materials present.

With these variables in mind, the difficulty in designing a test that will provide a basis for predicting performance under fire exposure becomes obvious. Equally obvious is the impracticality of designing tests to represent all fire conditions. A test designed to represent a "typical" fire situation or to expose materials to one set of "standard test" conditions may not provide a reliable basis for predicting "real-life" performance of all materials tested. Thus, there is a constant search for improved test methods having a numerical range of results, and for an adequate array of tests to suitably describe the behavior of the various materials available.

Although testing procedures cannot provide exact, "real-life" results, they do provide vital firesafety guidelines that should be considered by architects and builders when designing and constructing buildings.

CONFINEMENT OF FIRE AND SMOKE

To date, little has been done to build dwellings of materials and assemblies that confine a fire to the room or even the floor of origin. However, many other types of buildings are provided with some degree of protection for the occupants. Today, if a building is so constructed that it can neither confine a fire to a given area nor restrict the products of combustion from spreading throughout the building, it is likely that proper firesafety safeguards have not been incorporated. Currently it is relatively easy to ensure confinement of a fire through responsible designs, good construction, interior fire protection equipment, and fire-resistive interior materials.

FIRE LOADING

Estimates of the maximum heat that would be released if all combustibles in a given fire are burned can be made if the fire loading is determined. The fire load is expressed as the weight of combustible material per square foot of fire area, and includes the combustible structural elements, interior finish, floor finish, and combustible contents. The fire load also takes into account the kind and quantity of the material or materials involved. In addition to fire load, the way the building and its contents are arranged must be considered in order to determine how rapid the spread of a fire might be. Table 6.4 shows how the severity of a fire increases dramatically as the fire loading becomes heavier.

The Standard Time-Temperature Curve

The standard time-temperature curve is widely accepted and used by most of the standards and testing agencies. It is based on the maximum indication of the severity of a fire completely burning out an ordinary brick, wood-joisted building loaded with combustible contents. The use of this curve, together with information on the fire loading, is used to estimate the severity of a fire. Table 6.5 indicates the fire severity for typical occupancies.

Fire Loads by Occupancy

One of the difficulties faced by fire protection authorities is change of occupancy. The original occupancy of the designed building may have had only a slight fire severity possibility. If new owners or occupants change to another type of business using large amounts of combustible material, the building or area may become unsafe. Changes of occupancy demand the attention of the appropriate building and fire protection authorities. Building codes define the degree of protection required for walls, ceilings, and floors

Table 6.4   Estimated Fire Severity for Offices and Light Commercial Occupancies

Data applying to fire-resistive buildings with combustible furniture and shelving

Combustible Content Total, including finish, floor, and trim psf	Heat Potential Assumed* Btu per sq ft†	Equivalent Fire Severity Approximately Equivalent to That of Test under Standard Curve for the Following Periods:
5	40,000	30 min
10	80,000	1 hr
15	120,000	1½ hrs
20	160,000	2 hrs
30	240,000	3 hrs
40	320,000	4½ hrs
50	380,000	7 hrs
60	432,000	8 hrs
70	500,000	9 hrs

*Heat of combustion of contents taken at 8,000 Btu per lb up to 40 psf: 7,600 Btu per lb for 50 lb, and 7,200 Btu for 60 lb and more to allow for relatively greater proportion of paper. The weights contemplated by the table are those of ordinary combustible materials, such as wood, paper, or textiles.
†SI units: 1 psf = 4.9 kg/m²; 1 Btu/ft² = 1.14 J/m².

Table 6.5   Fire Severity Expected by Occupancy*

Temperature Curve A (Slight)
Well-arranged office, metal furniture, noncombustible building
Welding areas containing slight combustibles
Noncombustible power house
Noncombustible buildings, slight amount of combustible
    occupancy

Temperature Curve B (Moderate)
Cotton and waste paper storage (baled) and well-arranged,
    noncombustible building
Paper-making processes, noncombustible building
Noncombustible institutional buildings with combustible
    occupancy

Temperature Curve C (Moderately Severe)
Well-arranged combustible storage, e.g., wooden patterns,
    noncombustible buildings
Machine shop having noncombustible floors

Temperature Curve D (Severe)
Manufacturing areas, combustible products, noncombustible
    building
Congested combustible storage areas, noncombustible
    building

Temperature Curve E (Standard Fire Exposure—Severe)
Flammable liquids
Woodworking areas
Office, combustible furniture, and buildings
Paper working, printing, etc.
Furniture manufacturing and finishing
Machine shop having combustible floors

*See Figure 6.3 for the temperature curves identified in this table.

in accordance with the degree of hazard of the occupancy. If the occupancy is changed, a different degree of protection may be indicated and warranted.

FIRE DOORS

Fire doors are the most widely used and accepted means of protecting both vertical and horizontal openings in fire-rated walls and floors. Suitability of fire doors is determined by testing laboratories; doors not tested cannot be relied upon for effective protection. The doors are tested as they are installed in the field, i.e., with the frame, hardware, wired-glass panels, and other accessories necessary to complete the installation.

Fire doors may be classified by an hourly rating designation, an alphabetical letter designation, or a combination. Current practice is to use the hourly rating designation; formerly, the alphabetical letter designation was used. Appendix F of NFPA 80, *Standard for Fire Doors and Windows*, refers to openings as A, B, C, D, and E in accordance with the character and location of the wall. The alphabetical classification of the opening does not apply to the closure; however, in actual practice, the distinction between opening classification and door classification was rarely maintained. A three-hour door for use in a Class A opening was commonly called a Class A door.

Building codes and NFPA 80 commonly specify a door by its fire protection rating instead of by the classification of opening to be protected. Fire doors are now available with one-half and one-third hour ratings, which do not fall within the specified protection criteria for the classes of opening. The following paragraphs summarize current applications of doors, windows, and shutters for openings in fire-resistive walls.

Three-Hour Fire Doors: Openings in walls separating buildings or dividing a building into different fire areas are protected by three-hour fire doors. This can include the protection of openings in walls enclosing hazardous spaces, such as flammable liquid storage rooms. However, many codes only require three-hour doors in separation walls required to have a fire resistance of three hours or more. In addition, some authorities require two three-hour fire doors, one on each side of the opening, whenever a wall is required to have a fire resistance of four hours.

One-and-One-Half-Hour Fire Doors: Openings in two-hour enclosures of vertical openings in buildings are protected by one-and-one-half-hour fire doors. Many codes also permit the use of one-and-one-half-hour fire doors to protect openings in walls separating buildings or dividing buildings into different fire areas when the wall is only required to have a fire resistance of two hours.

One-and-One-Half-Hour Fire Doors and Shutters: Openings in exterior walls that can be subjected to severe fire exposure from outside the building are protected by one-and one-half-hour fire doors and shutters.

One-Hour Fire Doors: Openings in one-hour enclosures of vertical openings in buildings are protected by one-hour fire doors.

Three-Quarter-Hour Fire Doors: Openings in corridor and room partitions are protected by three-quarter-hour fire doors. Sometimes they are also permitted in partitions that subdivide floors of a building. Although three-quarter-hour fire-rated doors are for use in one-hour corridor partitions, many codes permit installation of other types of doors, such as one-half- or one-third-hour fire-rated or 1¾-in. (445-mm) solid bonded wood-core doors. Either self- or automatic closers are generally required on corridor doors. Since smoke control is also necessary to protect a corridor as an access to an exit, automatic closers are generally actuated by smoke detectors.

Three-Quarter-Hour Fire Doors and Shutters: Openings in exterior walls subject to a moderate or light fire exposure from outside the building are protected by three-quarter-hour fire doors and shutters.

Three-Quarter-Hour Fire Windows: Openings in corridor or room partitions or in exterior walls subject to a moderate or light external fire exposure can be protected by three-quarter-hour fire windows.

One-Half-Hour (30-Minute) and One-Third-Hour (20-Minute) Fire Doors: Doors with these ratings are for use where smoke control is a primary consideration. They are also used for the protection of openings in partitions between a habitable room and a corridor when the wall is constructed to have a fire-resistance rating of not more than one hour or across corridors where a smoke partition is required. Some codes permit the use of 1¾-in. (445-mm) solid bonded wood-core doors in corridor, room, and smoke-stop partitions that are required to have a fire resistance of not more than one hour. These doors cannot be used for corridor doors to hazardous areas or to protect openings in enclosures of vertical openings in buildings.

Classification of Openings in Walls

To protect openings in walls, nearly all building codes reference NFPA 80, *Standard for Fire Doors and Windows*.[8] The alphabetical letter designations used in NFPA 80 to classify openings in walls are as follows:[8]

1. Class A—Openings in fire walls and in walls that divide a single building into fire areas.
2. Class B—Openings in enclosures of vertical communications through buildings and in 2-hour rated partitions providing horizontal fire separations.
3. Class C—Openings in walls or partitions between rooms and corridors having a fire-resistance rating of 1 hour or less.
4. Class D—Openings in exterior walls subject to severe exposure from the outside of the building.
5. Class E—Openings in exterior walls subject to moderate or light fire exposure from outside of the building.

It is important to note that this classification applies to the various types of openings and not to the fire door itself. A fire door is not a Class A fire door. It is a door that is suitable for a Class A opening.

Types of Doors

Fire doors are manufactured in a wide variety of constructions. Some of the more common types are tin-clad doors, composite doors, hollow metal doors, metal-clad doors, rolling steel doors, and wood-core doors. The type of door and the way it is installed are important to secure the desired degree of protection. For severe exposures, double doors may be called for in Class A openings, tight-fitting doors are needed in hospitals and nursing homes for protection of occupants in rooms and corridors, and self-closing doors are indicated for rooms and corridors in hotels and apartments.

The following excerpt from the NFPA *Fire Protection Handbook* is a description of the various types of construction for fire doors:[2]

Composite Doors: These are of the flush design and consist of a manufactured core material with chemically impregnated wood edge banding and untreated wood face veneers, or laminated plastic faces, or surrounded by and encased in steel.

Hollow Metal Doors: These are of formed steel of the flush and paneled designs of No. 20 gage or heavier steel.

Metal-clad (Kalamein) Doors: These are of flush and paneled design consisting of metal-covered wood cores or stiles and rails and insulated panels covered with steel of No. 24 gage or lighter.

Sheet-Metal Doors: These are of formed No. 22 gage or lighter steel and of the corrugated, flush, and paneled designs.

Rolling Steel Doors: These are of the interlocking steel slat design or plate-steel construction.

Tin-clad Doors: These are two- or three-ply wood-core construction, covered with No. 30 gage galvanized steel or terneplate [maximum size 14 by 20 in. (36 × 51 cm)]; or No. 24 gage galvanized steel sheets not more than 48 in. wide (122 cm).

Curtain-type Doors: These consist of interlocking steel blades or a continuous formed spring steel curtain in a steel frame.

Wood-Core Doors: These doors consist of wood, hardboard, or plastic-face sheets bonded to a wood block or wood particle board core material with untreated wood edges.

Detailed information on fire doors can be found in NFPA 80, *Standard for Fire Doors and Windows*.[8] Fire doors are classified based on their performance when tested in accordance with NFPA 252, *Standard Methods of Fire Tests of Door Assemblies*.[9] Fire door classifications range from one-third hour, commonly used to limit smoke movement, to three hour, commonly used to provide protection for Class A openings. Other classified doors are used to protect vertical openings, such as stairs and elevators.

SMOKE MANAGEMENT

Control of the smoke that develops in a smoldering fire or in a fire not extinguished in its early stages is important both for reducing the danger of death and injury to people and for the efficiency of fire fighting. Smoke, including the gaseous products of combustion, and airborne particulate matter are the leading causes of death in fires.

In most dwelling fires and other small-building fires, smoke is vented by chopping a hole in the roof. The modern, large-area, single-story factory building is usually provided with roof vents and smoke curtains to vent the smoke. Smoke removal is a difficult problem in high-rise buildings and in windowless and below-ground structures. Because the contents of these buildings cannot be controlled, it is virtually impossible to dictate to a building owner or tenant what can be put into the building. Automatic sprinklers are, at present, the best answer to controlling the products of combustion and may be considered as the most fundamental form of smoke management.

Methods of smoke management should accomplish one or more of the following:
1. Maintain a tenable environment in the means of egress.
2. Contain the smoke to the area of fire origin.
3. Maintain a condition outside of the fire area which will assist manual suppression personnel with search/rescue and fire attack.
4. Assist with protection of human life and reduce property damage.

In order to obtain these goals, two design approaches should be considered. The first and most effective approach establishes an air pressure differential across barriers. The second, less-effective approach, is to establish high-volume air flows between barriers. This method is less effective since very high velocities are necessary. Dilution of smoke is not considered to be a means of smoke management because the control of smoke involves much more than supplying fresh air into the fire area and exhausting 100 percent of the contaminated air from the fire area.

Smoke management systems should account for a number of parameters which include but are not limited to: type and arrangement of combustible contents, size of fire areas, anticipated effects of automatic extinguishing systems, number and size of openings into and out of the fire areas, stack effect of the building's "natural" ventilation system, leakage of the building's construction materials, and the anticipated effect of the buoyant gases from the fire.

A properly designed and installed smoke management system can maintain a clear path in certain egress system components, such as corridors and stairs. In addition, horizontal evacuation to areas of refuge may be enhanced by a smoke management system. This "defend-in-place" concept is vital to health care occupancies. A thorough acceptance test of the system and a followup maintenance program should also be considered.

One of the most critical steps to be taken to ensure life safety in buildings is, of course, to educate designers and builders to life safety factors that must be incorporated into buildings.

Too often the best building designs are predicated on appearance and cost. True concern for life safety of the building occupants from fire is missing. A building arrangement which may be visually pleasing to the eye may become deadly in the event of a fire. The construction materials, egress system arrangement, and building systems must be coordinated so that they complement one another under fire conditions. Part of the reason for an attitude of indifference toward firesafe design may stem from a lack of awareness and education of many individuals about fire protection.

Currently, only two secondary educational programs in the U.S. involve degree-oriented fire study programs. The University of Maryland offers a Bachelor of Science degree in fire protection engineering, and Worcester Polytechnic Institute offers a graduate program in fire protection engineering which leads to a masters degree. As more students complete these programs, they are finding themselves in new areas that have traditionally not involved fire protection.

Summary

In terms of fire protection, building design and construction practices have improved over the years, but far too many buildings—old and new—are still not firesafe. Many building designers, either through ignorance or for reasons of economy, do not take the necessary precautions to ensure that buildings are firesafe.

Highly combustible contents, even in a well-designed building, can cause severe fire damage. Combustible interior finish and combustible furnishings have also caused the rapid spread of many fires. Unprotected steel can fail quickly in a fire. Tests can be made to determine the ability of the structural elements of a building to resist fire. Estimates of fire loading can determine the severity of a fire. The installation of automatic suppression systems has so far been the best life safety provision available to building owners and designers. The installation of fire doors is important to contain a fire to a limited area. Smoke and toxic gases from a fire present serious problems, and smoke management measures must be incorporated as a life safety factor.

Firesafety is taken for granted by many individuals in the U.S. until a serious fire occurs. Firesafety is the main topic of discussion for several months following a major fire and then silence ensues until the next major fire. Education of building designers, building owners, and the general public can lead to truly firesafe building design.

References

[1]"America Burning" 1973. The National Commission on Fire Prevention and Control, Washington, DC.

[2]*Fire Protection Handbook*, 16th ed. 1986. National Fire Protection Association, Quincy, MA.

[3]*Operation Skyline* 1975. National Fire Protection Association, Boston.

[4]NFPA 80A-1987. *Recommended Practice for Protection of Buildings from Exterior Fire Exposures*, National Fire Protection Association, Quincy, MA.

[5]NFPA *101*-1988. *Life Safety Code*, National Fire Protection Association, Quincy, MA.

[6]UL *Fire Resistance Directory*. Underwriters Laboratories Inc., Northbrook, IL.

[7]UL *Building Materials Directory*. Underwriters Laboratories Inc., Northbrook, IL.

[8]NFPA 80-1986. *Standard for Fire Doors and Windows*, National Fire Protection Association, Quincy, MA.

[9]NFPA 252-1984. *Standard Methods of Fire Tests of Door Assemblies*, National Fire Protection Association, Quincy, MA.

Additional Reading

Brannigan, Francis L. 1982. *Building Construction for the Fire Service*, 2nd Ed., National Fire Protection Association, Quincy, MA.

Butcher, E. G., and Parnell, A.C., *Smoke Control in Fire Safety Design* 1979. E & F.N. SPOM, London.

Chapter 7

Water-Based Fire Protection Systems and Equipment

Water has traditionally been the most commonly used fire extinguishing agent. Each water-based system works in a specialized manner, helping to ensure fire protection to life and property.

WATER AS AN EXTINGUISHING AGENT

The great majority of fires are extinguished by use of water—from a hose delivering a solid stream or a spray; from a sprinkler system or a water spray system; or from a pump, tank, or bucket. Water, which is usually available at or near the fire scene, has special physical properties particularly suited for fire fighting. These physical characteristics of water are pertinent to its extinguishing ability and its limitations as an effective extinguishing agent.

PHYSICAL PROPERTIES OF WATER

The physical properties that make water a good extinguishing agent are:

1. At ordinary temperature, water is a heavy, relatively stable liquid.
2. The melting of 1 lb (0.45 kg) of ice into water at $32°F$ ($0°C$) absorbs 143.4 Btu (151.3 kJ), which is the heat fusion of ice.
3. One Btu (1 kJ) is required to raise the temperature of 1 lb (0.45 kg) of water $1°F$ ($-17°C$), which is the specific heat of water. Therefore, raising the temperature of 1 lb (0.45 kg) of water from $32°F$ to $212°F$ ($0°C$ to $100°C$) requires 180 Btu (189.9 kJ). (In S.I. units, the specific heat capacity of water is 4.186 kJ/kg K.)
4. The latent heat of vaporization of water, i.e., converting 1 lb (0.45 kg) of water to steam at a constant temperature, is 970.3 Btu per lb (2254.8 kJ/kg) at atmospheric pressure.
5. When water is converted from liquid to vapor, its volume at atmospheric pressure increases about 1,600 times. This large volume of water

163

(saturated steam) displaces an equal volume of air surrounding a fire, thus reducing the volume of air (oxygen) available to sustain combustion.

Other than water, there is no material easily available which has all these characteristics. Of course, water applied in the form of ice or snow would cool even better than plain water, because it would take 143.4 Btu/lb (333.2 kJ/kg) to convert the ice or snow to water. However, to date there is no practical way to do this.

EXTINGUISHING PROPERTIES OF WATER

A fire can be extinguished only if an effective agent is applied at the point where combustion is occurring. Traditionally, the principal method of extinguishing fires has been to direct a solid stream of water (from a safe distance) into the base of the fire; this method still is widely used today. A more efficient method, however, is to apply water in spray form over the fire, rather than at the base.

Extinguishment by Cooling

The amount of water required to extinguish a fire depends on how hot the fire is. How quickly a fire is extinguished depends on how quickly the water is applied, how much of it is applied, and what form of water is applied. It is best to apply water so that the maximum amount of heat will be absorbed. Water absorbs the most heat when it is converted into steam, and it will be converted into steam more easily from droplets than from a solid stream.

Much theoretical information is available on the factors that affect the rates of heat absorption and vaporization of water droplets. Because these factors cannot be closely controlled under most actual fire conditions, they cannot be used for accurate fireground calculations.

Water spray cools a fire according to the following principles:

1. The rate of heat transfer between the water and burning surface is proportional to the exposed surface area of the water. For a given quantity of water, the surface area is greatly increased by conversion from droplets to steam.
2. The rate of heat transfer depends on the temperature difference between the water and the surrounding air or burning material.
3. The rate of heat transfer also depends on the vapor content of the air; increased water vapor decreases fire spread.
4. The heat-absorbing capacity of water depends upon the distance it traveled and its velocity in the combustion zone, both of which increase the temperature of the water thus reducing its heat-absorptive capacity. (This factor must take into account the necessity for projecting a suitable volume of water to extinguish the fire.)

Droplet Size: Calculations show that the optimum diameter of a water droplet is in the range of 0.01 to 0.04 in. (0.3 to 1.0 mm), and that the best fire-extinguishment results are obtained when the droplets are fairly uniform in size. At present there is no discharge device capable of producing completely uniform droplets, although many devices spray droplets that are fairly uniform within a broad range of discharge pressures. The droplets must be large enough to have sufficient energy to reach the point of combustion despite air resistance, the opposing force of gravity, and any air currents.

Wetting combustible materials is a method often employed to prevent ignition of unburned materials. If combustibles absorb water, it takes longer to ignite them because the water must be evaporated before the materials can get hot enough to burn.

Surface cooling is not usually effective on gaseous products and flammable liquids that have flash points below the temperature of the applied water. Water is generally not recommended for flammable liquids with a flash point below 100°F (37.8°C).

Extinguishment by Smothering

Fires in ordinary combustibles normally are extinguished by the cooling effect of water—not by the smothering effect created by the generation of steam. Although steam might suppress flames, it usually cannot extinguish such fires.

If enough steam is generated, air can be displaced or excluded. Fires in certain materials can be extinguished by this smothering action, which is speedier if the steam generated can be confined to the combustion zone. The process of heat absorption by steam ends when the steam starts to condense, a change which requires heat release from the steam. When this happens, visible clouds of water vapor form. Such condensation occurring above the fire has no cooling effect on the burning material. However, the steam may continue to carry heat away from the fire if it can effectively dissipate itself into clouds of water vapor above the fire.

Water might be used to smother a burning flammable liquid when the liquid has a flash point above 100°F (37.8°C), a specific gravity of 1.1 or heavier, and is not water-soluble. To achieve smothering most effectively, a foaming agent normally is added to the water. The water then must be applied gently to the surface of the liquid.

In cases where oxygen is produced while a burning material decomposes, smothering by any agent is not possible.

Extinguishment by Emulsification

An emulsion is formed when immiscible liquids—incapable of blending or mixing—are agitated together and one of the liquids is dispersed throughout the other. Extinguishment through emulsification can be achieved by applying water to certain viscous flammable liquids, which then cools the surfaces of such liquids and prevents the release of flammable vapors. With some viscous liquids (such as No. 6 fuel oil), the emulsification is a "froth" which retards the release of flammable vapors. Care must be used on liquids of appreciable depth, however, because frothing may spread the burning liquids over the sides of the container. A relatively strong coarse water spray normally is used for emulsification. Avoid a solid stream of water as it will cause violent frothing.

Extinguishment by Dilution

Fires in water-soluble flammable materials may, in some instances, be extinguished by dilution. The percentage of dilution necessary varies greatly, as does the volume of water and the time necessary for extinguishment. For example, dilution can be used successfully in a fire involving an ethyl or methyl alcohol spill if it is possible to get an adequate mixture of water and alcohol; however, dilution is not a common practice if tanks are involved. The danger of overflow because of the large amount of water required, and the danger of frothing should the mixture become heated to the boiling point of water, usually makes this form of extinguishment impractical.

ELECTRICAL CONDUCTIVITY OF WATER

Water in its natural state contains impurities that make it conductive. If water is applied to fires involving live electrical equipment, a continuous circuit might be formed which would conduct electricity back to the applier and cause a shock, especially if there are high voltages or potentials. Water-based foam-type extinguishing agents are very conductive. The amount of current rather than the voltage determines the extent of the shock. Principal variables, assuming contact with a live electrical charge, are:

1. The voltage and amount of current flowing.
2. The "breakup" of the stream as a result of the nozzle design, the pressures used, and the wind conditions. This breakup influences the conductivity of the stream, because the air spaces formed between the droplets interrupt the electrical path to ground. The electrical conductivity hazards of droplets are less than the hazards of solid streams of water. Modern water spray nozzles and combination straight-stream spray nozzles—the latter in the spray position—provide for effective dispersion of the water droplets.
3. The purity of the water and the relative resistivity of the water.
4. The length and cross-sectional area of the water stream.

5. The resistance to ground through a person's body as influenced by location (whether on wet ground or not), skin moisture, the amount of current the body can endure, the length of exposure to the current, and other factors, such as protective clothing.
6. The resistance to ground through the hose.

USE OF WATER ON SPECIAL HAZARDS

While water generally is a universal extinguishing agent, certain prohibitions and precautions must be observed when water is applied manually on some burning materials that react either chemically or explosively on contact with water. In other instances, the mechanical action of applying water must be carefully monitored in order to avoid creating conditions that intensify the hazard rather than control it. The following paragraphs give summary guidance on using water on different materials. The aim is to prevent problems caused when water is used indiscriminately as an extinguishing agent. A good reference to consult for recommendations on use of water on specific materials where problems could be encountered is the NFPA *Fire Protection Guide on Hazardous Materials*.[1]

Chemicals

As a general rule, water should not be used on chemicals, such as carbides, peroxides, etc., because the reaction might release flammable gases and heat. When wet, certain materials, such as unslaked lime, will heat spontaneously over a period of time if heat cannot be dissipated due to storage conditions.

Combustible Metals

Generally, water should not be used on fires involving combustible metals, such as magnesium, titanium, metallic sodium, hafnium, or on metals that are combustible under certain conditions, such as calcium, zinc, and aluminum.

Radioactive Metals

Water should not be used continuously on radioactive metals. The requirements for fire protection for radioactive metals generally are consistent with their nonradioactive counterparts. (For all practical purposes, radioactivity does not influence, nor is it influenced by, the fire properties of a metal.) Control of contaminated runoff water is a complicating factor if water is applied to radioactive metals.

Gases

Water used on gas fire emergencies generally is intended to control heat from the fire while efforts are made to stop the flow of escaping gas. Water in the form of spray, applied from hose lines or monitor nozzles or by fixed water spray systems, is commonly used to disburse or dilute concentrations of flammable gases, if the gases are water-soluble.

Combustible and Flammable Liquids

Heavy fuel oil, lubricating oil, asphalt, and other liquids with high flash points do not produce flammable vapors unless heated. Once such liquids are ignited, the heat of the fire will cause enough vaporization for continued burning. If water in spray form is applied to the surface of high-flash-point burning liquids, cooling will slow down the rate of vaporization—possibly enough to extinguish the fire. If water is applied to these burning liquids by means of a coarse spray, extinguishment may be achieved by emulsification.

The ability of water without additives (foaming agents) to put out a fire is limited on low-flash-point flammable liquids. These include the Class I flammable liquids [flash points below 100°F (37.8°C)], as defined in NFPA 30, *Flammable and Combustible Liquids Code*.[2] Any water that reaches the surface of a burning low-flash-point flammable liquid in a tank probably will sink and can cause the tank to overflow. In the case of a spill fire, water probably will cause the fire to spread. However, professional handling of certain types of water spray nozzles can result in extinguishment of fires in these liquids or, at a minimum, effective fire control.

The uses of water on petroleum product fires can be summarized as follows:[3]

1. As a cooling agent, water can be used to:
    (a) Cut off the release of vapor from the surface of a high-flash-point oil, thus extinguishing the fire.
    (b) Protect fire fighters from flame and radiant heat when closing a valve or doing other work requiring close approach to the fire.
    (c) Protect flame-exposed surfaces; most effective when the surface is above 212°F (100°C).
2. As a mechanical tool, a water stream can do work at a distance to:
    (a) Control leaks.
    (b) Direct the flow of the petroleum product to prevent its ignition, or to move the fire to an area where it will do less damage.
3. As a displacing medium, water may be used to:
    (a) Float oil above a leak in a tank either before or during a fire.
    (b) Cut off fuel escape by pumping it into a leaking pipe ahead of a leak.

AUTOMATIC SPRINKLER SYSTEMS

Automatic sprinklers are devices for automatically distributing water upon a fire in sufficient quantity to: (1) extinguish the fire entirely, (2) prevent its spread if the initial fire is out of range of the sprinklers, or (3) contain the fire if it is of a type that cannot be completely extinguished by water discharged from sprinklers.

Water is fed to the sprinklers through a system of piping, ordinarily suspended from the ceiling, with the sprinklers placed at intervals along the pipes. The orifice of the fusible-link automatic sprinkler is normally closed by a disk or cap held in place by a temperature-sensitive releasing element. Figure 7.1 shows in stop-action sequence the operation of a typical fusible-link, upright automatic sprinkler.

Development of Automatic Sprinklers

The forerunners of the automatic sprinkler were the perforated pipe and the open sprinkler. These were installed in a number of American mill

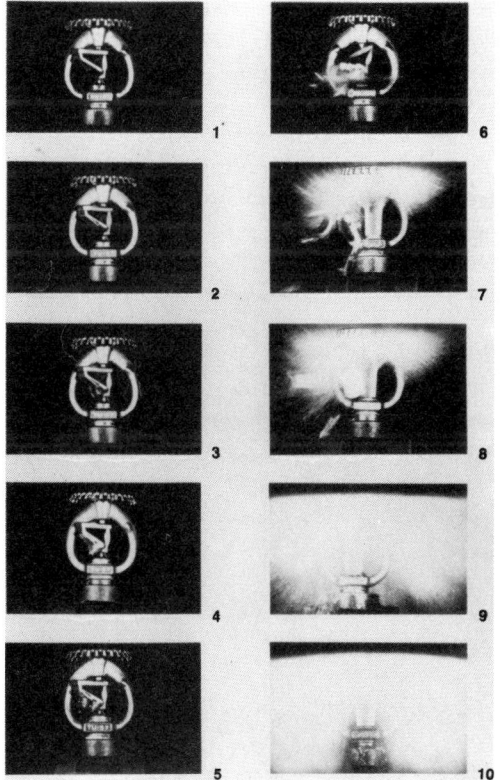

Fig. 7.1 Operation of a typical fusible-link automatic sprinkler is shown in this sequence of photos. As heat melts the solder, separation of members of the soldered link (the sloping side of the triangle in photos 1 through 5) is followed by complete separation of the link and lever arrangement (photo 6), which releases the cap over the sprinkler orifice, allowing water to escape and strike the deflector (photos 7 through 10).

properties from 1850 to 1880. These systems were not automatic, the discharge openings in the pipes often clogged with rust and foreign materials, and water distribution was poor. Open sprinklers, a slight improvement over perforated pipes, consisted of metal bulbs with numerous perforations. The bulbs were attached to piping and were intended to give better water distribution.

The idea of automatic sprinkler protection, whereby heat from a fire opens one or more sprinklers and allows the water to flow, dates back to about 1860. Its practical application in the United States, however, began about 1878 when the Parmelee sprinkler was first installed. This sprinkler, while very crude when compared with modern devices, gave generally good results and proved conclusively that automatic sprinkler protection was both practical and invaluable. (See Figure 7.2.)

Value of Automatic Sprinkler Protection

Automatic sprinkler protection helped develop modern industrial, commercial, and mercantile practices. Large areas, high buildings, hazardous occupancies, large values, or many people in one fire area all tend to develop conditions which cannot be tolerated without automatic fixed fire protection.

Automatic sprinklers are particularly effective for life safety because they give warning of the existence of fire and at the same time apply water to the burning area. Sprinklers can usually gain access to the seat of the fire as well as diminish smoke's interference with visibility for fire fighting. While the downward force of the water discharged from sprinklers can lower the smoke level in a room where a fire is burning, the sprinklers also serve to cool the smoke and make it possible for people to remain in the area much longer than

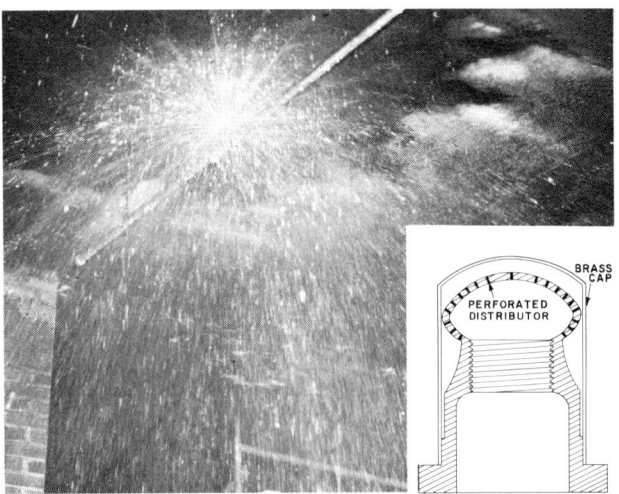

Fig. 7.2   An early automatic sprinkler: water is shown discharging from a Parmelee No. 3 upright sprinkler, which was first used in 1875. It consisted of a brass cap soldered over a perforated distributor and was designed to screw onto a nipple. A cross-sectional view of the sprinkler is at right.

they could if the room were without sprinklers. Through on-going research and development, automatic sprinklers are now practicable for dwellings and other small properties. In areas where water supplies are limited, a pressure tank can be provided; it should have sufficient capacity to control the fire during evacuation. NFPA 13D, *Standard for the Installation of Sprinkler Systems in One- and Two-Family Dwellings and Mobile Homes*,[4] gives guidance on some residential sprinkler systems.

NFPA *101*, the *Life Safety Code*,[5] recognizes sprinklers in numerous ways, particularly to offset deficiencies in existing buildings. For example, longer travel distances to exits, and interior finish of a higher combustibility than would otherwise be permitted, are allowed where sprinklers exist.

Automatic sprinklers, properly installed and maintained, provide a highly effective safeguard against the loss of life and property from fire. NFPA has no record of a multiple-death fire (a fire which kills three or more people) in a completely sprinklered building where the system was properly operating, except where an explosion occurred or flash fire killed victims prior to the system's operation. In most cases, victims of fatal fires in sprinklered properties were involved in the ignition of the fire and received their injuries prior to the operation of the sprinklers, or were unable to escape due to a physical or mental impairment.

Sprinkler Performance: Records of automatic sprinkler system performance were kept by the National Fire Protection Association from the time the Association was organized in 1896 until 1970. These remarkably comprehensive records show that in 95 percent of the some 117,770 fires in sprinklered buildings, where the Association had reliable data, the sprinklers performed satisfactorily. During the same period, there undoubtedly were numerous unreported small fires extinguished by one or two sprinkler heads, rather than the entire system. If these fires could be included in the records, the efficiency of sprinkler performance would be closer to 100 percent. Proof of this has come from Australia. Harry Marryatt, in his book *Fire: Automatic Sprinkler Performance in Australia and New Zealand, 1886-1968*,[6] reports that an 84-year record of all fires involving sprinkler systems shows 99.7 percent satisfactory performance.

The reasons for unsatisfactory performance of sprinkler systems from 1925-1969 are shown in Figure 7.3. Note that by far the largest reason has resulted from human error—the sprinklers were shut off at the time of the fire, so no water was being delivered to them. Also, partial sprinkler system protection is an important reason for sprinkler failure. If there are no sprinkler heads over the fire, the fire probably will get out of control.

Rarely do full-working automatic sprinkler systems fail to control fires. Failures are seldom due to the sprinklers themselves, but rather to inadequate water supply. Even with older types of sprinklers that are no longer approved, the failure of the sprinkler itself has been very infrequent. Under normal conditions, failure of the modern types is practically unknown.

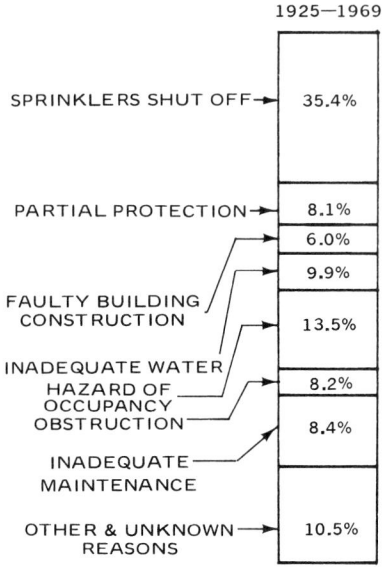

Fig. 7.3   *Reasons for unsatisfactory sprinkler performance—1925 to 1969.*

NFPA's practice of publishing summaries of sprinkler performance was stopped after 1970 when it became apparent that the data being used were biased by collection criteria which concentrated on fires causing large dollar loss. This bias led to an apparent but misleading decline in sprinkler effectiveness. Information from insurers on sprinkler effectiveness is also biased toward larger property loss due to absence of claims because of the widespread use of deductibles. Many fires which are extinguished by just a few sprinkler heads and result in small property loss are never reported to the insurer.

Thus, while current comprehensive statistics on sprinkler effectiveness are not available, ample documentation of sprinklers' effectiveness can be found in the files of the National Fire Protection Association, United States Fire Administration, National Fire Sprinkler Association, American Fire Sprinkler Association, and insurers.

In many situations, sprinkler protection is required by law for only specific parts of a building. Partial systems generally are not cost-effective. Should the fire start remote to the system, sprinklers will have no effect on the growing fire. A fire burning into the protected area generally will have developed sufficient intensity to overpower the sprinklers, thereby only wasting water needed by the fire service to fight the fire.

Minimizing Business Interruption and Water Damage

In addition to the saving in direct fire losses due to sprinkler protection, there is a saving represented by the freedom from business interruption.

There also is an undetermined but possibly even greater reduction in conflagration and exposure losses—a reduction which reasonably could be attributed to automatic sprinkler protection. The destruction of property and its adverse, sometimes permanent effect upon business can be a great hardship not only to the owner, tenants, and employees but also to the community as a whole. Safeguarding a business from serious interruption by fire is often a determining factor in a decision to install sprinkler protection.

Standard sprinkler systems have devices which automatically give an alarm in case of sprinkler operation. Thus, the systems not only apply water at the point most needed, but also give an audible signal on the premises. In many cases, they also give an alarm at a remote location, such as the local fire department or a central station. This permits immediate check of fire conditions, enhances life safety, and minimizes water damage.

A properly installed sprinkler system operating in a timely manner will generate less water damage than the later application of hose streams by the fire service. Sprinklers are not hampered in their operation by smoke or heat, as is the fire service. Sprinklers can apply water efficiently and promptly to the seat of the fire.

A common misconception is that *all* sprinklers in the sprinkler system discharge water at the time of fire. This is not the case; most fires are controlled by a few automatic sprinklers in the immediate vicinity of the fire. Fear of water damage sometimes is offered as an objection to the installation of automatic sprinkler protection. This fear comes in part from the emphasis, often in ignorance, placed upon water damage in news reports of fires. Statements that a fire was of insignificant size, but that water damage was severe, have been frequent. The probability of loss of life and severe destruction by fire in the absence of automatic sprinkler protection is seldom mentioned in these news accounts.

Rarely is there accidental discharge of water from an automatic sprinkler system or other parts of a fire protection water system due to defects in sprinklers, water control devices, piping, or associated equipment. Precautions must be taken, however, to prevent unnecessary discharge of water as a result of mechanical injury, freezing or overheating, or corrosion.

Economics of Sprinkler Protection

In addition to protection against destruction of property values and interruption of business, the savings in insurance costs often make the expenditure for automatic sprinkler protection a sound investment.

Many buildings do not have automatic sprinkler protection because the dollar cost of the protection has appeared unjustifiably high to the building owners in relation to the value of the structure. However, savings in insurance premiums alone could, in numerous cases, be adequate to finance, over a few years' time, the installation of automatic sprinkler protection. Of equal importance are the many building code "trade-offs" that are allowed when sprinklers are installed. These trade-offs often permit an increase in undivided area and less built-in fire resistance, resulting in less construction cost. In addition, no value can be placed on the life safety aspects of total sprinkler protection or the security that occupants feel when such systems are in place.

STANDARDIZING SPRINKLER INSTALLATIONS

The terms sprinkler protection, sprinkler installations, and sprinkler systems usually signify a combination of water discharge devices (sprinklers), one or more sources of water under pressure, waterflow-controlling devices (valves), distribution piping to supply the water to the discharge devices, and auxiliary equipment, such as alarms and supervisory devices. Outdoor hydrants, indoor hose standpipes, and hand hose connections also frequently are part of the system that provides protection. Figure 7.4 is an illustration of a typical sprinkler installation with all common water supplies, outdoor hydrants, and underground piping.

When considering water supply problems, the performance of sprinklers, dry pipe or wet pipe systems, or special arrangements of sprinkler protection, the designation "sprinkler system" applies to the sprinklers controlled by a single water supply valve. Under this definition, large buildings require several sprinkler systems, and a single water system may supply a number of sprinkler systems.

The fundamentals of sprinkler protection revolve around the principle of the automatic discharge of water in sufficient density to control or

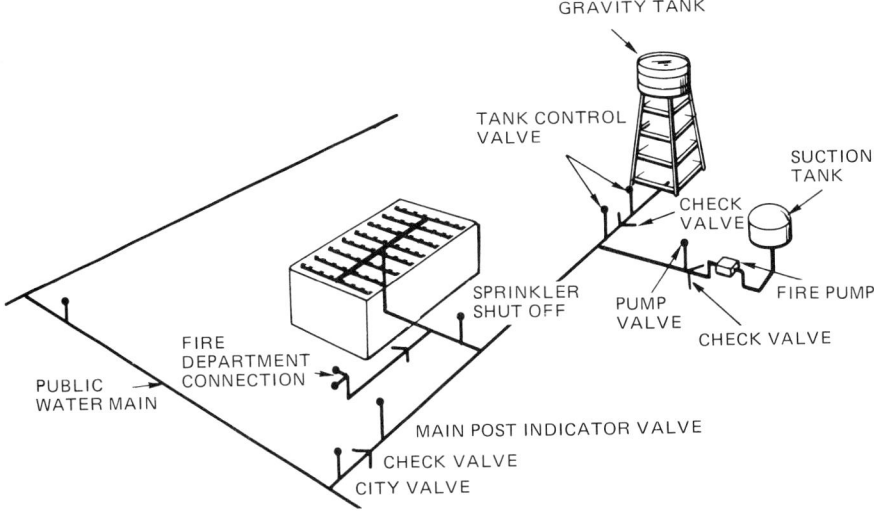

Fig. 7.4   A typical sprinkler installation showing all common water supplies, outdoor hydrants, and underground piping.

extinguish a fire in its incipiency. In planning for a system that fulfills this objective, many factors must be considered. They can, however, be broadly grouped into four categories: (1) the sprinkler system itself, (2) features of building construction, (3) hazards of occupancy, and (4) water supplies.

Automatic sprinkler systems of one type or another have been designed to extinguish or control practically every known type of fire in practically all materials in use today. It is essential, however, that the proper system be used for a given hazard. A sprinkler system designed to control and extinguish fire in an office occupancy with a relatively light amount of combustibles cannot be expected to have the same effectiveness in protecting a hazardous process involving considerable amounts of combustible materials, or a storage area where the fire loading is severe. On the other hand, it is not economical to overprotect by installing sprinkler equipment capable of controlling and extinguishing fire of a magnitude beyond any conceivable situation that could arise in the lifetime of a building.

The NFPA Sprinkler Systems Standard

NFPA 13, *Standard for the Installation of Sprinkler Systems,*[7] covers the planning and design of sprinkler protection, the type of materials and components used in systems, and the operations carried on in making the installation. Compliance with this nationally recognized sprinkler systems standard is often required by enforcement agencies, and it is used by insurance companies and insurance rating organizations. Property owners themselves often specify compliance with this sprinkler systems standard so the protection provided will be in accordance with the best-known practices.

While NFPA 13 is the primary document for guidance on installation of sprinklers, other NFPA standards, recommended practices, and guides also have a direct bearing on certain phases of sprinkler protection. They should be referred to during design and construction of sprinkler systems.

Listing of sprinkler system devices by a testing laboratory is a separate procedure. The use of devices and equipment listed by a laboratory on the basis of rigorous tests may be required by an authority having jurisdiction, or the authority itself may approve equipment.

The NFPA Residential Sprinkler Standard

In 1973, in response to recommendations in the Presidential Commission report, "America Burning,"[8] the NFPA Committee on Automatic Sprinklers appointed a Subcommittee on Residential and Light Hazard Occupancies to prepare a residential sprinkler standard. The first edition of NFPA 13D, *Standard for the Installation of Sprinkler Systems in One- and Two-Family Dwellings and Mobile Homes,*[4] published in 1975, was based on expert judgment and the best available information to date.

The purpose of NFPA 13D was "to provide a sprinkler system that will aid in the detection and control of dwelling fires and thus provide improved protection against injury, life loss, and property damage."[4] The standard permitted sprinklers to be omitted from certain areas where the incidence of life loss from fires was shown statistically to be low. In contrast, NFPA 13 required complete sprinkler protection in order to properly safeguard property. In departing from this ideal, the 1975 edition of NFPA 13D became the first attempt at a "life safety" sprinkler standard. In spite of these concessions, installations based on this standard were rare at first, primarily due to cost.

Beginning in 1976, the NFPCA (National Fire Prevention and Control Administration), later renamed the U.S. Fire Administration, acted on its mandate to reduce the nation's fire losses. It funded research programs focusing on the residential fire problem in general and residential sprinkler protection in particular.

Research showed that a more sensitive sprinkler was needed to respond faster to both smoldering and fast-developing residential fires if they were to be controlled with the water supplies typically available in residences, i.e., 20 to 30 gpm (76 to 114 L/min), and if low costs were to be achieved.

Full-scale tests conducted by Factory Mutual Research Corporation resulted in development of a prototype quick-response sprinkler which could control or suppress typical residential fires with the operation of not more than two sprinklers. It could also operate fast enough to maintain survivable conditions within the room of fire origin.[9] Survivable conditions were established as follows:

1. Maximum gas temperature at eye level—200°F (93°C).
2. Maximum ceiling surface temperature—500°F (260°C).
3. Maximum carbon monoxide concentration—1500 parts per million.

Thus, the concept changed from the traditional one of property protection to one of life safety. Full-scale field tests were then conducted in Los Angeles to establish system design parameters using the new prototype "quick-response" residential sprinkler developed by Grinnell Fire Protection Systems Company.[10-13] Data from these tests were studied by the NFPA Technical Committee on Automatic Sprinklers, and were used to establish the criteria for the 1980 edition of NFPA 13D.[4]

Residential Sprinklers in Other Occupancies

Residential-type sprinklers may be installed in buildings other than one- and two-family dwellings and mobile homes under certain specified condi- tions as covered in NFPA 13, *Standard for the Installation of Sprinkler Systems.*[7] Essentially, NFPA 13 allows residential-type sprinklers in dwelling units located in any occupancy, provided they are installed in conformance with the requirements of their listing and the positioning requirements of NFPA 13D. A dwelling unit is defined as one or more rooms arranged for use of one or more individuals living together, as in a single housekeeping unit normally having cooking, living, sanitary, and sleeping facilities. Dwelling units include hotel and motel rooms, dormitory rooms, sleeping rooms in nursing homes, and similar living units. Occupancies encompassing dwelling units include apartment buildings, board-and-care facilities, dormitories, condominiums, lodging and rooming houses, and other multiple-family dwellings.

TYPES OF SPRINKLER SYSTEMS

Automatic sprinkler systems fall into six major classifications. Each type of system includes piping for carrying water from a source of supply to the sprinklers in the area under protection. The six major classifications of systems are defined as follows:

1. *Wet Pipe Systems*: These systems employ automatic sprinklers attached to a piping system containing water under pressure at all times. When a fire occurs, individual sprinklers are activated by the heat, and water flows through those sprinklers immediately.
2. *Regular Dry Pipe Systems*: These systems have automatic sprinklers attached to piping which contains air or nitrogen under pressure. When a sprinkler is opened by heat from a fire, the pressure is reduced to the point where water pressure on the supply side of the dry pipe valve can force open the valve. Then water flows into the system and out any opened sprinklers.
3. *Preaction Systems*: These systems contain air in the piping that may or may not be under pressure. When a fire occurs, a supplementary fire detecting device in the protected area is activated. This opens a water control valve which permits water to flow into the piping system before a sprinkler is activated. When sprinklers are subsequently opened by the heat of the fire, water flows through the sprinklers immediately—the same as in a wet pipe system.
4. *Deluge Systems*: These systems have all sprinklers open at all times. When heat from a fire activates the fire detecting device, the deluge valve opens and water flows to, and is discharged from, all sprinklers on the piping system, thus deluging the protected areas.
5. *Combined Dry Pipe and Preaction Systems*: These systems include the essential features of both types of systems. The piping system contains air under pressure. A supplementary heat detecting device opens the water control valve and an air exhauster at the end of the unheated feed main. The system then fills with water and operates as a wet pipe system. If the supplementary heat detecting system should fail, the system will operate as a conventional dry pipe system.
6. *Special Types*: These systems depart from requirements of NFPA 13[7] in such areas as special water supplies and reduced pipe sizes. They are installed according to the instructions that accompany their listing by a testing laboratory.

Wet Pipe Sprinkler Systems

This type of system generally is used wherever there is no danger of the water in the pipes freezing, and wherever there are no special conditions requiring one of the other types of systems.

Where subject to temperatures below freezing, even for short periods, the ordinary wet pipe system cannot be used because the system contains water under pressure at all times. (See Figure 7.5.) There are, however, two recognized methods of maintaining automatic sprinkler protection where freezing is possible. One is to use systems where water enters the sprinkler piping only after operation of a control valve (dry pipe, preaction, etc.), and the other is to use antifreeze solution in a portion of the wet pipe system.

Antifreeze Solutions: When a recommended antifreeze solution is maintained in the piping from the riser, the normal water supply does not

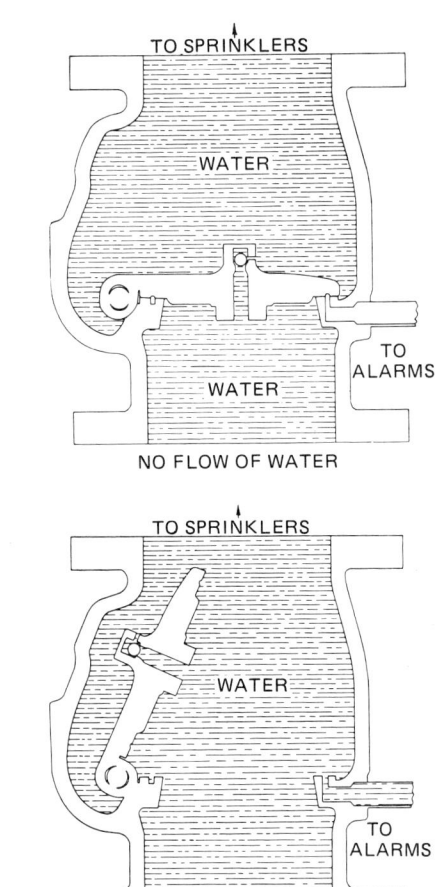

Fig. 7.5 A wet pipe sprinkler system is under water pressure at all times so that water will be discharged immediately when an automatic sprinkler operates. The automatic alarm valve shown causes a warning signal to sound when water flows through the sprinkler piping.

TO SPRINKLERS

WATER

TO ALARMS

WATER

NO FLOW OF WATER

TO SPRINKLERS

WATER

TO ALARMS

WATER FLOWING TO SPRINKLERS AND TO ALARMS

flow except when the solution is discharged from an opened sprinkler. Because antifreeze solutions are costly and may be difficult to maintain, their use usually is limited to small, unheated areas served by a wet pipe system where the volume of the section involved is not more than 40 gal (150 L) and where the piping otherwise would have to be shut off and drained during cold weather. Where the section involved is more than 40 gal (150 L), the cost of refilling the system or even of replenishing losses from small leaks makes it advisable to use small dry pipe valves. In any event, antifreeze solutions should be used only in accordance with applicable local health regulations.

Cold-Weather Valves: Automatic sprinkler piping should not be shut off and drained as a regular practice to avoid freezing during cold weather. However, where the fire hazard is not severe, permission may be given to shut off not more than 10 sprinklers on a wet pipe system. Such shutoff valves commonly are referred to as cold-weather valves.

Regular Dry Pipe Sprinkler Systems

Dry pipe sprinkler systems, in which the piping contains air under pressure until the dry pipe valve operates, are used only in locations that cannot be properly heated. However, dry pipe systems often are converted to wet pipe systems when adequate heat is provided. The principle of a dry pipe system is illustrated in Figure 7.6.

Efficiency of Dry Pipe Systems: According to fire records, more sprinklers open, on the average, at fires with dry pipe than with wet pipe systems. This tends to indicate that the control of fire is not as prompt with dry pipe as with wet pipe systems. However, in most classes of occupancy, and especially those of light and moderate hazard, dry pipe systems have shown generally good results and, when properly maintained, can be relied upon to satisfactorily extinguish or control fires.

Dry Pipe Valve Designs: Most dry pipe valves are designed so that a moderate air pressure in a dry pipe system will hold back a much greater water pressure. The difference between the air pressure and the water pressure, expressed as the ratio of these pressures when the air pressure is reduced to the value at which the valve opens, is called the differential. If the differential is obtained by having a large-diameter air clapper in a valve bear directly upon a smaller water clapper, the valve often is referred to as a differential-type dry pipe valve.

Quick-Opening Devices: One characteristic of a dry pipe system is a delay in time between the opening of a sprinkler and the discharge of water. This delay could allow the fire to spread and more sprinklers to open. The delay is due to the time required to exhaust the air from the sprinkler piping. The difficulty can be partly overcome by installation of quick-opening devices which either increase the rate of discharge of air from the piping or

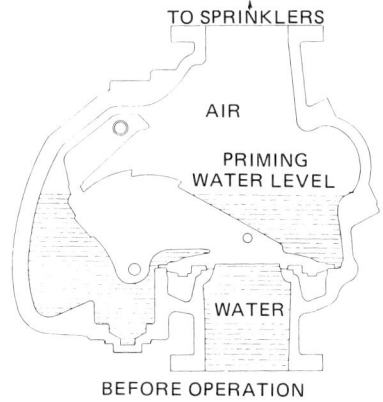

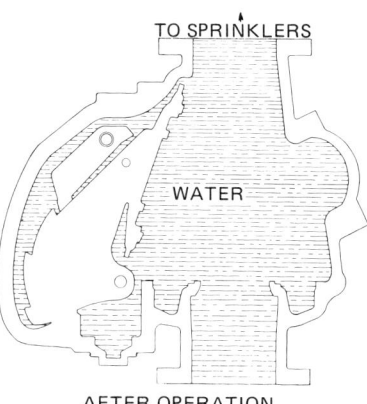

Fig. 7.6   The principle of a dry pipe system is illustrated by these simplified drawings of a dry pipe valve. Compressed air in the sprinkler system holds the dry valve closed, preventing water from entering the sprinkler piping until the air pressure has dropped below a predetermined point.

accelerate opening of the dry pipe valve when one or more sprinklers operate, depending upon the type of devices used.

Location of Dry Pipe Valves: The dry pipe valve should be located in an accessible place as near as practicable to the sprinkler system it supplies. It should be protected from mechanical injury. When exposed to cold, it must be housed in a well-constructed, lighted, and heated enclosure which will allow ready access to the valve. The water supply pipe below the dry pipe valve contains water at all times, and must be properly protected from freezing.

Preaction Sprinkler Systems

Preaction systems are designed primarily to protect properties where there is danger of serious water damage as a result of broken automatic sprinklers or piping.

The principal difference between a preaction system and a standard dry pipe system is this: in the preaction system, the water supply valve is activated independently of the opening of sprinklers. That is, the water supply valve is opened by the operation of an automatic fire detection system and not by the fusing of a sprinkler. The valve also can be operated manually.

The preaction system has several advantages over a dry pipe system. The valve is opened sooner because the system's fire detectors have less thermal lag than sprinklers. The detection system also automatically rings an alarm. Fire and water damage is decreased because water is on the fire more quickly and the alarm is given when the valve is opened. Because the sprinkler piping normally is dry, preaction systems are nonfreezing and, therefore, applicable to dry pipe service.

The same heat-responsive devices and release mechanisms used in preaction systems also can be used to operate water spray and foam extinguishing systems, as well as to actuate alarm and supervisory systems (protective signaling systems).

Preaction System with a Recycling Feature: A further refinement of the preaction system is a recycling system for controlling sprinklers. It shuts off the water when the fire has been extinguished, reactivates itself if the fire rekindles, and continues cycling as long as fire persists. Automatic sprinklers are used in the conventional manner. Supply water is held back by a flow control valve kept closed by water pressure. Operation of the flow control valve is controlled by an electrical panel activated by a system of heat detectors. The detectors are located in much the same way as sprinklers, with a specific ratio of detectors to sprinklers that depends on the building type and use and the degree of hazard.

Deluge Sprinkler Systems

The purpose of a deluge system is to wet down an entire fire area by admitting water to sprinklers that are open at all times. By using sensitive detectors operating on the rate-of-rise or fixed temperature principle, or controls designed for individual hazards, it is possible to apply water to a fire more quickly and with wider distribution than with systems whose operation depends on opening of sprinklers only as the fire spreads.

Deluge systems are suitable for various extra-hazard occupancies in which flammable liquids or other hazardous materials are handled or stored, and where there is a possibility that fire may flash ahead of the operation of ordinary automatic sprinklers. Deluge systems also are used often in aircraft hangars and assembly plants where ceilings are unusually high and where there is a likelihood of drafts, such as from open hangar doors. Drafts might deflect the direct rise of heat from an incipient fire so that ordinary sprinklers directly over the fire would not open promptly, although others at some distance would open without effect on the fire. Deluge systems also may be used to automatically control the water supply to outside open sprinklers for protection against exposure fires.

Open sprinklers and closed sprinklers can be combined in a single system where deluge protection is not needed over the entire area.

Combined Dry Pipe and Preaction Systems

The intent of a combined dry pipe and preaction system is to provide an acceptable means of supplying water through two dry pipe valves. The valves are connected in parallel to a sprinkler system of larger size than is permitted by NFPA 13[7] for a single dry pipe valve.

NFPA 13[7] does not restrict the use of combined systems to any particular classes of property. However, such systems were developed for protection of piers where long lines of supply piping could have been subject to freezing if a number of conventional dry pipe systems had been installed along the length of the pier. Due to the complications of combined dry pipe and preaction systems and the increased possibility of delayed water discharge, it is general practice to install them only in situations where it is difficult to protect a long supply main from freezing.

Special Types of Systems

In many situations, installation of sprinklers is advisable—especially for life safety—even though it is economically or otherwise impractical to meet all the requirements of NFPA 13.[7] These types of installations, commonly called nonstandard, involve features that depart from generally accepted practices. However, this does not necessarily imply questionable reliability or capacity to handle the specific fire problems for which the installations are intended. Their use does require evaluation by qualified individuals to determine their suitability.

Nonstandard sprinkler installations can involve water supplies of limited capacity, reduced pipe sizes, partial protection, sprinklers with orifice sizes different from those generally used, and other features not typical of standard installations.

Small Capacity Pressure Tanks: A sprinkler system might have a single water supply from a pressurized tank that is of less capacity than is recognized by NFPA 13[7] for limited water supply systems. It also might depart from the standard by having reduced pipe sizes, small orifice sprinklers, or increased sprinkler spacing. These special systems can employ a water supply tank pressurized by air or by compressed inert gas, such as nitrogen or carbon dioxide, from cylinders. Manufacturers of fire protection equipment have supplied, and laboratories have tested, systems of the latter type which have the advantage of using the full water capacity of a tank, with the water being discharged at a preselected, nearly constant pressure.

Substandard Water Supplies: Automatic sprinklers having water supplies from public mains, domestic or industrial systems, or other sources not meeting NFPA 13[7] requirements sometimes are installed to advantage. Such water supplies having pressure and capacity to effectively supply a few automatic sprinklers can furnish valuable protection for light fire hazards in limited areas, provided the supply is continuously available. Public water supplies of limited capacity are not likely to be dependable due to varying supply-and-demand relations, small sizes of mains, and frequently long pipe lines. Occasionally, an automatic limited service fire pump can strengthen the supply pressure if sufficient water volume is available.

Limited Water Supply Systems: Limited water supply systems are used where a public water supply or other conventional type of supply, such as a gravity tank or a fire pump, is not available for sprinklers with sufficient volume or pressure to satisfy the water supply requirements of NFPA 13.[7]

A pressure tank of limited capacity is one source of supply in this type of system which, in other respects, is the same as a conventional system because standard sprinkler system piping and standard sprinklers are used. The minimum sizes of pressure tanks recognized by NFPA 13[7] for supplemental supply for limited supply systems contain 2,000 gal (7570 L) of water for light hazard occupancies, and 3,000 gal (11 355 L) for ordinary hazard occupancies. Approval of plans for all proposed limited supply systems, including the amount of water available for pressure tanks, should be obtained from the appropriate authority having jurisdiction.

Outside Sprinkler Systems: Use of a water curtain on the outside wall of a building probably antedates automatic sprinklers. In the early years of sprinkler protection, ordinary sprinklers with the caps removed were used at the peaks of combustible roofs and at the eaves of buildings, particularly those of wood. Since then, special types of open sprinklers have been designed to protect window openings in brick walls; others protect combustible cornices. These sprinklers are placed near the top of the window or under the cornice. Water is discharged against the glass and frame or cornice to provide the desired protection.

Partial Installations: For complete protection to life and property, installation of sprinklers throughout the premises is necessary. However, sometimes partial sprinkler installations covering hazardous sections and other areas are specified in codes or standards for limited protection. This is permitted in the belief that partial installations provide opportunity for safe exit from the building, help reduce fire spread, and improve access for manual fire control.

When using partial sprinkler protection, reliance cannot be placed on these sprinklers to prevent the spread of fire originating in an unsprinklered area. NFPA fire records contain many case histories where partial systems have been overtaxed by fires originating in unsprinklered portions of buildings.

Circulating Closed Loop Systems: Sprinkler system piping in some instances can be used to circulate water for heating and cooling purposes. The closed loop piping is used only to circulate the water; none is removed for manufacturing processes or other nonfire uses.

A prominent feature of a closed loop system is that water for sprinklers is not required to pass through any heating or cooling equipment when sprinklers operate. The system provides for water to flow from the sprinkler water supply to each sprinkler without compromising the pressure or causing any loss due to outflow of water from the system because of operation of heating or cooling equipment.

AUTOMATIC SPRINKLERS: HOW THEY OPERATE

Automatic sprinklers are thermosensitive devices designed to react at predetermined temperatures by automatically releasing a stream of water and distributing it in specified patterns and quantities over designated areas. As stated earlier, the automatic distribution of water is intended to extinguish a fire or to prevent its spread if the initial fire is out of range of the sprinklers or is of a type that cannot be extinguished by water discharged from sprinklers. Water is fed to the sprinklers through a system of piping, ordinarily overhead, with the sprinklers placed at intervals along the pipes.

Since they were introduced in the latter part of the 19th century, the performance and the reliability of automatic sprinklers have been improved continually through experience and the efforts of manufacturers and testing organizations. In 1952 and 1953, a radical change which considerably improved its effectiveness was made in the pattern of the sprinkler's water discharge. Originally, this improved sprinkler was called the spray sprinkler. In 1958, it became the standard sprinkler, and sprinklers of the older design became known as old-style sprinklers.

Standard Automatic Sprinklers

Standard sprinklers generally are similar in appearance to old-style sprinklers and utilize the same style of frame and linkage or other release mechanism. The essential difference is in the deflector; seemingly minor differences in the deflector design make major differences in discharge characteristics.

On the assumption that discharge of water against the ceiling was essential to fire extinguishment, previous research on automatic sprinklers had been concerned largely with securing reasonably uniform distribution of water over the area protected by one sprinkler, and with wetting the ceiling. Later research showed that more effective extinguishment and a larger area of coverage could be achieved by directing all the water downward and horizontally. Research further showed that, with this pattern, discharge is effective even in controlling fires on the ceiling above the sprinklers. This is true because of the improved cooling effect of the spray, better high-level water distribution, and decreased exposure to the ceiling because of more effective direct discharge of water on burning materials below.

Due to the design of the deflector, the solid stream of water issuing from the orifice of a standard sprinkler is broken up to form an umbrella-shaped spray. The pattern is roughly that of a half-sphere filled with spray. Relatively uniform distribution of the water at all levels below the sprinklers is characteristic of a standard sprinkler. At a distance of 4 ft (1.2 m) below the deflector, the spray covers a circular area having a diameter of approximately 16 ft (4.9 m) when the sprinkler is discharging 15 gpm (58 L/min).

Standard sprinklers are made for installation in an upright or pendent position and must be installed in the position for which they are designed. (See Figure 7.7.) It is customary to replace old-style sprinklers with standard sprinklers in existing installations, although NFPA 13[7] permits replacing old-style sprinklers with similar devices. Most manufacturers, however, have discontinued producing old-style sprinklers. The general patterns of water discharge from the old-style and the standard sprinklers are shown in Figures 7.8 and 7.9.

Experimentation, engineering judgment, and experience determined that for pipe schedule systems a favorable rate of water discharge from an automatic sprinkler would be that of a ½-in. (12.7-mm) diameter orifice. This is often not the case with hydraulically designed systems. Therefore, sprinklers of various orifice sizes are utilized. Standard automatic sprinklers have a nominal ½-in. (12.7-mm) orifice.

Large Drop Sprinklers: Large drop sprinklers are special sprinklers designed to produce large drops that penetrate the strong updrafts generated by high-challenge fires. They have a k factor between 11 and 15 [ordinary ½-in. (12.7-mm) orifice sprinklers have k factors ranging from 5.3 to 5.8, roughly half of that for large drop sprinklers].

Small- and Large-Orifice Sprinklers: Small-orifice sprinklers with rates of discharge approximately one-half [⅜-in. (9.5-mm) orifice] and

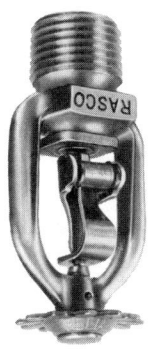

Fig. 7.7 A listed sprinkler showing upright (right) and pendent (left) models of the same issue; note the difference in design of the deflectors. The sprinklers shown are Reliable Model G.

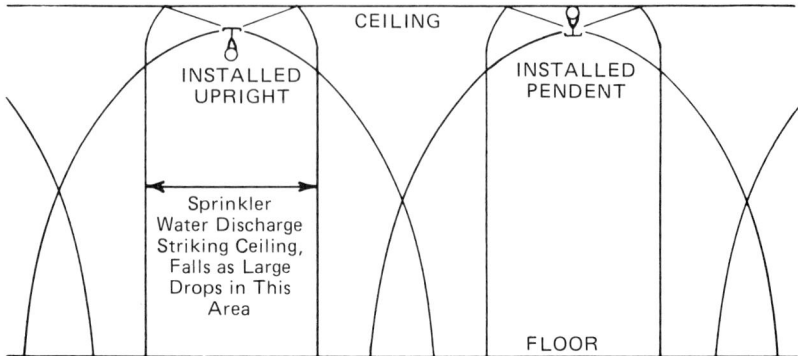

Fig. 7.8   Principal distribution pattern of water from old-style sprinklers (previous to 1953).

one-quarter [¼-in. (6.4-mm) orifice] of that of the ½-in. (12.7-mm) sprinkler are listed for special service. They are used in small enclosures or for other special conditions for which a reduced density of discharge is effective.

The established discharge rate of approved large-orifice [¹⁷⁄₃₂-in. (13.5-mm)] sprinklers is 140 percent of that of the ½-in. (12.7-mm) sprinkler. Intended for use where high density of water discharge is needed, large orifice sprinklers require a special engineering study of spacing and pipe size.

Fusible Sprinklers: A common fusible-style automatic sprinkler operates when a metal alloy fuses at a predetermined melting point. Various combinations of levers, struts, and links or other soldered members are used to reduce the force acting upon the solder. This permits the sprinkler to be held closed with the smallest practical amount of metal and solder, minimizing the time of operation by reducing the mass of fusible metal to be heated.

The solders used with automatic sprinklers are alloys of optimum fusibility composed principally of tin, lead, cadmium, and bismuth; all have

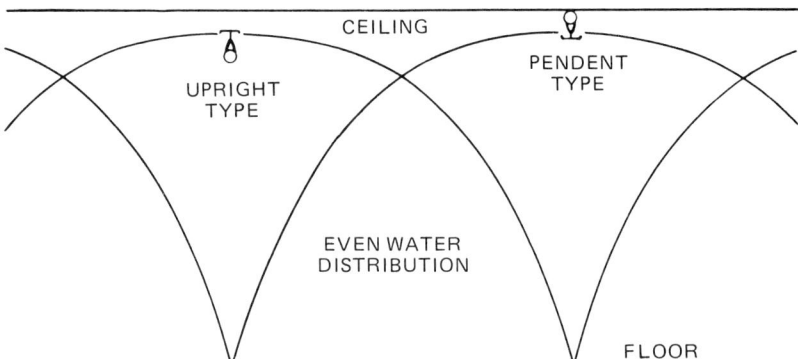

Fig. 7.9   Principal distribution pattern of water from standard sprinklers (in use since 1953).

sharply defined melting points. Alloys of two or more metals can have a melting point that is lower than that of the individual metal having the lowest melting point. The mixture of two or more metals that gives the lowest possible melting point is called a eutectic alloy.

Temperature Ratings: Today's sprinkler heads are designed with temperature ratings ranging from 135°F (57°C) to as high as 500°F (260°C). Ratings of 165°F (74°C) are common for use in buildings maintained at normal, constant temperatures.

Listed Automatic Sprinklers: In order to obtain acceptance or approval of their sprinklers, manufacturers submit them to fire testing organizations. After extensive tests and verification of the manufacturer's ability to properly manufacture the product, sprinklers found satisfactory are "listed." Acceptance of a sprinkler by inspection departments or other regulatory agencies is based on such a listing.

Standard sprinklers are designed to be installed and operated in their proper position. This is indicated by a stamping on the deflector bearing the appropriate word (upright or pendent) or the letters "SSU" (Standard Sprinkler Upright) or "SSP" (Standard Sprinkler Pendent).

Listed Quick-Response Residential Sprinklers

The design criteria in the 1980 edition of NFPA 13D[4] included for the first time the requirement that only *listed residential* sprinklers be used.

Sprinkler response time as a function of the temperature rating of the fusible element or link is well understood, i.e., a 165°F (74°C) rated sprinkler will operate when its temperature becomes 165°F (74°C), plus or minus 5°F (15°C). However, because of thermal lag of the link mass, the air temperature may be as high as 1,000°F (538°C) before the element operates.

One of the major differences between standard sprinklers and residential sprinklers is their response or sensitivity. Residential sprinklers are designed for quick response and operate much faster than standard sprinklers because they have less of a thermal lag.

Residential Sprinkler Sensitivity

Sensitivity requirements of residential sprinklers were arrived at somewhat by trial and error during the developmental test work on NFPA 13D.[4] To measure sensitivity, Factory Mutual researchers first developed the concept of the time constant, tau (τ), and later the Response Time Index (RTI).

Both the time constant, tau, and RTI refer to the performance of a sprinkler or link in a standardized air oven tunnel test. The test is known as a plunge test because a sprinkler at room temperature is plunged into a heated air stream.[14,15] The time constant, tau, is the time when the excess temperature of the sensing element of the sprinkler is approximately 63 percent of the excess gas temperature; in other words, when the temperature of the sprinkler link has risen 63 percent of the way to the higher temperature of the heated air. The smaller the time constant, tau, the faster the sprinkler sensing element heats up and operates.

The time constant, tau, is independent of the air temperature used in the plunge test, but is inversely proportional to the square root of the air velocity. During development of the 1980 edition of NFPA 13D,[4] a tau of 21 sec was considered to indicate the needed level of sensitivity, but this was associated with a specific velocity (5 ft/sec or 1.52 m/sec) used in the Factory Mutual plunge test. Since the time constant, tau, changes with the velocity of heated air moving past the sprinkler, it is a fairly inconvenient measure of sprinkler sensitivity.

The RTI now has replaced tau as the measure of sprinkler sensitivity. The RTI is determined by multiplying tau by the square root of the air velocity in the area of fire origin. The RTI therefore is practically independent of both air temperature and air velocity. Comparisons of RTI give a good indication of relative sprinkler sensitivity.

The smaller the RTI, the faster the sprinkler operation. Standard sprinklers have RTIs in the range of 225 to 700 $sec^{1/2}/ft$ ($100^{1/2}$ to 400 $sec^{1/2}/m^{1/2}$), while the RTI for residential sprinklers is about 50 $sec^{1/2}/ft^{1/2}$ (28 $sec^{1/2}/m^{1/2}$).

The tau of 21 sec at 5 ft/sec converts to an RTI of 25.9 $sec^{1/2}/m^{1/2}$. Factory Mutual has set the maximum RTI for residential sprinklers at 55 $sec^{1/2}/ft^{1/2}$. (In English units, an RTI is 1.81 times its metric value.)

Residential Sprinkler Distribution: In addition to their increased sensitivity, residential sprinklers differ from standard sprinklers in other ways. The most crucial probably is the distribution pattern.[9] Effective control of residential fires often depends on a single sprinkler in the room of fire origin. Thus, distribution of residential sprinklers must be more uniform than that of standard sprinklers, which in large areas can rely upon the overlapping patterns of several sprinklers to make up for voids. Additionally, residential sprinklers are required to protect sofas, drapes, and similar furnishings around the periphery of the room. The sprinklers' discharge spray patterns, therefore, must be capable not only of throwing water to the walls of their assigned areas, but must be high enough on the walls to prevent the fire from getting above them. Water delivered close to the ceiling not only protects the portion of the wall close to the ceiling, but also enhances the capacity of the spray to cool gases at the ceiling level, thus reducing the likelihood of excessive sprinkler openings.[9]

Residential Sprinkler Listing: Because of their differences, residential and standard sprinklers are not listed by product evaluation organizations under the same product standards. Product approval standards for residential sprinklers include a plunge test with specific sensitivity requirements and a distribution test that checks the spray pattern in the vertical as well as the horizontal plane. Product standards for standard sprinklers contain neither test.

Quick-Response Conventional Sprinklers

In addition to standard conventional sprinklers and residential quick-response sprinklers, a third type of sprinkler is the "quick-response standard sprinkler."

Development and listing of quick-response standard sprinklers have been accomplished in two different ways. One was is to replace the actuating mechanism of standard sprinklers with the more sensitive heat-responsive element used in residential sprinklers. This "quick-response" sprinkler then is resubmitted to a testing laboratory for testing and listing under a product standard for standard sprinklers. The other way is simply to submit a residential sprinkler to a testing laboratory for testing and listing under provisions of a product standard for standard sprinklers.

These quick-response standard sprinklers are designed to be used in hotels, motels, office buildings, and other large occupancies where faster operation could enhance life safety. The design for sprinkler systems using these quick-response standard sprinklers, including water supply requirements, is required to be in accordance with NFPA 13.[7]

STANDPIPE AND HOSE SYSTEMS

Standpipe and hose systems provide a means for manual application of water to fires in buildings. They do not take the place of automatic extinguishing systems which are the generally preferred form of protection. They always are needed where automatic protection is not provided and in areas of buildings not readily accessible to hose lines from outside hydrants.

Standpipe systems are designed for fire department use as a quick and convenient means of obtaining effective water streams for fire in large low buildings or the upper stories of high-rise buildings. Many jurisdictions have discontinued the requirement for occupant-use hose systems in buildings that are completely protected by automatic sprinklers. The most effective use of standpipe systems is by fire departments or personnel who are trained in the use of 2½-in. (64-mm) hose streams at high pressures.

NFPA 14, *Standard for the Installation of Standpipe and Hose Systems*,[16] contains specific requirements for these systems and should be consulted for installation details. NFPA 13[7] is another essential reference for installation of combined sprinkler and standpipe systems.

Standpipe and Hose System Classification

Class I Systems: Class I systems [2½-in. (64-mm) hose connections] are provided for use by fire departments and others trained in handling heavy water streams. In nonsprinklered high-rise buildings beyond the reach of fire department ladders, Class I systems can provide water supply for the primary means of fire fighting, i.e., manual attack on the fire.

Class II Systems: Class II systems [1½-in. (38-mm) hose lines] are provided for use by building occupants until the fire department arrives. The hose is connected to ⅜- or ½-in. (9.5- or 12.7-mm) open nozzles or combination spray/straight stream nozzles with shutoff valves. Shutoff or spray nozzles seldom are provided unless the occupancy is one where hand hose would be used frequently. Normally the hose is kept attached to the shutoff valves at the outlets. Where the hose streams used by occupants can be properly supplied by connections to the risers of wet-pipe automatic sprinkler systems, separate standpipes for these smaller streams are not required.

Class III Systems: Class III systems are provided for use either by fire departments and those trained in handling heavy hose streams or by the building occupants. Because of the multiple use, this type of system is provided with both 2½-in. (64-mm) hose connections (for use by fire departments or those trained in handling heavy hose streams) and 1½-in. (38-mm) hose connections (for use by the building occupants). One method for accommodating this multiple use is by means of a 2½-in. (64-mm) hose valve with an easily removable 2½-in. (64-mm) by 1½-in. (38-mm) adapter, permanently attached to the standpipe.

The use of hose smaller than 1½ in. (38 mm) in diameter is permitted by NFPA 14[16] in Class II service in light hazard occupancies when listed for this service and approved by the authority having jurisdiction. The reasoning is that untrained building occupants might not be able to handle 100 ft (30 m) of 1½-in. (38-mm) hose with a residual pressure of 65 psi (448 kPa) at the outlet and a flow of 100 gpm (378 L/min). If smaller diameter hose mounted on a reel with a flow smaller than 100 gpm (378 L/min) is provided, an untrained person might be less hesistant and more capable of using the equipment under fire conditions. Hard rubber ¾-in. (19-mm) and 1-in. (25-mm) fire hose is being used successfully in many foreign countries.

Water Supplies

The water supply for standpipe systems depends on the size and number of required streams, the length of time the systems may have to be operated, and the demands of automatic sprinklers using the same riser. The probable number of streams required should be ascertained before the water supply is determined. Water supplies for combined sprinkler and standpipe systems do not require that the water demand for sprinklers be added to the demand for standpipes if the occupancy is completely sprinklered; the greater amount that is required—whether for the standpipe or the sprinkler system—is sufficient. For a combined system in a partially sprinklered occupancy, the demands for each system must be added. Standpipe and hose systems should have water pressure maintained at all times. Where this is impractical, as in unheated buildings, the system should be arranged to admit water automatically by means of a dry-pipe valve or other approved device.

Acceptable water supplies include the following:

1. City waterworks systems where pressure is adequate.
2. Automatic fire pumps.
3. Manually controlled fire pumps with pressure tanks.
4. Pressure tanks.
5. Gravity tanks.
6. Manually controlled fire pumps operated by remote-control devices at each hose station.

Table 7.1 summarizes the principal specifications for the three classes of standpipe systems. Fire department connections to standpipe systems should be readily accessible on the outsides of buildings so fire department pumpers can supply water to the standpipe systems. A standpipe system is primarily designed to save time for fire department personnel when placing hose streams in service on the upper floors of buildings.

Types of Standpipe Systems

The four generally recognized types of standpipe systems are:

1. A wet standpipe system, in which the supply valve is open and water pressure is maintained at all times. This is the most desirable type of system.
2. A dry standpipe system arranged to admit water through manual operation of approved remote-control devices located at each hose station. The water supply control mechanism introduces an inherent question of reliability which must be considered.
3. A dry standpipe system in an unheated building arranged to admit water automatically by means of a dry-pipe valve or other approved device. The depletion of system air at the time of use introduces a delay in the application of water to the fire. It also increases the level of competency required to control the pressurized hose and nozzle assembly during the charging period.
4. A dry standpipe system having no permanent water supply used to reduce the time required for fire departments to put hose lines into action on upper floors of tall buildings. A dry standpipe system also might be used in buildings during construction when allowed in lieu of the wet standpipe in unheated areas.

Table 7.1   Summary of National Fire Protection Association Standpipe Standards*

Type	Intended Use	Size Hose and Distribution	Minimum Size Pipe	Minimum Water Supply
Class I	Heavy streams	2½-in. connections	4 in. up to 100 ft	500 gpm 1st standpipe
	Fire department	All portions of each story or section within 30 ft of nozzle with 100 ft of hose	6 in. above 100 ft	250 gpm each additional (2,500 gpm maximum)
	Trained personnel			
	Advanced stages of fire		(275 ft maximum unless pressure regulated)	30-minute duration
				65 psi at top outlet with 500 gpm flow
Class II	Small streams	1½-in. connections (Distribution same as Class I)	2 in. up to 50 ft	100 gpm per building
	Building occupants			30-minute duration
			2½ in. above 50 ft	
	Incipient fire			65 psi at top outlet with 100 gpm flowing
Class III	Both of above	Same as Class I with added 1½-in. outlets or 1½-in. adapters and 1½-in. hose	Same as Class I	Same as Class I

*From NFPA 14, *Standard for the Installation of Standpipe and Hose Systems.*[16]

Combined Sprinkler and Standpipe Systems

In a combined sprinkler and standpipe system, the sprinkler risers can be used for feeding both the sprinkler system and the hose outlets. The outlets are 2½ in. (64 mm) in diameter. If the building is completely sprinklered, 1½-in. (38-mm) hose for occupant use should be permitted.

Piping must comply with the requirements of NFPA 13[7] for the automatic sprinkler portions of the system, and with NFPA 14[16] for sizing of vertical risers and water supplies.

Outside Hose Systems

Where hose is kept connected to hydrants in hose houses, fire lines can be laid and water turned on in about half a minute. This is compared with the two or three minutes required where a hose cart must be run up, hose coupled to hydrants, run out, and nozzle attached before water is turned on. In addition to the advantage of accessibility, hose kept in dry hose houses lasts longer than hose kept in heated buildings. In large plants having a fire department with trained fire fighters and hose-carrying vehicles, hose houses may not be needed, but for most plants they provide the best means for storing hose.

Hose and Hydrant Houses and Equipment: NFPA 24, *Standard for the Installation of Private Fire Service Mains and Their Appurtenances*,[17] gives requirements for construction and equipment for outside hose and hydrant houses.

WATER SPRAY PROTECTION

The term "water spray" refers to the use of water that has a predetermined pattern, particle size, velocity, and density, and that is discharged from specially designed nozzles or devices. Water spray for fire protection has been called water fog, fog, or by trade name designations applied by equipment manufacturers. (The use of trade name designations cannot be taken as indicative of any specific discharge pattern or spray characteristics of the nozzles so marketed, and has been discouraged.)

No sharp line of demarcation separates water spray protection from sprinkler protection. The discharge from nozzles or sprinklers producing a spray pattern differs from water spray discharge only in the particular form of the spray and the other variables indicated in the above paragraph. In some cases, the same device can serve both purposes.

NFPA 15, *Standard for Water Spray Fixed Systems for Fire Protection*,[18] is the standard applicable to water spray systems and should be consulted on details of design and installation of the systems not covered in NFPA 13.[7]

Fixed Water Spray Systems

A water spray system is a special fixed-pipe system connected to a reliable supply of fire protection water. The system is equipped with water spray nozzles for specific water discharge and distribution over the surface or area to be protected. The piping system is connected to the water supply through an automatically or manually actuated valve which initiates the flow of water.

Automatic water control valves for spray systems can be actuated electrically by operation of automatic detection equipment, such as heat detectors, relay circuits, or gas detectors, or mechanically by hydraulic or pneumatic systems, depending upon the operating mode of the individual valves. Generally, each manufacturer of valves—most of which can do dual service in deluge systems—provides its own particular combination of water control valve, releasing mechanism, heat detection system, and supervisory service.

Application of Systems: Fixed water spray systems most commonly are used to protect flammable liquid and gas tankage, piping, and equipment; electrical equipment, such as transformers, oil switches, and rotating electrical machinery; and openings in firewalls and floors through which conveyors pass. The type of water spray required for any particular hazard will depend on the nature of the hazard and the purpose for which the protection is provided.

Uses for Water Spray Protection

Water spray can be used effectively for any one or a combination of the following purposes: (1) extinguishment of fire, (2) control of fire, (3) exposure protection, and (4) prevention of fire.

Extinguishment: Fire extinguishment by water spray is accomplished by cooling, smothering by the steam produced, emulsification of some liquids, dilution in some cases, or a combination of these factors.

Controlled Burning: With its consequent limitation of fire spread, controlled burning can be used if the burning combustible materials are not susceptible to extinguishment by water spray, or if extinguishment is not desirable.

Exposure Protection: To accomplish exposure protection, water is applied directly to the exposed structures or equipment to remove or reduce the heat transferred to them from the exposing fire. Water spray curtains mounted at a distance from the exposed surface are less effective than direct water spray application.

Prevention of Fire: It sometimes is possible to use water spray to dissolve, dilute, disperse, or cool flammable or combustible materials before they can ignite from an exposing ignition source.

Application of Water Spray Systems

Water spray protection is advantageous in meeting the previously listed purposes when it is applied to the following types of materials or equipment.

1. Ordinary combustible materials, such as paper, wood, and textiles, particularly to extinguish fires rather than as a control measure.
2. Electrical equipment installations, such as transformers, oil switches, and rotating electrical machinery.
3. Flammable gases and liquids, particularly to control fires in these materials and to extinguish certain types of fires involving combustible liquids.
4. Flammable liquid and gas tanks, processing equipment, and structures, as protection of those installations against exposure fires.
5. Open cable trays and runs containing electrical cables or tubing.

Fixed water spray systems are designed specifically to provide optimum control, extinguishment, or exposure protection for special fire protection problems. They are not intended to replace automatic sprinkler systems, but they may be independent of, or supplementary to, other forms of protection. Limitations to the use of water spray should be recognized. Such limitations involve the nature of the equipment to be protected, the physical and chemical properties of the materials involved, and the environment of the hazard.

Summary

Water is the principal fire extinguishing agent because it is readily available, is inexpensive, and has excellent cooling effects. Water discharged through automatic sprinkler systems provides valuable protection for lives and property in places of public assembly as well as in factories and commercial occupancies. The fire record of completely sprinklered properties is remarkably good, the most common cause of failure being the result of human error: the sprinklers were shut off at the time of the fire. Standpipe and hose systems are used in many buildings to provide fire fighters with water at each floor level. In addition, water fixed-spray systems provide protection for special hazards, such as flammable liquid tanks and electrical transformers.

References

[1]*Fire Protection Guide on Hazardous Materials,* 9th ed. 1986. National Fire Protection Association, Quincy, MA.

[2]NFPA 30-1987. *Flammable and Combustible Liquids Code*, National Fire Protection Association, Quincy, MA.

[3]Johnson, Oliver W. 1961. "Water on Oil Fires." *NFPA Quarterly*, Vol. 55, No. 2.

[4]NFPA 13D-1984. *Standard for the Installation of Sprinkler Systems in One- and Two-Family Dwellings and Mobile Homes*, National Fire Protection Association, Quincy, MA.

[5]NFPA *101*-1985. *Life Safety Code*, National Fire Protection Association, Quincy, MA.

[6]Marryatt, H.W. 1971. *Fire: Automatic Sprinkler Performance in Australia and New Zealand, 1886-1968*, Australian Fire Protection Association, Melbourne, Australia.

[7]NFPA 13-1987. *Standard for the Installation of Sprinkler Systems*, National Fire Protection Association, Quincy, MA.

[8]"America Burning" 1973. The National Commission on Fire Prevention and Control, Washington, DC.

[9]Kung, H.C., *et al* 1980. *Sprinkler Performance in Residential Fire Tests*, Factory Mutual Research Corp., Norwood, MA.

[10]Cote, A.E., and Moore, D. 1980. *Field Test and Evaluation of Residential Sprinkler Systems*, Los Angeles Test Series (a report for the NFPA 13D Subcommittee), National Fire Protection Association, Quincy, MA.

[11]Moore, D. 1980. *Data Summary of the North Carolina Test Series of USFA Grant 79027 Field Test and Evaluation of Residential Sprinkler Systems* (a report for the NFPA 13D Subcommittee), National Fire Protection Association, Quincy, MA.

[12]Kung, H.C., *et al* 1982. "Field Evaluation of Residential Prototype Sprinkler: Los Angeles Fire Test Program," *FMRC J.I. OEOR3.RA(1)*, Factory Mutual Research Corporation, Norwood, MA.

[13]Cote, A.E. 1982. *Final Report on Field Test and Evaluation of Residential Sprinkler Systems*, National Fire Protection Association, Quincy, MA.

[14]Heskestad, G., and Smith, H.F. 1976. "Investigation of a New Sprinkler Sensitivity Approval Test: The Plunge Test," *FMRC Technical Report 22485*, Factory Mutual Research Corporation, Norwood, MA.

[15]Heskestad, Gunner, and Smith, Herbert 1980. "Plunge Test for Determination of Sprinkler Sensitivity," *FMRC J.I.3AIE 2.RR*, Factory Mutual Research Corporation, Norwood, MA.

[16]NFPA 14-1986. *Standard for the Installation of Standpipe and Hose Systems*, National Fire Protection Association, Quincy, MA.

[17]NFPA 24-1984. *Standard for the Installation of Private Fire Service Mains and Their Appurtenances*, National Fire Protection Association, Quincy, MA.

[18]NFPA 15-1985. *Standard for Water Spray Fixed Systems for Fire Protection*, National Fire Protection Association, Quincy, MA.

Chapter 8

Nonwater-Based Fire Protection Systems and Equipment

Chemical and mechanical extinguishing methods have been developed to provide wider latitude for fire control. Each works in a specialized manner, ensuring greater fire protection to life and property.

FOAM EXTINGUISHING AGENTS AND SYSTEMS

Fire fighting foam is an aggregate of gas-filled bubbles made from aqueous solutions of specially formulated concentrated liquid foaming agents. The gas used is normally air, but in certain applications can be an inert gas. Since foam is lighter than the aqueous solutions from which it is formed, and lighter than flammable liquids, it floats on all flammable or combustible liquids. This produces an air-excluding, cooling, continuous layer of vapor-sealing, water-bearing material that halts or prevents combustion.

Foam is produced by mixing a foam concentrate with water at the appropriate concentration, and then aerating and agitating the solution to form the bubble structure. Some foams are thick and viscous and create tough, heat-resistant blankets over burning liquid surfaces and vertical areas; other foams are thinner and spread more rapidly. Some foams are capable of producing a vapor-sealing film of surface-active water solution on a liquid surface. Others, such as medium- or high-expansion foam, are meant to be used as large volumes of wet-gas cells for inundating surfaces and filling cavities.

Fire fighting foams are defined by their expansion ratio—the ratio of final foam volume to original foam solution volume before adding air. The foams are subdivided arbitrarily into three ranges: (1) low-expansion foam—expansion up to 20:1, (2) medium-expansion foam—expansion 20 to 200:1, and (3) high-expansion foam—expansion to 200 to 1,000:1.

Uses and Limitations of Fire Fighting Foams

Low-expansion foam is used principally to extinguish burning flammable or combustible liquid spills or tank fires by application to develop a cooling,

197

coherent blanket. Foam is the only permanent extinguishing agent used for fires of this type. Its application allows fire fighters to extinguish fires progressively. A foam blanket covering a tank's liquid surface can prevent vapor transmission for some time, depending upon the stability and depth of the foam. Fuel spills quickly are rendered safe by foam blanketing. The blanket can be removed after a suitable period of time; often it has no detrimental effect on the product with which it comes in contact.

Foams can be used to diminish or halt the generation of flammable vapors from nonburning liquids or solids, and to fill cavities or enclosures where toxic or flammable gases might collect.

Where aircraft are fueled and operated, foam is of great importance. Sudden large fuel spills resulting from aircraft accidents or malfunctions require rapid foam application. Hangar fire protection is best accomplished by foam-water sprinkler systems and portable foam equipment.

Foams of the medium- or high-expansion type (20 to 1,000:1) might be used to fill enclosures such as basement areas or holds of ships, where fires are difficult or impossible to reach. Here foams act to halt convection and access to air for combustion. Their water content also cools and diminishes oxygen by steam displacement. Some high-expansion foams (with expansion ratios of 400 to 500:1) can be used to control liquefied natural gas (LNG) spill fires and to help disperse the resulting vapor cloud.

Many foams are generated from solutions with very low surface tension and penetration characteristics. Foams of this type are useful where Class A combustible materials are present. In such instances, the water solution draining from the foam wets and cools the solid combustibles.

Foam breaks down and vaporizes its water content under attack by heat and flame. Therefore, it must be applied to a burning liquid surface in sufficient volume and rate to compensate for this water loss, with an additional amount applied to guarantee a residual foam layer over the extinguished liquid to guard against post-fire rekindling. Foam is unstable and can be easily broken down by a physical or mechanical force, such as a water hose stream. Certain chemical vapors or fluids also quickly can destroy foam. When certain other extinguishing agents are used in conjunction with foam, severe breakdown of the foam can occur. Turbulent air or violently uprising combustion gases from fires also can divert foam from the burning area.

Foam solutions are conductive and therefore not recommended for use on electrical fires. If foam is used, a spray is less conductive than a straight stream. However, because foam is cohesive and contains materials that allow water to conduct electricity, foam spray is more conductive than water spray.

Engineering design requirements and recommended application methods must be followed for successful use of foams. These requirements can be found in NFPA 11, *Standard for Low-Expansion Foam and Combined Agent Systems;*[1] NFPA 11A, *Standard for Medium- and High-Expansion Foam Systems;*[2] NFPA 11C, *Standard for Mobile Foam Apparatus;*[3] NFPA 16, *Standard for the*

Installation of Deluge Foam-Water Sprinkler Systems and Foam-Water Spray Systems;[4] and NFPA 403, *Recommended Practice for Aircraft Rescue and Fire Fighting Services at Airports and Heliports.*[5]

Foam is fully effective for hazardous liquid fires when the following general criteria are met:

1. The liquid must be below its boiling point at the ambient conditions of temperature and pressure.
2. Care must be taken in application of foam to liquids with a bulk temperature higher than 212°F (100°C). At and above these fuel temperatures, foam forms an emulsion of steam, air, and fuel. This can produce a four-fold increase in foam volume when applied to a tank fire, possibly creating dangerous frothing or slopover of the burning liquid.
3. The liquid must not be unduly destructive to the foam used, or the foam must not be highly soluble in the liquid to be protected.
4. The liquid must not be water reactive.
5. The fire must be a horizontal surface fire. Three-dimensional (falling fuel) or pressure fires cannot be extinguished by foam unless the hazard has a relatively high flash point and can be cooled to extinguishment by the water in the foam.

Types of Foam

A number of types of foaming agents are available. Some are known as foam concentrates, which are designed for specific applications. Others are suitable for extinguishing all types of flammable liquids, including water-soluble and foam-destructive liquids. Descriptions of the common types of foam are shown in Table 8.1.

CARBON DIOXIDE SYSTEMS

Carbon dioxide (CO_2) has been used for many years in the extinguishment of flammable liquid fires, gas fires, fires involving electrically energized equipment, and—to a lesser extent—fires in ordinary combustibles, such as paper, cloth, and other cellulosic materials. CO_2 will suppress fire in most combustible materials; exceptions are a few active metals, such as magnesium, and metal hydrides, and materials such as cellulose nitrate that contain available oxygen. Further practical limitations of CO_2 are related to the method of application and to restrictions imposed by the hazard itself.

NFPA 12, *Standard on Carbon Dioxide Extinguishing Systems,*[6] provides guidance to those responsible for the purchase, design, installation, testing, inspection, operation, and maintenance of carbon dioxide systems.

Table 8.1　Fire Fighting Foams

Type	Company	Trade Name	Effective on Types of Fuels
PROTEIN	Nat'l Foam Ansul Angus Rockwood Rockwood	Aero foam Nicerol Regular 6% Dbl-Strength-3%	Hydrocarbon
Fluoroprotein	Nat'l Foam Angus Ansul Rockwood	XL-3 FP70 Fluoroprotein Super Pro	Hydrocarbon FP3
Synthetic	MSA Nat'l Foam Angus Rockwood Ansul	Syndet Expandol Jet X Full EX	Hydrocarbon Class A
AFFF	3M Nat'l Foam Ansul Angus Rockwood	Light Water Aero Water Ansul Light Tridol AFFF	Hydrocarbon
AFFF/Fluoroprotein (AFFF P)	Angus Nat'l Foam	Petroseal Aerofilm 3	Hydrocarbon
Fluoroprotein Polar Solvent	Angus	Fluoro-Polydol	Hydrocarbons & Polar Solvents
AFFF/Alcohol Resistant	3M Nat'l Foam Ansul MSA Angus Rockwood	ATC Universal ARC Polar Compound AFFF Alcoseal Aqua Foam	Polar Solvents and Hydrocarbons

*May be used on some types of hazardous materials.
Source: Industrial Fire World.

Properties of Carbon Dioxide

Carbon dioxide has a number of properties that make it a desirable fire extinguishing agent: it is noncombustible, it does not react with most substances, and it provides its own pressure for discharge from the storage container. Also, since carbon dioxide is a gas, it can penetrate and spread to all parts of the fire area. As a gas, or as a finely divided solid called "snow" or "dry ice," it will not conduct electricity and, therefore, can be used on energized electrical equipment. It leaves no residue, thus eliminating cleanup due to the agent itself.

Toxicity: Although carbon dioxide is only mildly toxic, it can produce unconsciousness and death when present in fire extinguishing concentrations.

Use on Hazmat	Nominal Expansion Ratio	Recommended Storage Construction	Sub- Surface	% Available
No	(7-10)-1	Mild Steel	No	3% 6%
No	(7-10)-1	Mild Steel	Yes	3% 6%
Yes	500-1 to 1500-1	Mild Steel Plastic	No	1%-10%
*Yes	(7-10)-1	Stainless or Fiberglass	Yes	1%-2%-3% 6%
No	(7-10)-1	Mild Steel	Yes	3% or 6% 3%
Yes	(7-10)-1	Mild Steel	Yes	3% Hydrocarbons 6% Polar Solvents
Yes	(7-10)-1	Stainless Steel Fiberglass	Yes	6% Polar Solvents 3% Hydrocarbons

These reactions are due more to suffocation than to any toxic effect of the carbon dioxide itself. A concentration in air of nine percent is about all most people can withstand without losing consciousness within a few minutes. Breathing a higher concentration of carbon dioxide could render a person helpless almost immediately.

Extinguishing Properties of CO_2

Carbon dioxide is effective as an extinguishing agent primarily because it reduces the oxygen content of the atmosphere by dilution to a point where the atmosphere no longer will support combustion. Under suitable conditions of control and application, the available cooling effect also is helpful, especially where carbon dioxide is applied directly on the burning material.

Extinguishment by Smothering: In any fire, heat is generated by rapid oxidation of a combustible material. Some of this heat raises the unburned fuel to its ignition temperature, while a large part of the heat is lost through radiation and convection—especially in the case of surface-burning materials. If the atmosphere that supplies oxygen to the fire is diluted with carbon dioxide vapor, the rate of heat generation (oxidation) is reduced until it is below the rate of heat loss. When the fuel is cooled below its ignition temperature, the fire dies out and is completely extinguished.

The minimum concentration of carbon dioxide needed to extinguish surface-burning materials, such as liquid fuels, can be accurately determined, since the rate of heat loss by radiation and convection is reasonably constant. Table 8.2 lists the minimum concentrations of CO_2 necessary for some common liquid and gaseous fuel fires as determined by the U.S. Bureau of Mines. It is difficult to obtain similar data for solid materials, because the rate of heat loss through radiation and convection can vary widely, depending upon shielding effects caused by the physical arrangement of the burning material.

Extinguishment by Cooling: The value of rapid cooling is more apparent where the CO_2 agent is discharged directly on the burning material, such as a liquid-filled dip tank. A massive application quickly covering the entire surface area prevents reignition when the discharge ends and normal air again contacts the fuel area. The presence of dry-ice particles in the discharge stream helps to promote fast surface cooling.

Limitations of CO_2 as an Extinguishing Agent

The use of carbon dioxide on general Class A fires is limited mostly by (1) its low cooling capacity (particles of dry ice do not wet or penetrate), and (2) enclosures incapable of retaining an extinguishing atmosphere. True surface burning fires are extinguished easily because natural cooling takes place quickly. On the other hand, if the fire penetrates below the surface, or under materials that provide thermal insulation that slows down the rate of heat loss (generally referred to as "deep-seated burning"), a higher concentration of carbon dioxide and a much longer holding time are needed for complete extinguishment.

Liquid fuel fires frequently are extinguished by discharging CO_2 directly on the burning material. No enclosure is needed and a 30-second discharge usually is adequate to cool everything below the reignition temperature of the fuel. This use of carbon dioxide is limited mainly to situations where there is serious overheating of massive metal objects or a substantial quantity of glowing embers from carbonaceous materials. A much longer discharge time may be needed to prevent reignition.

Table 8.2   Minimum Carbon Dioxide Concentrations for Extinguishment

Material	Theoretical Min. CO_2 Concentration (%)
Acetylene	55
Acetone	26*
Benzol, Benzene	31
Butadiene	34
Butane	28
Carbon Disulfide	55
Carbon Monoxide	53
Coal Gas or Natural Gas	31*
Cyclopropane	31
Dowtherm	38*
Ethane	33
Ethyl Ether	38*
Ethyl Alcohol	36
Ethylene	41
Ethylene Dichloride	21
Ethylene Oxide	44
Gasoline	28
Hexane	29
Hydrogen	62
Isobutane	30*
Kerosene	28
Methane	25
Methyl Alcohol	26
Pentane	29
Propane	30
Propylene	30
Quench, Lubricating Oils	28

Note: The theoretical minimum extinguishing concentrations in air for the above materials were obtained from U.S. Bureau of Mines, Bulletin 503 (Coward and Jones, 1952). Those marked * were calculated from accepted residual oxygen values.

Oxygen-Containing Materials and Reactive Chemicals: Carbon dioxide is not an effective extinguishing agent for fires involving chemicals, such as cellulose nitrate, that contain their own oxygen supply. Fires involving reactive metals, such as sodium, potassium, magnesium, titanium, zirconium, and the metal hydrides, cannot be extinguished by carbon dioxide because the metals and hydrides decompose CO_2.

Life Safety Considerations: Carbon dioxide should not be used in normally occupied spaces unless arrangements can be made to assure evacuation before discharge. The same restriction applies to spaces not normally occupied but where personnel might be present for maintenance or other purposes. It can be difficult to assure evacuation if the space is large or if egress is in any way impeded by obstacles or complicated passageways.

Escape is even more difficult after the discharge starts, because of possible confusion due to noise and greatly reduced visibility.

Consideration also should be given to any possibility of large volumes of carbon dioxide vapor leaking or flowing into unprotected lower, possibly occupied, levels, such as cellars, tunnels, or pits. In this event, the suffocating atmosphere would not be visible and might not be detected until too late to save lives.

Methods of Application

Two basic methods are used to apply carbon dioxide in extinguishing fires. One method is to discharge a sufficient amount of the agent into an enclosure to create an extinguishing atmosphere. This is called "total flooding." The second method is to discharge the agent directly on the burning material without relying on an enclosure to retain the carbon dioxide. This is called "local application."

Hand Hose Lines: Carbon dioxide systems can consist of hand hose lines permanently connected by means of fixed piping to a fixed supply of CO_2. Such systems frequently are provided for manual protection of small localized hazards. Although not a substitute for a fixed system, a hose line can be used to supplement a fixed system where the hazard is accessible for manual fire fighting, as well as to supplement portable equipment.

HALOGENATED AGENTS AND SYSTEMS

Halogenated extinguishing agents are hydrocarbons in which one or more hydrogen atoms have been replaced by atoms from the halogen series: fluorine, chlorine, bromine, or iodine. This substitution confers non-flammability as well as flame extinguishment properties to many of the resulting compounds. Halogenated agents are used both in portable fire extinguishers and in extinguishing systems.

Prior to 1945, three halogenated fire extinguishing agents were widely used: carbon tetrachloride (Halon 104), methyl bromide (Halon 1001), and chlorobromomethane (Halon 1011). The earliest, carbon tetrachloride, became available in the early 1900s and found immediate wide use in portable hand-pump extinguishers. Its main advantages were electrical non-conductivity and lack of residue following application.[7]

In the late 1920s, methyl bromide (Halon 1001) was found to have greater extinguishing potential than carbon tetrachloride. It was used extensively in German aircraft and ships and in British aircraft during World War II, but because of its high vapor toxicity it was never widely used in portable extinguishers. Chlorobromomethane (Halon 1011) was developed in

Germany from 1939 to 1940 as a replacement for methyl bromide, but its use did not become widespread until after World War II.[8]

For toxicological reasons, however, concern about using these three early halogenated agents gained significant momentum during the early 1960s. Except for a few Halon 1001 and Halon 1011 systems that may still be in service on several older models of European and United States military aircraft, and except for certain explosion suppression applications, systems containing these three early halogenated agents have been removed from service.

In 1947, the Purdue Research Foundation performed a systematic evaluation of more than 60 new extinguishing agents. Simultaneously, the U.S. Army Chemical Center undertook toxicological investigations of these same compounds. From these tests, four halogenated agents— bromotrifluoromethane (Halon 1301), bromochlorodifluoromethane (Halon 1211), dibromodifluoromethane (Halon 1202), and dibromotetrafluoromethane (Halon 2402)—were selected for further evaluations in specific applications.

From these further tests, Halon 1301 was determined to be the second most effective and least toxic of the group, while Halon 1202 was the most effective but also the most toxic. As a result, Halon 1301 was selected by the U.S. Army for use in portable extinguishers and by the Federal Aviation Administration for use in commercial aircraft engine nacelles. Halon 1202 was selected by the U.S. Air Force to protect military aircraft engines. In a similar evaluation program in England, Halon 1211 was selected for military and civilian aircraft systems and for portable fire extinguishers.

The concept of using halogenated agents in commercial total-flooding systems seems to have originated between 1962 and 1964. From 1964 through 1968, a number of Halon 1301 total-flooding systems using carbon dioxide equipment and technology were installed in the U.S. Such a system designed to protect the Winterthur Museum was described in the November 1969 issue of *Fire Journal*.[9]

In 1966, NFPA organized the Technical Committee on Halogenated Fire Extinguishing Agent Systems to develop standards covering installation, maintenance, and use of such systems. NFPA 12A, *Standard on Halon 1301 Fire Extinguishing Systems*,[10] and NFPA 12B, *Standard on Halon 1211 Fire Extinguishing Systems*,[11] were approved in the early 1970s.

During 1966, attention began to focus on the use of Halon 1301 to protect computer rooms and electronic data processing (EDP) equipment. In 1972, following extensive testing by several major companies on the effects of Halon 1301 decomposition products on electronic equipment,[12] the NFPA Committee on Electronic Computer/Data Processing Equipment recognized Halon 1301 total-flooding systems as suitable for protection of electronic computer/data processing equipment. In Europe, Halon 1211 total-flooding systems now are used with precautions, such as time delays, to allow for evacuation.

Chemical Compositions and Classification

Halogenated extinguishing agents currently are known simply as halons. The halon system for naming the halogenated hydrocarbons was devised by the U.S. Army Corps of Engineers. This simplified system of nomenclature describes the chemical composition of the materials without the use of chemical names or possibly confusing abbreviations. Examples of this system are shown in Table 8.3. The first digit of the number represents the number of carbon atoms in the compound molecule; the second digit, the number of fluorine atoms; the third digit, the number of chlorine atoms; the fourth digit, the number of bromine atoms; and the fifth digit, the number of iodine atoms (if any). If the fifth digit is zero, it is not expressed; bromo-trifluoromethane ($BrCF_3$), for example, is referred to as Halon 1301 (not 13010), although its chemical formula shows one carbon atom, three fluorine atoms, no chlorine atoms, one bromine atom, and no iodine atoms.

Because they are either gases or liquids that rapidly vaporize in fire, halons leave no corrosive or abrasive residue. They are nonconductors of electricity, and their high liquid densities permit use of compact storage containers. Halon is used primarily for the protection of electrical and electronic equipment, petroleum production facilities, engine compartments (e.g., those of ships, military vehicles, and aircraft), and other areas where rapid extinguishment is important. Halons also are used where damage to equipment or materials or post-fire cleanup must be minimized.

Extinguishing Characteristics

The extinguishing mechanism of the halogenated agents is not clearly understood. However, a chemical reaction undoubtedly occurs which interferes with the combustion process. The agents act by removing the active chemical species involved in the flame chain reactions (a process known as "chain breaking"). While all the halogens are active in this way, bromine-based halon is much more effective than chlorine- or fluorine-based halon.

Table 8.3   Sample Halon Numbers for Various Halogenated Fire Extinguishing Agents

Chemical Name	Formula	Halon No.
Methyl bromide	CH_3Br	1001
Methyl iodide	CH_3I	10001
Bromochloromethane	CH_2BrCl	1011
Dibromodifluoromethane	CF_2Br_2	1202
Bromochlorodifluoromethane	CF_2BrCl	1211
Bromotrifluoromethane	CF_3Br	1301
Carbon tetrachloride	CCl_4	104
Dibromotetrafluoroethane	$C_2F_4Br_2$	2402

Fire Extinguishing Effectiveness: In total-flooding systems, the effectiveness of the halogenated agents on flammable liquid and vapor fires can be dramatic. Rapid and complete extinguishment can be achieved with low concentrations of agent. On a world-wide basis, systems using Halon 1301 and Halon 1211, together with realistic test methods, have been developed. Applied to both of these halon agents and recognized by NFPA, tests have shown that in total-flooding application for flame extinguishment or inerting, Halon 1301 requires an average of 10 percent less material on a gas volume basis than does Halon 1211 for any given fuel. Table 8.4 shows a comparison of flame extinguishment values included in NFPA 12A[10] and NFPA 12B.[11] It generally is recognized that on a weight-of-agent basis, both agents are approximately two and one-half times more effective than carbon dioxide.

The flame extinguishing ability of Halon 2402 vapors is very good and quite similar to that for Halon 1301 and Halon 1211. However, because it is a liquid at room temperature and is intended mainly for local application situations, its performance is not directly comparable to the total-flooding data obtained for Halon 1301 and Halon 1211.

The effectiveness of halogenated agents in Class A fires is less predictable and depends to a large extent upon the specific burning material, its configuration, and how early in the combustion cycle the agent is applied. Most plastics behave as flammable liquids—they can be extinguished rapidly and completely with 4 to 6 percent concentrations of Halon 1211 or Halon 1301. Other materials, particularly cellulosic products, can in certain forms develop deep-seated fires in addition to flaming combustion. The flaming portion of such fires can be extinguished with low 4 to 6 percent concentrations of agent, but the deep-seated portion can continue glowing under certain circumstances.[13] Even so, the deep-seated fire will be controlled to some extent: its rate of burning and consequent heat release will be reduced. Considerably higher agent concentrations (18 to 30 percent) are required to achieve complete extinguishment, but these levels are seldom economical to

Table 8.4   Comparison of Flame Extinguishment Values for Halon 1301 and Halon 1211

Fuel	Average Percent by Volume of Agent in Air Required for Flame Extinguishment	
	Halon 1301	Halon 1211
Methane	3.1	3.5
Propane	4.3	4.8
n-Heptane	4.1	4.1
Ethylene	6.8	7.2
Benzene	3.3	2.9
Ethanol	3.8	4.2
Acetone	3.3	3.6

Note: Design flame extinguishment concentrations for Halon 1301 and 1211 total flooding systems are calculated from the tested value by adding a 20 percent safety factor. However, they are never less than 5 percent.

apply. However, the concept of controlling deep-seated fires with halogenated agents has been accepted in the respective NFPA standards.

Toxic and Irritant Effects

The toxicity of Halon 1301, Halon 1211, and Halon 2402 has been studied extensively in both animals and humans. As a result, safety guidelines for these agents have been developed. From the extensive medical data available, exposure guidelines have been produced for use of Halon 1301, Halon 1211, and Halon 2402. (See Table 8.5.)

NFPA 12A[10] permits Halon 1301 design concentration up to 10 percent in normally occupied areas, and up to 15 percent in areas not normally occupied. Because the required extinguishing concentration of Halon 1211 is near or above its limit for safe exposures, Halon 1211 systems are not recognized in NFPA 12B[11] for use in normally occupied areas.

Decomposition Products of Halon

Consideration of life safety during use of halogenated agents also must include the effects of breakdown products, which have a relatively higher toxicity to humans. Decomposition of halogenated agents takes place on exposure to flame, or at surface temperatures above approximately 900°F (482°C). In the presence of available hydrogen (from water vapor or the combustion process itself), the main decomposition products of Halon 1301 are hydrogen fluoride (HF), hydrogen bromide (HBr), and free bromine (Br_2). Although small amounts of carbonyl halides (COF_2, $COBr_2$) were reported in early tests, more recent studies have failed to confirm the presence of these compounds. The decomposition products of Halon 1211 and Halon 2402 are similar, but in the case of Halon 1211 also include hydrogen chloride (HCl) and free chlorine (Cl_2).

Halon's Effect on the Ozone Layer

In the mid-1980s, it was determined that halon emissions were contributing to reductions in the protective ozone layer in the earth's upper atmosphere. The Montréal Protocol on Substances that Deplete the Ozone Layer, adopted September 16, 1987, became the first global agreement to deal specifically with a worldwide environmental concern.

Under terms of the Montréal Protocol, production and use of the fire fighting halons—Halon 1211, 1301, and 2402—are allowed to continue at 1986 levels. While the Montréal Protocol calls for terms of the agreement to go into effect in 1989, in the United States the effective date will be set by the Environmental Protection Agency (EPA). EPA regulations are due to be announced in August 1988.

**Table 8.5   Permitted Exposure Times to Halon
1301, Halon 1211, and Halon 2402**

	Concentration Percent by Volume	Permitted Time of Exposure
Halon 1301	Up to 7	15 min.
	7–10	1 min.
	10–15	30 sec
	Above 15	Prevent exposure
Halon 1211	Up to 4	5 min.
	4–5	1 min.
	Above 5	Prevent exposure
Halon 2402	0.05	10 min.
	0.10	1 min.

Although halon fire protection will become a controlled industry with the implementation of domestic regulations, there is recognition internationally as well as in the U.S. of the following:

1. Fire fighting halons are essential to the modern economy.
2. At the present time, there are no alternatives for halons in many applications.
3. Any regulations for reduction in the use of halons will depend upon further scientific review of the negative effects of halons upon the ozone layer.

INERTING GAS SUPPRESSION SYSTEMS

For many years a range of gases, including carbon dioxide, nitrogen, and helium, has been used successfully to prevent ignition of potentially flammable mixtures or to extinguish fires—particularly those involving flammable liquids or gases. These methods have been based on the knowledge that fires can be extinguished if a sufficient volume of an inert gas is introduced into an enclosed space where a fire is burning. The concentration must be maintained in the space long enough to prevent rekindling of the burned material.

Traditionally, this method of extinguishment has been explained as a reduction of the fuel or oxygen concentration to the point where combustion is prevented. Now, however, modern fire dynamics show that the mechanism is related to the fact that the added inerting gas acts as "thermal ballast" to reduce the temperature of the flame/vapor mixture below the limiting adiabatic flame temperature necessary to sustain combustion.[14] The gaseous extinguishing agent actually absorbs the combustion energy, and its thermal capacity is the important factor in inerting explosive or burning materials.

The thermal capacities of the more commonly used inerting agents are shown in Table 8.6. Table 8.6 shows why carbon dioxide is an important gaseous extinguishing agent and why helium, with its lower heat-absorbing

**Table 8.6   Thermal Capacities of the More
Commonly Used Inerting Agents.**

Agent	Symbol	Thermal Capacity at 340°F (727°C)	
		Btu/mol °F	J/mol-K
Carbon Dioxide	CO_2	12.9	54.3
Steam	$H_2O(g)$	9.8	41.2
Nitrogen	N_2	7.8	32.7
Helium	He	5.0	20.8

capacity, is not widely used. Halon extinguishing agents have not been included in the table. While they have even higher thermal capacity than carbon dioxide, their extinguishment mechanism is more related to chemical interaction in the combustion process.

Nitrogen

Nitrogen has been used in a number of inerting applications, mostly those related to aircraft fire suppression systems. It has been considered for aircraft engine nacelle fires, but halons are generally far superior for this application.[15] Nitrogen fire fighting systems also have been considered for cargo compartment underfloor areas, wheel wells, and wing areas in military and civil aircraft.[15] This technique has been used on U.S. Air Force C-5A transport aircraft. Nitrogen also might be used in fires of unattended parked aircraft to reduce oxygen concentrations to below 10 percent by volume and to eliminate cabin fires.

Steam Inerting Systems

Steam can be used to smother fires in the same manner as other inert gases. It effectively reduces the concentrations of fuel vapor and oxygen and absorbs sufficient heat to stop the combustion process. Although steam systems for fire extinguishment preceded other modern smothering systems, such as those using carbon dioxide and foam, steam is rarely used today. It clearly is an impractical method except where a large steam supply is continuously available, and this supply can be tapped effectively and efficiently when a fire emergency arises. Also, the possible burn hazard to personnel always must be considered.

Steam extinguishing systems are not recommended for fire protection purposes in any current NFPA standards. However, the appendix to NFPA 86, *Standard for Ovens and Furnaces*,[16] does offer suggestions for fire protection with steam systems which can be followed "where·steam flooding is the only alternative" after automatic sprinklers or water spray systems and approved types of supplementary fire protection (carbon dioxide, foam, or dry chemical systems) have already been considered.[16]

Steam smothering systems used to be employed for the protection of cargo spaces and the holds of steamships. This method is no longer recommended. Tests indicating the relative inefficiency of such systems to control cotton cargo fires were conducted by the U.S. Coast Guard during "Operation Phobos."[17]

DRY CHEMICAL AGENTS AND APPLICATION SYSTEMS

Dry chemical is a powder mixture which is used as a fire extinguishing agent. It is intended for application by means of portable extinguishers, hand hoseline systems, or fixed systems. Borax- and sodium bicarbonate-based dry chemical were the first such agents developed. Sodium bicarbonate became the standard because of its greater effectiveness as a fire extinguishing agent. About 1960, sodium bicarbonate-based dry chemical was modified to render it compatible with protein-based low-expansion foams to permit a dual-agent attack. Multipurpose (monoammonium phosphate base) and "Purple-K" (potassium bicarbonate base) dry chemicals then were developed for fire extinguishing use. Shortly thereafter, "Super-K" (potassium chloride base) was developed to equal "Purple-K" in effectiveness. In the late 1960s, the British developed urea-potassium bicarbonate-based dry chemical. Currently, five basic varieties of dry chemical extinguishing agents are available.

The terms "regular dry chemical" and "ordinary dry chemical" generally refer to powders that are listed for use on Class B and Class C fires. "Multipurpose dry chemical" refers to powders listed for use on Class A, B, and C fires. The terms "regular dry chemical," "ordinary dry chemical," and "multipurpose dry chemical" should not be confused with "dry powder" or "dry compound" which are used to identify powdered extinguishing agents developed primarily for use on combustible metal fires.[25]

Dry chemical agents usually are efficient in extinguishing fires in flammable liquids. They also can be used on fires involving some types of electrical equipment. Although regular dry chemical has certain limited applications in extinguishment of flash surface fires with ordinary combustibles, the chemical requires the addition of water to put out deep-seated smoldering fires. Multipurpose dry chemical can be used on fires in flammable liquids, fires involving energized electrical equipment, and fires in ordinary combustible materials. Multipurpose dry chemical seldom needs the help of water to completely extinguish fires in Class A materials.

The principal base chemicals used in the production of currently available dry chemical extinguishing agents are sodium bicarbonate, potassium bicarbonate, potassium chloride, urea-potassium bicarbonate, and monoammonium phosphate. Various additives are mixed with these base materials to improve their storage, flow, and water-repellency characteristics. The most commonly used additives are metallic stearates, tricalcium phosphate, or silicones, which coat the particles of dry chemical to make them free flowing and resistant to the caking effects of moisture and vibration.

Toxicity: The ingredients presently used in dry chemicals are nontoxic. However, the discharge of large quantities can cause temporary breathing difficulty during and immediately after discharge and can seriously interfere with visibility.

Extinguishing Properties

Fire tests on flammable liquids have shown potassium bicarbonate-based dry chemical to be more effective than sodium bicarbonate-based dry chemical in extinguishment. Similarly, monoammonium phosphate has been found equal to or better than sodium bicarbonate in extinguishing fires.[18] The effectiveness of potassium chloride is about equivalent to potassium bicarbonate, and urea-potassium bicarbonate exhibits the greatest effectiveness of all the dry chemicals tested.

When introduced directly to the fire area, dry chemical causes the flame to go out almost at once. Smothering, cooling, and radiation shielding contribute to the extinguishing efficiency of dry chemical, but studies suggest that a chain-breaking reaction in the flame is the principal cause of extinguishment.[19]

Smothering Action: For many years, it was widely held that regular dry chemical extinguishing properties relied primarily on the smothering action of the carbon dioxide released when sodium bicarbonate was heated by fire. The carbon dioxide undoubtedly does contribute to the effectiveness of dry chemical, as does the like volume of water vapor released when dry chemical is heated. However, tests generally have disproved the belief that these gases are a major factor in extinguishment.

When multipurpose dry chemical is discharged into burning ordinary combustibles, the decomposed monoammonium phosphate leaves a sticky residue (metaphosphoric acid) on the burning material. This residue seals glowing material from oxygen, thus helping to extinguish the fire and prevent reignition.

Cooling Action: It cannot be substantiated that the cooling action of dry chemical is an important reason for its ability to extinguish fires promptly. More information on this subject is contained in a paper based on studies of the heat capacities of various powders tested for extinguishing effectiveness. The heat energy required to decompose dry chemicals plays an undeniable role in contributing toward their individual extinguishing abilities, but the effect, *per se*, is minor. To be effective, any dry chemical must be heat sensitive and, as such, absorbs heat in order to become chemically active.[20]

Radiation Shielding: Discharge of dry chemical produces a cloud of powder between the flame and the fuel; this cloud shields the fuel from some of the heat radiated by the flame. Tests to evaluate this factor concluded that the shielding is of some significance.[20]

Chain-Breaking Reaction: The preceding extinguishing actions each contribute to a certain degree to the total extinguishing action of dry chemical. However, studies reveal that still another factor makes a contribution even greater than that of the other factors combined.

The chain-reaction theory of combustion has been advanced by some investigators to provide the clue to the identity of this unknown extinguishing factor. This theory assumes that free radicals are present in the combustion zone and that the reactions of these particles with each other are necessary for continued burning. The discharge of dry chemical into the flames prevents reactive particles from coming together and continuing the combustion chain reaction. The explanation is referred to as the chain-breaking mechanism of extinguishment.[21,22]

Uses and Limitations

Dry chemical is used primarily to extinguish flammable liquid fires. Because it is electrically nonconductive, it also can be used on flammable liquid fires involving live electrical equipment. Regular dry chemical extinguishers have been tested by fire equipment testing laboratories, and have been found suitable for use on flammable liquid and electrical fires (Class B and C fires).

Due to the rapidity with which it extinguishes flame, dry chemical is used on surface fires involving ordinary combustible materials (Class A fires). In several areas in the textile industry, notably opener-picker rooms and carding rooms in cotton mills, regular dry chemical has been used effectively. However, wherever regular dry chemical is provided for use on surface-type Class A fires, it should be supplemented by water spray for extinguishing smoldering embers or in case the fire gets beneath the surface. In some baled cotton storage areas, the tops of bales can be covered with regular dry chemical to prevent surface spread should fire break out. However, this preventive measure does not eliminate the need for automatic sprinkler protection in such areas. Since multipurpose dry chemical becomes sticky when heated, it is not recommended for textile card rooms or other locations where removal of the residue from fine machine parts would be difficult.

Dry chemical does not produce a lasting inert atmosphere above the surface of a flammable liquid; consequently, its use will not result in permanent extinguishment if there are reignition sources, such as hot metal surfaces or persistent electrical arcing.

Dry chemical should not be used in installations where relays and delicate electrical contacts are located (e.g., in telephone exchanges and computer equipment rooms), since the insulating properties of dry chemical might render such equipment inoperative. Because some dry chemicals are slightly corrosive, they should be removed from all undamaged surfaces as soon as possible after fire extinguishment.

Regular dry chemical will not extinguish fires that penetrate beneath the surface, or fires in materials that release their own oxygen for continued combustion. Also, dry chemical may be incompatible with mechanical (air) foam unless the dry chemical has been specially prepared to be reasonably foam compatible.

Specifications have been established by fire equipment testing laboratories to assure the positive and consistent performance of dry chemical as an extinguishing agent. These specifications apply to moisture content, water repellency, electrical resistivity, storage at elevated temperatures, flow capability, caking resistivity, and abrasive action. The discharge characteristics of the device in which the dry chemical is to be used are also evaluated. Extinguishing effectiveness is determined by performance tests of the application to standard fires under conditions recommended by the manufacturer.

Dry Chemical Extinguishing Systems

Although dry chemical had been used for many years in fire extinguishers, it was not until 1954 that the first dry chemical extinguishing system was tested and listed by a testing laboratory. In 1952 the NFPA Committee on Dry Chemical Extinguishing Systems was established, and in 1957 the first edition of NFPA 17, *Standard for Dry Chemical Extinguishing Systems*,[23] was adopted.

Dry chemical extinguishing systems can be used where quick extinguishment is desired and where reignition sources are not present. Dry chemical systems are used primarily for flammable liquid fire hazards, such as dip tanks, flammable liquid storage rooms, and areas where flammable liquid spills might occur. Systems have been designed for kitchen range hoods, ducts, and associated rangetop hazards, such as deep-fat fryers. Where it is necessary to extinguish a flammable liquid or gas fire being fed by fuel under pressure, dry chemical hand hoseline systems can be used, followed by closure of fuel shutoff valves.

Since dry chemical is electrically nonconductive, extinguishing systems using this agent can be activated on electrical equipment that is subject to flammable liquid fire, such as oil-filled transformers and oil-filled circuit breakers. Dry chemical system protection is not recommended, however, for delicate electrical equipment, including telephone switchboards and computers. Such equipment is subject to damage by dry chemical deposit and, because of the insulating properties of the dry chemical, might require excessive cleaning to restore operation.

Hand hoseline systems containing regular dry chemical have been used to a limited extent for quick-spreading surface fires on ordinary combustible material. In such applications, the dry chemical system only stops or prevents a rapid surface spread; it must be supplemented by a water-type extinguishing device to put out deep-seated smoldering fires. Fixed systems containing multipurpose dry chemical are also suitable for protection of ordinary combustibles, provided the dry chemical can reach all burning surfaces.

Methods of Application

The two basic types of dry chemical systems are referred to as fixed systems and hand hoseline systems. Other methods of applying dry chemical are by portable and wheeled extinguishers.

Fixed Systems: Fixed dry chemical systems consist of a supply of dry chemical, an expellent gas, an actuating method, fixed piping, and nozzles through which the dry chemical can be discharged into the hazard area. Fixed dry chemical systems are of two types: total flooding and local application. (See Figure 8.1.)

In *total flooding*, a predetermined amount of dry chemical is discharged through fixed piping and nozzles into an enclosed space or an enclosure around the hazard. Total flooding is applicable only when the hazard is totally

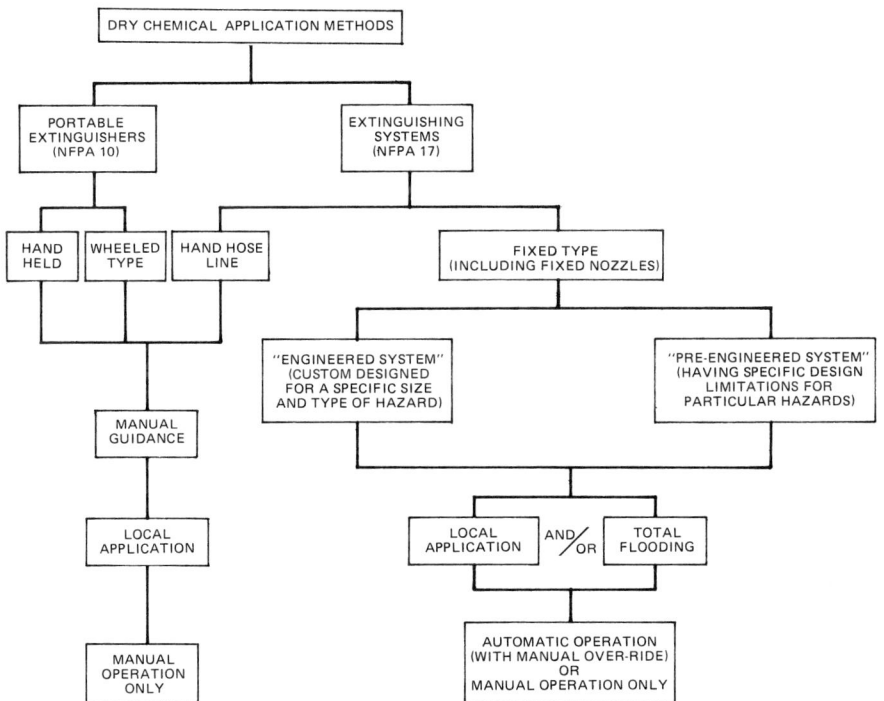

Fig. 8.1   Methods of dry chemical application.

enclosed or when all openings surrounding a hazard can be closed automatically when the system is discharged. Total flooding can be used only where no reignition is anticipated, because the extinguishing action is transient.

Local application differs from total flooding in that the nozzles are arranged to discharge directly into the fire. Local application is practical in those situations where the hazard can be isolated from other hazards so that fire will not spread beyond the area protected, and where the entire hazard can be protected. The principal use of local application systems is to protect open tanks of flammable liquids. As with total flooding systems, local application is ineffective unless extinguishment can be immediate and there are no reignition sources.

COMBUSTIBLE METAL AGENTS AND APPLICATION TECHNIQUES

A variety of metals burn, particularly those in finely divided form. Some metals burn when heated to high temperatures by friction or exposed to external heat; others burn from contact with moisture or in reaction with other materials. Because accidental fires can occur during the transportation of these materials, it is important to understand the nature of the various fires and their hazards.

Hazards involved in the control or complete extinguishment of metal fires include extremely high temperatures, steam explosions, hydrogen explosions, toxic products of combustion, explosive reaction with some common extinguishing agents, breakdown of some extinguishing agents with the liberation of combustible gases or toxic products of combustion, and, in the case of certain nuclear materials, dangerous radiation. Because some agents displace oxygen, especially in confined spaces, agents and methods for their specific application must be selected with care. Some metal fires should not be approached without suitable self-contained breathing apparatus and protective clothing, even if the fire is small. Other metal fires might be readily approached with minimum protection; still others should be fought only with unmanned fixed equipment.

Numerous agents have been developed to extinguish combustible metal (Class D) fires, but any one given agent does not necessarily control or extinguish all metal fires. Some agents are valuable in working with several metals, while others are useful in combating only one type of metal fire. Despite their use in industry, some of these agents provide only partial control and cannot be considered actual extinguishing agents. Certain agents suitable for other classes of fires should be avoided for metal fires because violent reactions could result (e.g., water on sodium; vaporizing liquids on petroleum fires).

Certain combustible metal extinguishing agents have been used for years, and their success in handling metal fires has led to the designation "approved extinguishing powder" and "dry powder." These designations have appeared in codes and other publications where it was not possible to employ the proprietary names of the powders. These terms have been accepted in describing extinguishing agents for metal fires and should not be confused with the name "dry chemical," which normally applies to an agent suitable for use on flammable liquid (Class B) and live electrical equipment (Class C) fires.

Successful control or extinguishment of metal fires depends to a considerable extent upon the method of application and the training and experience of the fire fighters. Practice drills should be held on the particular combustible metals on which the agent is expected to be used. Prior knowledge of the capabilities and limitations of agents and associated equipment is always useful in emergency situations. Fire control or extinguishment will be difficult if the burning metal is in a place or position where the extinguishing agent cannot be applied in the most effective manner. In industrial plant locations where work involves use of combustible metals, public fire departments and industrial fire brigades should have the advantage of fire control drills conducted under the guidance of knowledgeable individuals.

Transportation of combustible metals creates unique problems, in that fire could occur in a location where necessary fire fighting knowledge and suitable extinguishing agents are not readily available. The U.S. Department of Transportation, anticipating such situations, specifies cargo limitations, labeling, and placarding for the various means of transportation.

Approved Extinguishing Agents for Combustible Metals

A number of proprietary combustible metal extinguishing agents have been submitted to testing agencies for approval or listing. Others have not, particularly those agents developed for special metals in rather limited commercial use.

Two of the best-known extinguishing agents used to smother fires in combustible metals, such as aluminum and magnesium, are G-1 Powder and Met-L-X Powder. G-1 Powder is applied by spreading it over the surface of the burning metal. Met-L-X Powder is applied from extinguishers in portable and wheeled units and in fixed-pipe systems. A number of other commercially available dry powders can be used to extinguish fires in various metals. Graphite powder, talc, and sand all have been used to smother metal fires.

PORTABLE FIRE EXTINGUISHERS

Virtually all fires are small at first and might be easily extinguished if the proper type and amount of extinguishing agent were applied promptly.

Portable fire extinguishers are designed for this purpose, but their successful use depends on the following conditions:

1. The extinguisher must be properly located and in good working order.
2. The extinguisher must be the proper type for the fire.
3. The fire must be discovered while still small enough for the extinguisher to be effective.
4. The fire must be discovered by a person ready, willing, and able to use the extinguisher.

Fire extinguishers are the first line of defense against unfriendly fires, and should be installed regardless of other fire control measures. However, the fire department should be notified as soon as a fire is discovered; notification should never be delayed in hope that the extinguisher will be sufficient.

Historical Background

The first portable fire extinguishers were developed in the late 1800s. They contained glass bottles of acid which, when broken, dumped acid into a soda solution and produced a mixture with sufficient gas pressure to expel the solution. Cartridge-operated water extinguishers of the inverting type were introduced in the late 1920s. In 1928, a nonfreeze alkali metal-salt solution called "loaded stream" was developed for use in cartridge-operated extinguishers. Stored-pressure water extinguishers were developed in 1959, and over the next 10 years gradually replaced cartridge-operated models. (In 1969, the manufacture of all inverting extinguishers was discontinued in the United States, and these extinguishers are no longer listed or approved by testing laboratories.)

The first foam extinguisher was developed in 1917. It looked and worked much like the soda-acid extinguisher. The use of foam extinguishers steadily increased over the years until, during the 1950s, dry chemical extinguishers gained widespread acceptance.

Vaporizing Liquids: Carbon tetrachloride (CCl_4), in 1908, was one of the first chemicals used in portable fire extinguishers. However, it subsequently was found that its vapors were toxic; when used on a fire, CCl_4 could produce highly toxic hydrogen chloride and phosgene. Slightly less toxic chlorobromomethane (CH_2ClBr) was produced after World War II, when the term "vaporizing liquid" was first used to designate these extinguishers. However, some federal agencies banned vaporizing-liquid extinguishers in the 1950s because they were poisonous. By the mid 1960s, many states, cities, and industrial firms also had banned them. In the late 1960s, listings of vaporizing liquids were discontinued by testing laboratories.

Halogenated Agents: Although vaporizing liquids proved unacceptable, less-toxic halogenated hydrocarbon chemicals found use in the form of compressed or liquefied gases. Bromotrifluoromethane (Halon 1301) was introduced in 1954 in a high vapor pressure compressed gas extinguisher for use on fires in flammable liquids and live electrical equipment. A medium vapor pressure extinguisher using bromochlorodifluoromethane (Halon 1211) became available in 1973. In 1974, extensive testing was begun on extinguishers containing low vapor pressure dibromotetrafluoromethane (Halon 2402), which is a liquid at room temperature. Tests indicated that this agent could be used on all types of fires, but its use in portable extinguishers in the United States has been limited thus far due to high costs and potential toxicity problems.

Carbon Dioxide: The first carbon dioxide extinguishers were produced during World War I, and during World War II they were the leading extinguisher for flammable liquids fires. By 1950, however, dry chemical agents had replaced carbon dioxide as the preferable agent for flammable liquids fires. A further decline in the use of carbon dioxide occurred as the popularity of halogenated agents increased.

Dry Chemicals: Although the extinguishing ability of sodium bicarbonate was recognized as early as the late 1800s, it was not until 1928 that an effective cartridge-operated dry chemical extinguisher was developed. Research and development produced an improved, finely granulated agent in 1943 and a further improved model in 1947.

As the use of flammable liquids increased, so did development of effective dry chemical agents. In 1959, a potassium bicarbonate-based agent about twice as effective as the sodium bicarbonate (ordinary) based agent was introduced.

In 1961, manufacturers introduced a new kind of agent called "multipurpose dry chemical." It had the advantage of being 50 percent more effective than ordinary dry chemical on flammable liquid and electrical fires, and also was capable of extinguishing fires in ordinary combustibles. Originally, diammonium phosphate was used because it was relatively inexpensive, but monoammonium phosphate soon became preferred because it was considered less hygroscopic (moisture-absorbing).

In 1968, an agent with a potassium chloride base was introduced. It was 80 percent more effective than ordinary dry chemical, but more corrosive and hygroscopic than potassium bicarbonate. A urea-potassium bicarbonate-based agent (potassium carbamate) was developed in Europe in 1967 and brought to America in 1970. It has been judged as at least 2½ times as effective as ordinary dry chemical.

Dry Powder: Increased use of combustible metals (magnesium, sodium, lithium, etc.) established the need for a special agent to extinguish fires involving them. The designation "dry powder" was chosen to indicate an agent's suitability for use on Class D (combustible metal) fires; the term "dry chemical" was reserved for agents effective on Class A:B:C or B:C fires. In

1950, a dry powder extinguisher using sodium chloride as a base agent was first marketed.

Reliability and Design Safety of Fire Extinguishers

Portable fire extinguishers might remain idle for many years, but they must be capable of functioning at any time with maximum efficiency and without hazard to the user. Because most extinguishers are pressure vessels, they can rupture if not properly designed, constructed, and maintained. The initial responsibility for safe extinguishers belongs to the manufacturer, who is subject to the design standards and the testing, inspection, and labeling procedures of responsible fire testing laboratories. However, the owner must assume responsibility for maintaining an extinguisher once it is in service; many owners have a contract with a qualified service company.

The NFPA Extinguisher Standard: NFPA 10, *Standard for Portable Fire Extinguishers*,[24] contains provisions for selection, installation, inspection, maintenance, and testing extinguishers. These provisions have been widely adopted by property owners, sales/servicing agencies, and enforcing officials.

The first edition of NFPA 10[24] was published in 1921 from data assembled by the former NFPA Committee on Field Practices. This standard is continuously reviewed, and revised editions are published every few years. Starting with the 1978 edition, more detailed information, with specific examples, was added to give guidance on selection and distribution of portable extinguishers.

Relation of Extinguishers to Class of Fires

Because the effectiveness of various extinguishing agents is not uniform on different fires, NFPA 10[24] classifies fires into the following four types:

Class A: Fires in ordinary combustible materials (wood, cloth, paper, rubber, and many plastics. These fires require (1) the heat absorbing (cooling) effects of water or water solutions, (2) the coating effects of certain dry chemicals which retard combustion, or (3) the interrupting of the combustion chain reaction by halogenated agents (medium vapor pressure agents also have some cooling capability).

Class B: Fires in flammable or combustible liquids, flammable gases, greases, and similar materials. These fires must be put out by (1) excluding air (oxygen), (2) inhibiting the release of combustible vapors, or (3) interrupting the combustion chain reaction.

Class C: Fires in live electrical equipment when safety to the extinguisher operator requires use of electrically nonconductive extinguishing agents only. (Note: When electrical equipment is deenergized, extinguishers for Class A or B fires may be used.)

Class D: Fires in certain combustible metals (magnesium, titanium, zirconium, sodium, potassium, etc.) which require a heat-absorbing extinguishing medium that does not react with the burning metals.

Some portable extinguishers will put out only one class of fire; some are suitable for two or three classes; but none is suitable for all four classes. Most extinguishers are labeled to enable users to quickly identify the classes of fire for which they may be used. This classification is contained in NFPA 10,[24] which gives the applicable picture symbol or symbols. Color coding also is used. (See Figures 8.2 and 8.3.) The NFPA classification is required to appear on the extinguisher label in accordance with ANSI standards.

Rating numerals also are used on the labels of extinguishers for Class A and Class B fires; the rating numeral gives the relative extinguishing effectiveness of the extinguisher. For example, an extinguisher rated 4-A; 20-B:C indicates: (1) It should extinguish approximately twice as much Class A fire as a 2-A rated extinguisher. (2) It should extinguish approximately 20 times as much Class B fire as a 1-B rated extinguisher. (3) It is suitable for use on energized electrical equipment.

Extinguishers rated for Class B fires have numeral ratings based on the relative quantity of burning flammable liquid in a flat pan that can be extinguished during a laboratory test. Again, the point is that an extinguisher rated 20-B can put out much more fire than one rated 5-B.

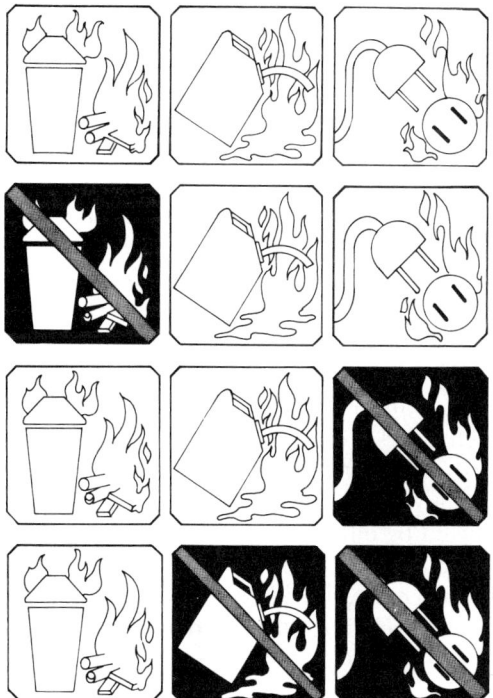

Fig. 8.2 These pictographs are designed so that their proper use may be determined at a glance. When an application is prohibited, the background is black and the slash is bright red. Otherwise the background is light blue. Top row of picture symbols indicates an extinguisher for Class A:B:C fires; second row indicates an extinguisher for Class B:C fires; third row indicates an extinguisher for Class A:B fires; and bottom row indicates an extinguisher for Class A fires.

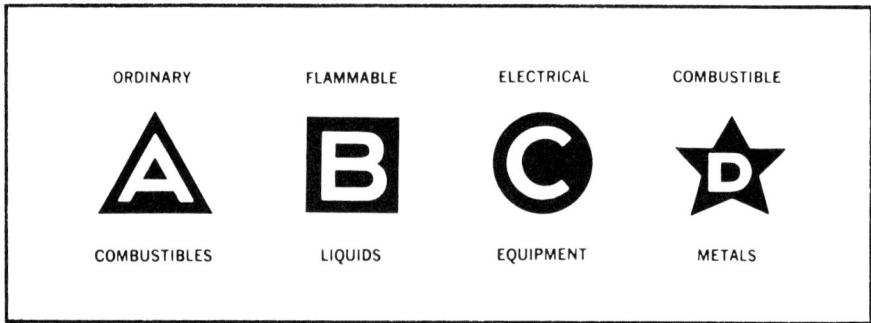

Fig. 8.3 Extinguisher markings that can be used until conversion to pictographs is complete. Color coding is part of the identification system, and the triangle (Class A) is colored green, the square (Class B) red, the circle (Class C) blue, and the five-pointed star (Class D) yellow.

No rating numerals are used for extinguishers labeled for Class C fires. Since electrical equipment has either ordinary combustibles or flammable liquids, or both, as part of its construction, an extinguisher for Class C fires should be chosen according to the nature of the combustibles in the immediate area.

Extinguishers for Class D fires contain different dry powders that are effective on fires in different kinds of combustible metals. An extinguisher for magnesium fires might not work on a sodium fire, or at least would not have the same effectiveness as one intended for use on sodium. For that reason, general numerical ratings are not used; instead, each extinguisher for Class D fires has a nameplate detailing the type of metal in which the particular agent will extinguish a fire.

Extinguishers which are effective on more than one class of fire have multiple "letter" and "numeral-letter" classifications and ratings. Fractional ratings are not recognized in NFPA 10[24] and are not included in requirements of the ANSI standards.

The most recently recommended marking system combines pictographs of both uses and nonuses on a single label. (See Figure 8.2.) Letter-shaped symbol markings, as shown in Figure 8.3, are recommended for use until full conversion to the newer pictographs is completed.

Sometimes extinguishers produced by different manufacturers (or even by the same manufacturer) have the same quantity of the same extinguishing agent, but have different ratings. In such cases, the difference in performance usually is due to differences of design, including rates of discharge, nozzle design, discharge patterns, etc.

Selection, Operation, and Distribution of Fire Extinguishers

Before choosing an extinguisher, it is important to know: (1) the nature of the fuels present, (2) who will use the extinguisher, (3) the physical environment in which the extinguisher will be placed, and (4) whether any chemicals in the area will react adversely with an extinguishing agent. When choosing from among various extinguishers, these points should be considered: (1) whether it is effective on the specific hazards present, (2) whether it is easy to operate, and (3) what maintenance and upkeep it requires.

Summary

Fire fighting foam is used to blanket and smother flammable liquid fires. Mechanical foams, chemical foams, and high-expansion foams are used for various purposes and conditions. Carbon dioxide gas is another useful extinguishing agent for flammable liquid and electrical fires; it is an inert gas and does not conduct electricity. Halons are relatively new extinguishing agents. Halon 1301 and 1211 are the recognized halogenated agents, used for protecting electrical equipment. Dry chemical extinguishing agents are effective on flammable liquid and electrical fires, and also can be used on ordinary combustibles.

Portable fire extinguishers are valuable as a first line of defense. They are classified and rated for various kinds of fire. Selection of the right kind and size of portable fire extinguisher is important for efficient and effective use.

References

[1]NFPA 11-1983. *Standard for Low Expansion Foam and Combined Agent Systems*, National Fire Protection Association, Quincy, MA.

[2]NFPA 11A-1983. *Standard for Medium and High Expansion Foam Systems*, National Fire Protection Association, Quincy, MA.

[3]NFPA 11C-1986. *Standard for Mobile Foam Apparatus*, National Fire Protection Association, Quincy, MA.

[4]NFPA 16-1986. *Standard on Deluge Foam-Water Sprinkler and Foam-Water Spray Systems*, National Fire Protection Association, Quincy, MA.

[5]NFPA 403-1978. *Recommended Practice for Aircraft Rescue and Fire Fighting Services at Airports and Heliports*, National Fire Protection Association, Quincy, MA.

[6]NFPA 12-1985. *Standard on Carbon Dioxide Extinguishing Systems*, National Fire Protection Association, Quincy, MA.

[7]Wharry, David, and Hirst, Ronald 1974. *Fire Technology: Chemistry and Combustion*, Institution of Fire Engineers, Leicester, UK.

[8]Strasiak, Raymond 1954. "The Development of Bromochloromethane (CB)," *WADC Technical Report 53-279*, Wright Air Development Center, OH.

[9]Dowling, John, and Ford, C. 1969. "Halon 1301 Total Flooding System for Winterthur Museum," *Fire Journal*, Vol. 63, No. 6.

[10]NFPA 12A-1985. *Standard on Halon 1301 Fire Extinguishing Systems*, National Fire Protection Association, Quincy, MA.

[11]NFPA 12B-1985. *Standard on Halon 1211 Fire Extinguishing Systems*, National Fire Protection Association, Quincy, MA.

[12]Ford, Charles 1972a. *Halon 1301 Computer Fire Test Program—Interim Report*, E.I. duPont de Nemours & Co., Inc., Wilmington, DE.

[13]Ford, Charles 1962. "Overview of Halon 1301 Systems," *Symposium on the Mechanism of Halogenated Extinguishing Agents*, ACS Symposia Series.

[14]Drysdale, D.D. 1985. *Fire Dynamics*, Wiley & Sons, Chichester, UK.

[15]Dyer, J.H., *et al* 1977. "The Extinction of Fires in Aircraft Jet Engines—Part III, Extinction of Fires at Low Airflows," *Fire Technology*, National Fire Protection Association, Boston, MA.

[16]NFPA 86-1985. *Standard for Ovens and Furnaces*, National Fire Protection Association, Quincy, MA.

[17]NFPA 1947. *Proceedings of the Fifty-First Annual Meeting of the National Fire Protection Association*, May 26-29, 1947, National Fire Protection Association, Quincy, MA.

[18]Guise, A.B. 1962. "Potassium Bicarbonate-Based Dry Chemical," *NFPA Quarterly*, National Fire Protection Association, Boston, MA.

[19]Haessler, W.M. 1974. *The Extinguishment of Fire*, Revised ed., National Fire Protection Association, Boston, MA.

[20]McCamy, C.S., *et al* 1956. "Fire Extinguishment by Means of Dry Powder," *Sixth Symposium on Combustion*, The Combustion Institute, Reinhold, NY.

[21]Guise, A.B. 1960. "The Chemical Aspects of Fire Extinguishment," *NFPA Quarterly*, National Fire Protection Association, Boston, MA.

[22]Haessler, W.M. 1962. "Fire and Its Extinguishment," *NFPA Quarterly*, National Fire Protection Association, Boston, MA.

[23]NFPA 17-1985. *Standard for Dry Chemical Extinguishing Systems*, National Fire Protection Association, Quincy, MA.

[24]NFPA 10-1984. *Standard for Portable Fire Extinguishers*, National Fire Protection Association, Quincy, MA.

[25]FMEC annually. *Factory Mutual Approval Guide*, Factory Mutual Research Corporation, Norwood, MA.

9

Alarm and Detection Systems and Devices

Communication devices, both manual and automatic, are of great value to fire protection. Different types of signaling systems have been developed that not only alert fire departments to the presence of a fire, but through mechanical detection of smoke, flames, or excessive heat, can warn building occupants of a situation of potential danger.

PUBLIC FIRE SERVICE COMMUNICATIONS

The fire alarm box on a pole on a street corner has been a familiar sight to city dwellers for several generations. The first such municipal fire alarm system was installed in the City of Boston more than a century ago. Most of today's school children have attended demonstrations, usually by fire fighters, on how to turn in an alarm from a fire alarm box; also, most of these same children have been lectured on the dangers of false alarms.

An effective fire alarm system fulfills two functions: (1) receiving alarms from the public through fire alarm boxes located on the street or on private property, and (2) transmitting the alarm to the fire companies and personnel who should respond to the emergency.

MUNICIPAL FIRE ALARM SYSTEMS

Scope, installation, maintenance, and use of all municipal fire alarm systems, regardless of the principles of operation, are covered in detail in NFPA 1221, *Standard for the Installation, Maintenance and Use of Public Fire Service Communications Systems*.[1] This standard makes no distinction among coded, voice, or code-voice alarm systems, because each must perform the same function, providing a means by which an alarm can be transmitted from a street alarm box to the communications center.

From the standpoint of the dispatcher responsible for receiving the alarm, the use of a public emergency reporting station (fire alarm box) eliminates the difficulty of determining the location from which the alarm is being transmitted. When actuated, each device must transmit a distinct numerical code in addition to any other functions or capabilities provided.

225

A general abandonment of jurisdictionally owned municipal fire alarm systems or more aptly, public emergency reporting systems, began to take place in many major metropolitan centers in the mid-1970s. The movement, based on economics and the efficiency of the service provided, has since spread to smaller suburban jurisdictions.

The average metropolitan system consisted of a mix of public reporting stations (street boxes) and auxiliary fire alarm systems interconnected to master boxes. In most cases, the operational jurisdictions reduced their obligations toward auxiliary interconnects by referring all risks, exclusive of schools, hospitals, churches, government buildings, and major hotels, to privately operated central stations. Public reporting stations were eliminated shortly thereafter. Those agencies who still maintain and operate such systems have essentially directed their applications toward auxiliary systems.

The expanding implementation of "911" systems has eliminated the necessity of public reporting stations (street boxes), at least in those areas where publicly accessible phones are well distributed and functionally usable. Public emergency reporting systems are electrically operated networks divided into three basic categories: (1) coded, (2) voice, and (3) code-voice combination.

A public emergency reporting system may be used for the transmission of other emergency signals or calls, provided such transmission does not interfere with the proper handling of fire alarms. For example, systems employing voice communications between the street alarm box and communications center can be used to transmit alarms of nonfire emergency nature. Fire alarm boxes in a radio-type system can be provided with pushbuttons for calling the police department, the ambulance service, or other emergency services, and signals can be transmitted directly to the service called. With parallel-type telephone systems (each box served by a separate circuit), the dispatcher can crossconnect to the proper emergency service, such as the police department, or systems can be arranged so authorized persons, such as law enforcement personnel, can be connected directly to their own departments.

Fire alarm systems traditionally have been grouped into two types, depending on whether retransmission of alarms is manual or automatic. These two types of systems—Type A and Type B—are described briefly in the following paragraphs, and are more fully detailed in the standards of the National Fire Protection Association.

Type A (Manual)

The Type A (manual) system (see Figure 9.1) is one in which operators are required to check the receipt of fire alarms and to retransmit them over alarm circuits to fire stations. All fire calls, whatever their origins, are considered to be alarms. One operator is required when more than 600 alarms per year are experienced; two where the number exceeds 1,500. In addition to handling the fire alarm system, operators may handle telephone calls for other emergencies or telephone business.

With Type A systems there is a sufficiently large number of circuits to use a considerable portion of an operator's time in daily tests of the circuit. One reason for having two operators is that if one suddenly becomes ill in an isolated fire alarm office, the system would not be out of service.

Type B (Automatic)

The Type B (automatic) system (see Figure 9.2) is one in which alarms are retransmitted automatically to fire stations. Type B systems are designed for small communities that have organized fire departments, but that do not have or need a Type A system.

Fire Alarm Boxes

The effectiveness of a municipal fire alarm system depends on a number of factors. For example, to provide proper protection to the community, fire alarm boxes must be well distributed and located, so they are readily visible and accessible for public use. As a general rule, boxes should be not more than 500 ft (152.4 m)—one block—apart in congested districts, or more than 800 ft (243.8 m)—two blocks—apart in residential areas. A fire alarm box should be at or near the entrance to every school, hospital, nursing home, and

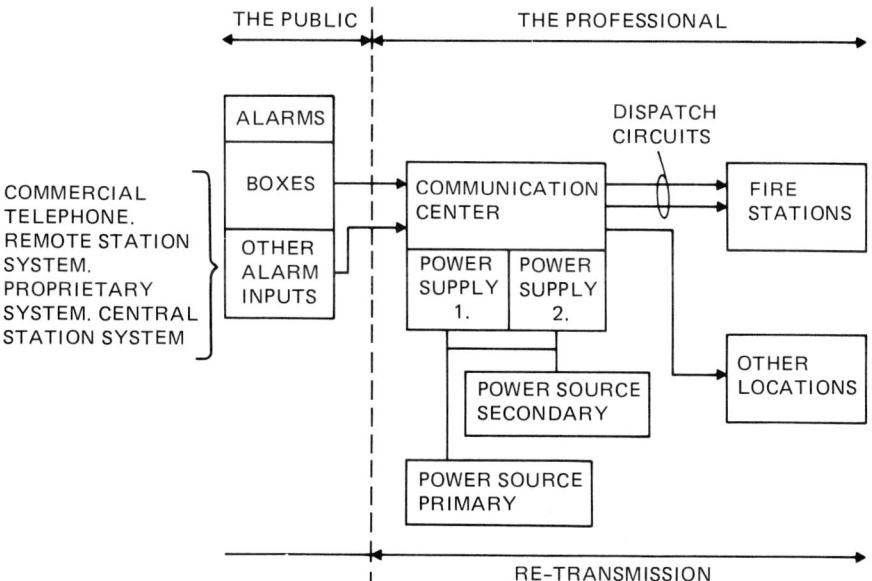

Fig. 9.1   Type A public reporting system.

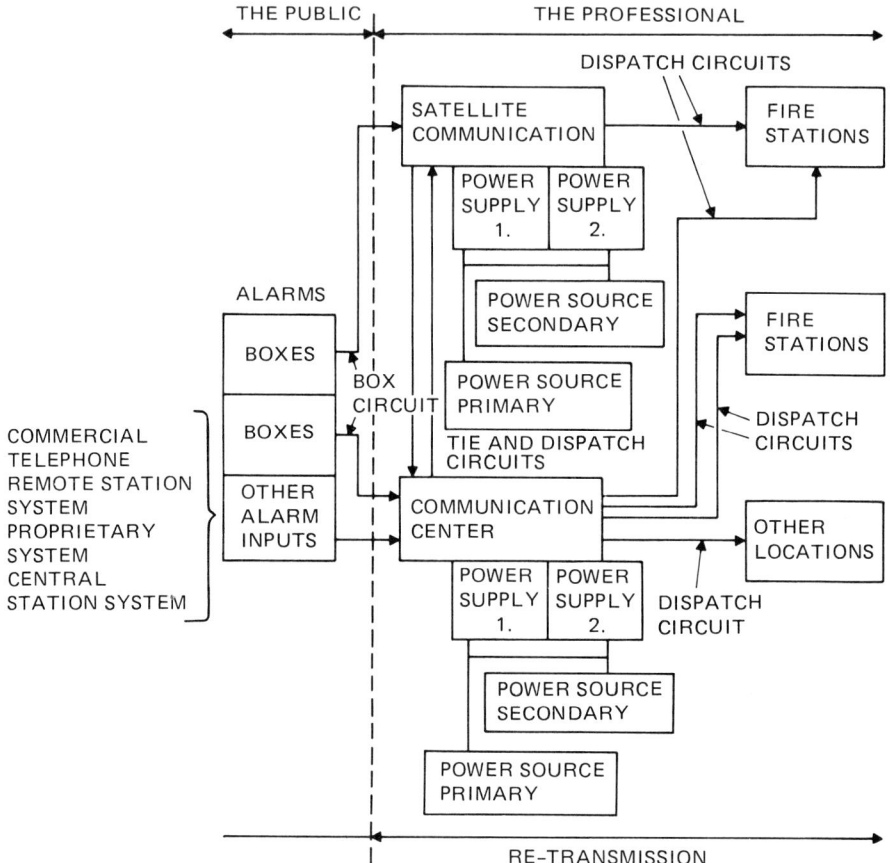

Fig. 9.2   Type B public reporting system.

place of public assembly. A fire alarm box also should be installed near the entrance to each fire station, since many people are more aware of the location of the nearest fire station than of the nearest fire alarm box. People often have gone to a fire station to report a fire only to find that the fire company was away from the station on another alarm and that there was no readily available means of transmitting an alarm to the communication center.

In addition to being located and marked so that they are readily visible, fire alarm boxes should be illuminated at night with a light of distinctive color.

The local fire department should repeatedly remind the public of the function and value of fire alarm boxes, location of the boxes nearest their home and place of work, and the fact that false alarms can jeopardize protection of life and property.

Coded Systems

The coded category is itself divided into two classifications: (1) the wired telegraphic system, and (2) the radio (wireless) system.

The wired telegraph-type box is actuated by depressing a lever which is accessible through a small door. (See Figure 9.3.) This starts a spring-wound clockwork mechanism and transmits a code number by the rotation of a code wheel that opens and closes the circuit. The transmitted code number of each box is different so the location of the box transmitting the alarm can be determined. If a circuit wire is broken, the telegraph-box system usually is designed to transmit the coded signal through a ground connection.

Radio boxes are similar to wired telegraphic boxes. (See Figure 9.4.) When activated, they transmit a coded signal, representative of their geographical location; however, in addition to the basic numerical identification code, each box must be equipped to send at least three specific messages identifying the actuation cause. A "test" message must be transmitted at least once in each 24-hour period, assuring the monitoring point (communications center) that the box is working. If a box is struck or physically battered, a specific "tamper" message must be transmitted. If the box is used for public reporting only, the third required message would be "fire." Radio boxes are available that allow transmission of up to 50 messages actuated either publicly (fire, police, medical aid, etc.), privately (through auxiliary connections), or combination of both.

Voice Systems

Voice systems are wired telephone networks. The fire alarm box is essentially a telephone handset installed in a specifically designed housing. (See Figure 9.5.) The location from which the alarm is transmitted is definitely established without dependence upon voice transmission and without interference from other boxes. Removing the handset from its cradle in the box lights a lamp on the communications center switchboard where the number of the box and time of message can be recorded.

Like the coded system category, voice system technology is itself divided into parallel and series classifications.

Code-Voice Combination: Boxes which also contain a telegraph mechanism are called combination telephone-telegraph type, and both voice and coded signals can be transmitted over the same circuit to the communications center. (See Figure 9.6.) The series telephone box may be intermixed with telegraph street boxes and master boxes.

THE COMMUNICATIONS CENTER

A communications center is the location at which alarms are received and from which appropriate signals are transmitted to the fire department to

Fig. 9.3 (Left) A coded telegraph-type box. (Right) An interior view of the telegraphic alarm box, showing the actuating lever (arrow) that normally protrudes through the hole shown on the inner door at right. When depressed, the handle releases the spring-wound mechanism and sends a coded signal from the code wheel shown at center. (Gamewell Corp.)

initiate response of apparatus and personnel. The communications center houses the equipment and personnel to perform the two functions of receiving and transmitting.

A communications center might serve a single municipality, several adjacent municipalities, an entire county, or a large political jurisdiction of another type. The combining of municipal police and fire communications into one communications center is becoming more common.

Location of the communications center preferably should be in a separate building in a park or open space. The building should be well protected against fire exposure. If the communications center is located in

Fig. 9.4 (Left and center) Typical battery/dc-powered coded radio-type alarm boxes equipped for multiple messages (fire, police, ambulance, etc.). (King Fisher Co.) (Right) A user-power-coded radio box. This type of device must incorporate an automatic test feature to meet standard requirements. (Signal Communications Corp.)

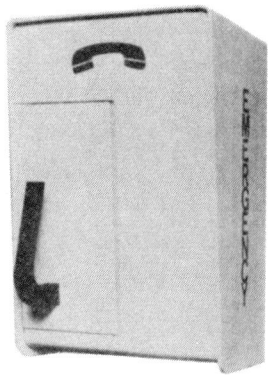

Fig. 9.5   A voice-telephone alarm box, series type, for handset operation only. (Gamewell Corp.)

the fire department headquarters, it must be situated to minimize the probability of interruption from any cause.

When an alarm is received at fire alarm headquarters: (1) it should be received and recorded automatically, (2) it should be indicated both visually and audibly, and (3) the time of receipt and the exact location of the alarm should be recorded automatically. Running cards defining the response required for alarms from each box are on file, so the operator can transmit information to the fire companies designated on the appropriate card.

Even the smallest fire department must have some means of receiving notice of a fire. In some communities, the communications center consists of only a telephone in a home or business where someone is always available, and where there is a switch to activate an audible device to call volunteers. Frequently, the local police station is set up to serve as the communications center for a community. As fire departments become larger, a room or area within a fire station often is provided to house the fire alarm dispatcher and alarm facilities.

In most communities, including those that maintain a complete and reliable municipal fire alarm system, the majority of fires are reported by telephone. Generally, reporting a fire by telephone is successful, although

Fig. 9.6   A combination telephone and coded telegraph-type alarm box. (Gamewell Corp.)

there have been situations where complete reliance on the telephone to report a fire has been disastrous. For instance, a person under the stress and excitement of a fire might not give the right information or enough details to the fire department; a telephone might not be readily accessible; the line may be busy; the fire may have disrupted the telephone circuit; or the caller might dial the wrong fire department. All such situations, as well as many others, have happened and have resulted in delayed alarms. While everyone should know how to use the telephone to summon the fire department, everyone also should know where the nearest fire alarm box is and how to use it.

Fire Department Radio

Most fire departments now make extensive use of two-way radio equipment for fireground operations. Nearly all of the larger departments have fire radio licenses and frequencies issued by the Federal Communications Commission (FCC). The rules of the FCC provide that fire radio be used only by authorized personnel of the fire department, and not used for sending nonfire messages. Most fire departments give training courses in radio procedures, and have specific rules for the operation of their fire radio systems.

Radio is particularly useful for alerting volunteer fire fighters. Many volunteer fire fighters have portable radio receivers in their automobiles, their homes, and where they work. In fully paid departments, it is now common for fire radios to be furnished to off-duty personnel so they can be alerted when needed for serious fires.

PROCESSING COMMUNICATIONS WITHIN THE DEPARTMENT

Regardless of the method by which an alarm is received, the location of the emergency is the minimum information required to respond to an emergency. The nature of the incident being reported is also obviously essential. However, one must bear in mind that thousands of responses to numerically coded (location) nonvoice notifications are made each year.

Data Collection Systems

To efficiently operate a fire department communications center, reliable data must be available. To be useful, this data must be both accurate and readily available. The data may take a wide variety of forms, from card indexes to sophisticated computer files, and may include such categories as the status of vehicles, the availability of resources, geographic information, and personnel information. The following categories outline the various types and uses of data that can be collected and maintained by a communications center.

Incident-Related Data: It is important (and often a legal responsibility) to keep accurate records of fire and emergency medical incidents to which the department responds. The type of data worth collecting include:
1. Time of dispatch.
2. Address.
3. Identification of responding units.
4. Arrival times at the scene.
5. Departure times from the scene.
6. Radio traffic tape logs.
7. Telephone tape logs.
8. Times of significant events at the scene.

Operational Related Data: This data consists of information which must be rapidly available at all times to maintain efficient operation of the communications center. The data include:
1. Geographic data.
2. Equipment data.
3. Response order data.
4. Additional resource data.

Administrative/Management Data: This category includes data which may be required for the general fire department management:
1. Personnel records.
2. Personnel scheduling data.
3. Personnel specialty skills information.
4. Occupancy information.

Reports: Many types of reports are necessary for overall fire department management. Some of these reports will be required at regular intervals, while others will be required on demand to fill a specific departmental information need. A list of probable reports are:
1. Response time reports.
2. Emergency activity reports.
3. Occupancy activity reports.
4. Specific incident reports.
5. Personnel staffing reports.
6. Reports for outside agencies.
7. Fire loss/injury reports.
8. Equipment-related reports.

These reports can be used to project future needs of the department, such as location of new fire stations, hiring of additional personnel, and relocation of existing equipment.

Receipt of Alarms

Most alarms or emergency notifications are received by telephone over a conventional commercial telephone network. Automatic recording devices must be interconnected to all types of voice reporting circuits, including radio systems. However, recording devices cannot assure the credibility or accuracy of the information received.

Correct telephone numbers are the only means of positively identifying the location of an informant. The cross-matching of telephone numbers to addresses is a telephone company service, provided on a select basis, to public safety agencies.

The 911 systems are designed to provide the answering dispatcher with an immediate reference to the number of the telephone in use and its specific location. The data is usually displayed on a cathode-ray tube (CRT) device. The data input is formulated for electronic data processing (EDP) usage and can be fed concurrently to a departmental computer-aided dispatch (CAD) system.

All public emergency reporting systems (municipal fire alarm systems) provide an individual numeric code for each unit installed, and the proper numeric code is transmitted when use of a specific unit is initiated. All incoming coded signals are permanently recorded on printers integral to the specific systems, and in some cases are also displayed on direct readout light-emitting diode [(LED), CRT etc.] devices.

The translation of the numerically coded data into a usable address is performed with the aid of a running card file, which, incidentally, is a necessary factor in the processing of all alarms and requests for emergency assistance.

Voice Recording and Reproducing Systems

The voice recording and reproducing system should consist of a multi-channel recording device which provides a permanent, time referable, unalterable, and court admissible recording of all telephonic/radio communications within a fire department.

The watch commander and one of the dispatchers should be totally familiar with the operation and maintenance of the multichannel recording device. Familiarizing other department personnel is advisable, but responsibility should lie with no more than two individuals per watch/shift.

A multichannel recording system will provide:

1. Verification of emergency messages.
2. A log of all communications within the department.
3. Protection of the department against civilian or other department complaints.

4. Substantiation or negation of claims against the department.
5. Added protection to the civilian population.
6. A training tool for dispatch personnel.
7. Validation of response time.

The importance of these various benefits will differ from department to department.

Running Card Files

A running card file contains a dispatch plan for every area of the jurisdiction served. The number of running cards in a given file is dependent upon the size of the jurisdiction and the manner in which it has been indexed for responses. The most common method of indexing is to assign numeric designations to all intersections or cadastral networks of horizontal and vertical crosspoints. There is a specific running card to each intersection or crosspoint. Public emergency reporting stations (alarm boxes) are normally assigned an identity code corresponding to the intersection or crosspoint site closest to their physical location.

When an emergency notification is received over a conventional commercial telephone network, through a 911 system, or by a two-way radio, the dispatcher utilizes the addresses and/or intersection data obtained from the informant to find the correct running card.

Status-Keeping Systems

The ability to maintain an accurate record of the current status of all emergency units in the system is essential to a successful fire dispatch system and will aid in: (1) minimizing response times, (2) ensuring appropriate response, and (3) collecting nonemergency unit activity information.

Status keeping may involve the tracking of various pieces of information, such as:

1. Availability of the unit for emergency response.
2. Service capabilities of the unit (paramedic service, extrication equipment, all-terrain capability, etc.).
3. Location of the unit.
4. Means of contacting the unit (radio, phone, etc.).
5. Miscellaneous items based on the particular department's operating philosophy.

The mechanism of a status-keeping system will vary, depending upon the number of units on which status must be kept and the actual dispatch process utilized by a department. A small department might utilize a manual system with a small "clipboard," while a large department may require a computer.

Dispatch Circuits and Equipment

A dispatch circuit is the means by which the file alarm dispatcher notifies fire companies to respond to an alarm. The location from which the alarm was received is the minimum information that must be transmitted.

NFPA 1221[1] provides requirements, some of which are:

1. Two separate means of transmitting alarms to fire stations shall be provided at the communications center. However, only one means of transmittal is required when less than 600 alarms per year are received.
2. Each alarm transmitted, date, and time shall be automatically recorded.
3. Devices for transmitting coded or other types of signals shall be arranged for manual setting and operation. (The automatic dispatching of fire companies by the use of computers is under development.)

Note: Computer-aided dispatch (CAD), unlike fully automated dispatching systems, is not a circuit. CAD is a method used to assess related response information. CADs identify the companies normally assigned, and the current status of each of the companies.

Computer-Aided Dispatch (CAD)

Computers are being used in all areas of the fire service. These areas include personnel management, budgeting, fire prevention, training, inventory, etc. One major use is the dispatching of fire units. The degree which a CAD system can be implemented depends upon the size of the department. A small fire department with one to three stations does not require a complex CAD system with mobile digital terminals, station terminals, and printers. Dispatch software packages are available for personal computers, and small departments can use them to rapidly recover the information needed to expeditiously dispatch fire units, including preplans, hazardous material information, lists of other responsible public safety agencies, names, addresses, and phone numbers of responsible parties—the list is endless. (See Figure 9.7.)

Caution must be used in both the selection and use of personal computers. Computer professionals should match the application to the proper computer. The dispatch computer should not be used for other department applications. It should be kept free of time-consuming processes to allow for rapid retrieval of dispatch information. CAD systems for medium to large departments are more complex and must be designed for the needs of an individual department.

AUTOMATIC AND MANUAL PROTECTIVE SIGNALING DEVICES

Signaling systems are installed in all kinds of buildings for various types of fire alarm, fire prevention, and fire protection purposes. In general, such systems fulfill the following six functions:

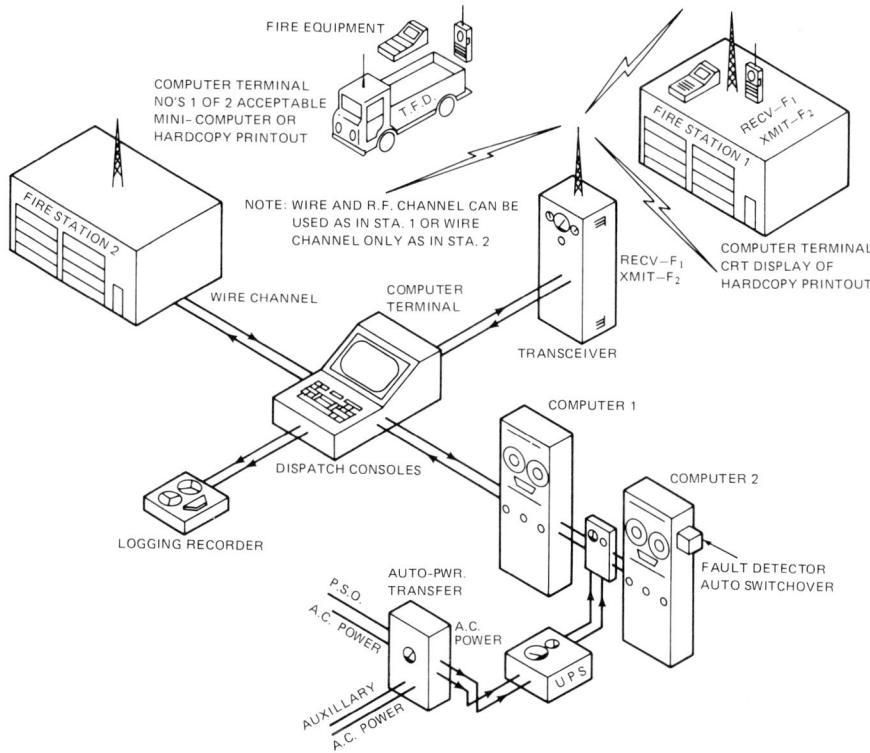

Fig. 9.7   Diagram of a Computer-Aided Dispatch (CAD) system.

1. Notify occupants so they can get out when a fire occurs.
2. Summon the appropriate people to fight the fire.
3. Supervise automatic sprinklers or other extinguishing systems to assure their operation when needed.
4. Supervise various industrial operations to warn of situations that might develop a fire hazard.
5. Supervise security guards or other personnel to assure their performance of their duties.
6. Actuate fire suppression equipment.

Classification of Signaling Systems

Six types of protective signaling systems are generally recognized. These systems, and the NFPA standards covering their installation, maintenance, and use, are as follows:

1. *Central Station Systems.* NFPA 71, *Standard for the Installation, Maintenance, and Use of Central Station Signaling Systems.*[2]

2. *Local Systems.* NFPA 72A, *Standard for the Installation, Maintenance and Use of Local Protective Signaling Systems for Guard's Tour, Fire Alarm and Supervisory Service.*[3]
3. *Auxiliary Systems.* NFPA 72B, *Standard for the Installation, Maintenance and Use of Auxiliary Protective Signaling Systems for Fire Alarm Service.*[4]
4. *Remote Station Systems.* NFPA 72C, *Standard for the Installation, Maintenance and Use of Remote Station Protective Signaling Systems.*[5]
5. *Proprietary Systems.* NFPA 72D, *Standard for the Installation, Maintenance and Use of Proprietary Protective Signaling Systems.*[6]
6. *Household Fire Warning Systems.* NFPA 74, *Standard for the Installation, Maintenance, and Use of Household Fire Warning Equipment.*[7]

Basic minimum requirements for performance of automatic fire detectors are contained in NFPA 72E, *Standard on Automatic Fire Detectors.*[8] Specific requirements for simple local alarm units for use with automatic sprinkler systems appear in NFPA 13, *Standard for the Installation of Sprinkler Systems.*[9] NFPA *101, Life Safety Code,*[10] suggests provisions for alarm and fire detection systems having particular application to the hazard of life. One provision recommends that fire detectors and the components of fire detection systems be tested and listed by appropriate testing laboratories.

Central Station Systems: Central station systems are operated by firms whose principal business is to furnish and maintain supervised signaling service. The central station serves various properties subscribing to the service. The alarm and signaling devices on the subscribers' properties are connected to the central station where trained operators are on hand to receive the signals and take appropriate action. Central station operators retransmit alarms to the fire department. In addition, when an alarm sounds, a runner usually is dispatched to the subscriber's premises to check on the situation.

In the central station system, signals register in the office of an independent agency, usually located at a distance from the protected properties. The agency has trained and experienced personnel continually on duty to receive signals, to retransmit fire alarms to the fire department, and to take whatever action supervisory signals indicate is necessary.

Central station systems customarily serve a number of properties of different ownership, and usually are operated under contract by an agency that has no direct monetary interest in the protected properties. To reduce expense, it is customary to connect the plants of several subscribers to a single transmission circuit; each such circuit terminates in a recording instrument, and each subscriber is distinguished by one or more coded signal numbers that are not otherwise repeated on that particular circuit.

In addition to transmission of fire alarms, the central station service includes supervision of sprinkler systems; supervision of other extinguishing systems; supervision of industrial processes, temperature, and humidity; and supervision of guard and "watchman" services. Central station systems have demonstrated a satisfactory performance record over the years.

Local Systems: Local protective signaling systems produce a signal at the protected premises for an alarm of fire and for supervisory services required. Such services include supervision of a security guard's rounds, supervision of sprinkler system waterflow alarm service and of sprinkler systems, and supervision of smoke alarm service. Local systems are used primarily for protection of life by indicating the necessity for evacuation, and secondarily for protection of property. In some cases and upon approval of the authority having jurisdiction, the local system can be arranged so a presignal will alert selected personnel responsible for sounding a general alarm.

In this system, the alarm or supervisory signal registers in the protected premises. From the relatively uncomplicated arrangements of wiring and bells in early systems, local signaling systems have been developed that embody a number of sophisticated techniques intended to assure operability of equipment and to initiate alarm devices.

A first essential of a satisfactory local system is a reliable primary (main) power supply, which should be either (1) a dependable commercial power service or (2) an engine-driven generator, where an operator is on duty at all times. A secondary (standby) power supply should be batteries, generator, and batteries of multiple generators. The second essential of a satisfactory local system is the use of wiring materials and methods suitable for the purpose. The requirements for protective signaling systems contained in NFPA 70, *National Electrical Code®*,[11] Article 760, are applicable.

The third essential of a satisfactory local system is for the equipment to activate a distinctive signal in case of trouble—such as a broken or grounded circuit or the failure of a main power source—where such trouble prevents the intended operation of the system. Generally, a local protective signaling system can be expected to operate where such disability exists. Trouble signals for this purpose are distinctive from alarm or supervisory signals and, as a rule, cannot easily be disabled or made inoperative.

Auxiliary Systems: The auxiliary-type system connects the appropriate devices in the protected facility with the municipal fire alarm system. Alarms from such a system are received at fire alarm headquarters by the same equipment and by the same alerting methods as alarms transmitted from municipal street boxes. In this system, signals are recorded at a municipal fire department. The facilities connecting the protected property and the fire department are part of the municipal fire alarm system. The devices in the protected plant customarily are owned and maintained by the municipality, or leased by it, as part of the municipal alarm system. The auxiliary-type of system is limited to alarm service only.

The three types of auxiliary systems in current use are: (1) local energy type, (2) shunt type, and (3) parallel telephone type. The local energy- and the shunt-type auxiliary systems make use of the receiving equipment and the interconnecting wires of an established municipal fire alarm telegraph system. The parallel telephone auxiliary system is one in which the alarms are transmitted over a circuit, usually leased lines, directly connected to the annunciating switchboard at an alarm communication center and terminated at the protected property by an end-of-line resistor or its equivalent.

Remote Station Systems: Remote station systems usually are used to protect premises frequently unoccupied. The alarm or supervisory signal is received at fire alarm headquarters or at the office of a communications agency, usually located at a distance from the protected property. Signals are transmitted and received on privately owned equipment. The agency receiving the signals may be a municipal fire department or a communications agency capable of receiving the signals and acting upon them. For fire alarms only, the remote station system should be connected with municipal fire alarm headquarters. Of prime importance is the choice of the remote station. If supervisory signals are to be included, the remote station should be located in a commercial signaling agency with personnel trained to provide the proper response to the signals received. Systems of this type generally are leased by the occupant of the protected premises and maintained under contract by the leasing company.

For purposes of registering fire alarms, the desirable agency is the municipal fire alarm headquarters—if it has personnel constantly in attendance. Supervisory remote station signals, however, require action different from the dispatching of fire department equipment. The preferred location for the remote station is in the quarters of a commercial agency that is open continuously, and that has personnel in attendance who are trained to provide proper response to the signals received. Such an agency also might serve in areas where fire department quarters are not continuously staffed for the reception of fire alarm signals. In such cases, it is the function of the agency personnel to do whatever is needed to ensure prompt dispatching of fire department equipment.

Proprietary Systems: A proprietary protective signaling system is for individual properties where the system is under constant supervision by competent experienced personnel in a central supervisory station at the property protected. Such systems almost always are found in large industrial plants. In this system, operators are trained to take whatever action the signals indicate. The central supervisory station is under the control of the owner or occupant of the protected property. Operation of the central supervisory system and the signaling systems connected to it is a function of the personnel employed for this purpose by the owner of the property. The equipment usually is purchased by the user and is not subject to outside control.

Proprietary systems can be considered elaborate local systems to which recording devices at an in-plant central supervising station have been added. This type of system is under the supervision of operators trained to investigate the situations that the system reports and to take whatever steps are necessary. Steps might be to summon the fire department or to call for in-plant assistance to adjust abnormalities. The distinguishing features of a proprietary system are its ownership by the occupant of the protected plant and the presence of a central supervising station within the confines of that plant. Proprietary systems basically are utilized for the supervision of patrolling guards and the reception of emergency signals from them when a fire or other unusual situation is discovered. In addition, alarm signals should have precedence over all other signals.

Household Fire Warning Systems: Household fire warning systems should provide reasonable protection for residents in ordinary dwellings. Smoke and toxic gases that develop in most dwelling fires are more likely to cause death and injury than exposure to direct heat or flames; exceptions are where clothing is ignited and where a smoke detector would not be of any help. Many home fires are slow smoldering fires, such as those caused by a discarded cigarette in an overstuffed chair. In many cases, such fires start at night when the family is asleep; thus, a smoke detector located outside each sleeping area offers the most protection. Depending on the particular arrangement of the home, additional smoke detectors often are needed. When activated, the detector must give a signal that is distinctive and loud enough to wake sleeping persons. The source of power needed to operate the detector can come either from the house electrical supply, or from a battery that will last at least a year. A battery-operated detector must be capable of emitting a distinctive trouble signal to indicate the need for battery replacement.

The two extremes of fire to which household fire warning equipment must respond are the rapidly developing fire and the slow smoldering fire. Either type can produce smoke and toxic gases. The detection and alarm system devices in the household fire warning system category are for the sole benefit of occupants of the protected household. If the alarm is extended to any other location, such as the fire department, the system should be considered one of the aforementioned types as applicable. In either case, the requirements of detector location and spacing, as they apply to home warning systems, would continue to be followed. Home fire warning systems are concerned primarily with life safety—not with protection of property.

Automatic Fire Extinguishing Systems

In addition to the six types of protective signaling systems described above, various automatic fire extinguishing systems can cause an alarm to be sounded when the system is activated. For example, the fire detecting capability of automatic sprinklers is employed to advantage in protective signaling. Any of the several forms of electrical waterflow alarm switches commonly used in automatic sprinkler systems can be designed to provide an alarm during sprinkler system operation. Electrical waterflow alarm switches function when a sprinkler head fuses. Sprinkler systems frequently are equipped with additional devices to signal abnormal conditions that might hamper the effective operation of the sprinklers. A supervisory device which will indicate that a sprinkler supply valve has been shut is most useful.

All of the sprinkler manufacturers provide waterflow alarm and other supervisory signaling equipment for wet-pipe and dry-pipe sprinkler systems. Other types of automatic fire extinguishing systems also can serve as fire detection systems, including those employing foam, foam-water, high-expansion foam, a halogenated agent, carbon dioxide, dry chemical, water spray, etc.

Power Supplies for Protective Signaling Systems

Of major concern in the use of protective signaling devices is the dependability of the source of power—the power supply. Usually the electric utility furnishing commercial light and power is the most acceptable primary power supply. An exception is that a battery may be the power supply in the case of household fire warning equipment. For the other types of systems, batteries can be used as a secondary power supply. This secondary supply often consists of fully charged storage batteries that are automatically put into service if the commercial power fails, ensuring uninterrupted operation.

Experience has demonstrated the need for power supplies of top quality and with an excellent reputation for reliability. Distinction is made between primary and secondary supplies. The primary supply furnishes power for transmission and reception, or for sounding of the alarm or supervisory signal. The secondary supply either provides energy to the system for fire alarm signaling in case the main supply fails, or provides for trouble signals and other functions. Although not essential for alarm transmission, such functions are associated with the reliability of the system or can provide emergency signaling as well as trouble signals.

Selection of a primary source of power should stress the importance of continual reliability under most of the conditions likely to be encountered on the property. In recent years, commercial light and power supplies have been found sufficiently reliable, in most areas, to permit their use for primary system powering. This is true provided the commercial supplies are used in conjunction with fully charged storage batteries or a generator or a combination of the two that can be automatically thrown into service should the commercial power fail. A number of acceptable secondary supplies are suggested by the NFPA standards.

FIRE DETECTION MECHANISMS AND DEVICES

This section discusses the operating principles of fire detection mechanisms and devices, the spacing of detection devices, fusible links and releases, and the actuation of fire-controlling equipment.

Heat Detectors

Heat detectors are the oldest type of automatic fire detection device. They began with the development of automatic sprinklers in the 1960s and have continued to the present with a proliferation of various types of devices. A sprinkler can be considered a combined heat-activated fire detector and extinguishing device when the sprinkler system is provided with waterflow indicators connected to the fire alarm control system. Waterflow indicators detect either the flow of water through the pipes or the subsequent pressure change upon actuation of the system.

Heat detectors which only initiate an alarm and have no extinguishing function are still in use. Although they are the least expensive fire detectors and have the lowest false alarm rate of all automatic fire detector devices, they also are the slowest in detecting fires. A heat detector is best suited for fire detection in a small confined space where rapidly building high-heat-output fires are expected, in areas where ambient conditions would not allow the use of other fire detection devices, or where speed of detection is not a prime consideration.

Heat detectors are generally located on or near the ceiling and respond to the convected thermal energy of a fire. They respond either when the detecting element reaches a predetermined fixed temperature or to a specified rate of temperature change. In general, heat detectors are designed to operate when heat causes a prescribed change in a physical or electrical property of a material or gas contained within the operating element of the detector.

Operating Principles of Fixed-Temperature Heat Detectors

Fixed-temperature detectors are designed to alarm when the temperature of the operating element reaches a specified point. The air temperature at the time of alarm is usually considerably higher than the rated temperature, because it takes time for the air to raise the temperature of the operating element to its set point. This condition is called thermal lag. Fixed-temperature heat detectors are available to cover a wide range of operating temperatures ranging from about 135°F (57°C) and higher. Higher temperature detectors are necessary so that detection can be provided in areas normally subjected to high ambient (nonfire) temperatures, or in areas zoned so that only detectors in the immediate fire area operate.

Fusible Element Type: Eutectic metals—alloys of bismuth, lead, tin, and cadmium that melt rapidly at a predetermined temperature—can be used as operating elements for heat detection. The most common such use is the fusible element in an automatic sprinkler. Fusing of the element allows the cover on the orifice to fall away, water to flow in the system, and the alarm to be initiated.

A eutectic metal may also be used to actuate an electrical heat detector. The eutectic metal is often used as a solder to secure a spring under tension. When the element fuses, the spring action closes contacts and initiates an alarm. [See Figure 9.8(a), parts D, F, and G.] Devices using eutectic metals cannot be restored; either the device or its operating element must be replaced following operation.

Continuous Line Type: As alternatives to spot-type fixed-temperature detection, various methods of line-type detection have been developed. The detector shown in Figure 9.8(b) uses a pair of steel wires in a normally open circuit.

The conductors are held apart by a heat-sensitive insulation. The wires, under tension, are enclosed in a braided sheath to form a single cable assembly. When the design temperature is reached, the insulation melts, the two wires contact, and an alarm is initiated. Following an alarm, the fused section of the cable must be replaced to restore the system.

A similar alarm device, utilizing a semiconductor material and a stainless steel capillary tube, has been used where mechanical stability is also a factor. The capillary tube contains a coaxial center conductor separated from the tube wall by a temperature-sensitive glass semiconductor material. Under normal conditions, a small current (i.e., below alarm threshold) flows in the circuit. As the temperature rises, the resistance of the semiconductor decreases, allows more current flow, and initiates the alarm.

Bimetallic Type: When two metals with different coefficients of thermal expansion are bonded together and then heated, differential expansion causes bending or flexing toward the metal having the lower expansion rate. This action closes a normally open circuit. The low-expansion metal commonly used is invar, an alloy of 36 percent nickel and 64 percent iron. Several alloys of manganese-copper-nickel, nickel-chromium-iron, or stainless steel may also be used for the high-expansion component of a bimetal assembly. Bimetals are used for the operating elements of a variety of fixed-temperature detectors. These detectors are generally of two types—the bimetal strip and the bimetal snap disc.

As it is heated, a bimetal strip deforms in the direction of the contact point. With a given bimetal, the width of the gap between the contacts determines the operating temperature; the wider the gap, the higher the operating point. The operating element of a snap-disc device is a bimetal disc formed into a concave shape in its unstressed condition. [See Figure 9.8(c).] Generally, a heat collector is attached to the detector frame to speed the transfer of heat from the room air to the bimetal. As the disc is heated, the stresses developed cause it to suddenly reverse curvature and become convex. This provides a rapid positive action that closes the alarm contacts. The disc itself is not usually part of the electrical circuit.

All heat detectors using bimetal elements are automatically self-restoring after operation, when the ambient temperature drops below the operating point.

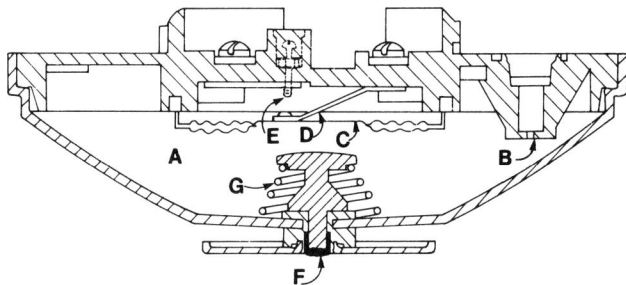

Fig. 9.8(a) A spot-type combination rate-of-rise, fixed-temperature device. The air in chamber A expands more rapidly than it can escape from vent B. This causes pressure to close electrical contact D between diaphragm C and contact screw E. Fixed-temperature operation occurs when fusible alloy F melts, releasing spring G which depresses the diaphragm closing contact points. (Edwards Company.)

Rate-Compensation Detectors

A rate-compensation detector is a device that responds when the temperature of the surrounding air reaches a predetermined level, regardless of the rate of temperature rise.

A typical example is a spot-type detector with a tubular casing of metal that tends to expand lengthwise as it is heated, and an associated contact mechanism that will close at a certain point in the elongation. A second metallic element inside the tube exerts an opposing force on the contacts,

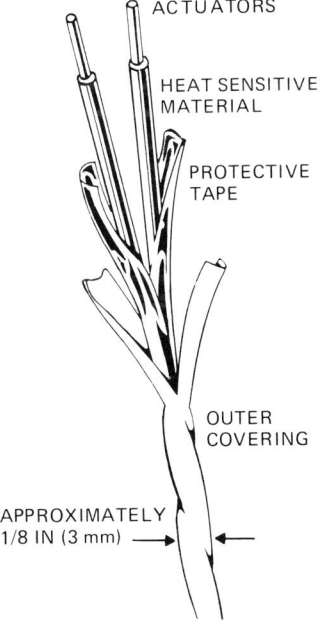

Fig. 9.8(b) Line-type heat detector.
(The Protectowire Company.)

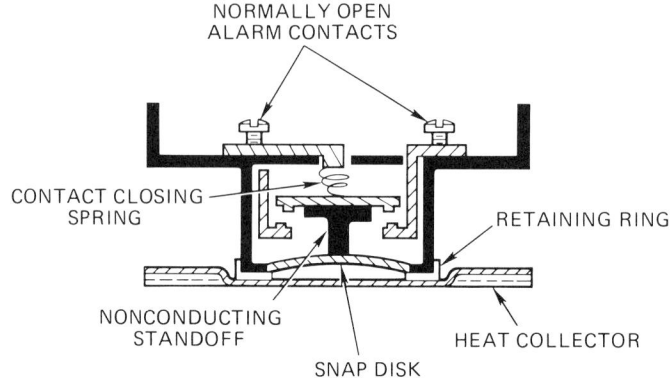

NORMALLY OPEN
ALARM CONTACTS

CONTACT CLOSING
SPRING

RETAINING RING

NONCONDUCTING
STANDOFF

HEAT COLLECTOR

SNAP DISK

Fig. 9.8(c)   Snap-disc device.

tending to hold them open. The forces are balanced so that with a slow rate of temperature rise, there is more time for heat to penetrate to the inner element. This inhibits contact closure until the total device has been heated to its rated temperature level. However, with a fast rate of temperature rise, there is less time for heat to penetrate to the inner element. The element therefore exerts less of an inhibiting effect, so contact closure is obtained when the total device has been heated to a lower level. This, in effect, compensates for thermal lag.

Thermal detectors using expanding metal elements are also automatically self-restoring after operation, when the ambient temperature drops to some point below the operating point.

Rate-of-Rise Detectors

One effect that a flaming fire has on the surrounding area is to rapidly increase air temperature in the space above the fire. Fixed-temperature heat detectors will not initiate an alarm until the air temperature near the ceiling exceeds the design operating point. The rate-of-rise detector, however, will function when the rate of temperature increase exceeds a predetermined value, typically around 12 to 15°F (7 to 8°C) per minute. Rate-of-rise detectors are designed to compensate for the normal changes in ambient temperature [less than 12°F (6.7°C) per minute] which are expected under nonfire conditions.

In a pneumatic fire detector, air heated in a tube or chamber expands, increasing the pressure in the tube or chamber. This exerts a mechanical force on a diaphragm that closes the alarm contacts. If the tube or chamber were hermetically sealed, slow increases in ambient temperture, a drop in the barometric pressure, or both, would cause the detector to initiate an alarm regardless of the rate of temperature change. To overcome this, pneumatic detectors have a small orifice to vent the higher pressure that builds up during slow increases in temperature or during a drop in barometric pressure. The vents are sized so that when the temperature changes rapidly, as in a fire situation, the rate of expansion exceeds the venting rate and the pressure rises. When the temperature rise exceeds 12 to 15°F (7 to 8°C) per minute, the pressure is converted to mechanical action by a flexible diaphragm. Pneumatic heat detectors are available for both line- and spot-type detectors. A schematic of a spot-type pneumatic heat detector is shown in Figure 9.8(a) and a line-type is shown in Figure 9.8(d).

Line-Type: The line-type [Figure 9.8(d)] consists of metal tubing, in a loop configuration, attached to the ceiling or side wall near the ceiling of the area to be protected. Lines of tubing are normally spaced not more than 30 ft (9.1 m) apart, not more than 15 ft (4.5 m) from a wall, and with no more than 1,000 ft (305 m) of tubing on each circuit. Also, a minimum of a least 5 percent of each tube circuit or 25 ft (7.6 m) of tube, whichever is greater, must be in each protected area. Without this minimum amount of tubing exposed to a fire condition, insufficient pressure would build up to achieve proper response.

In small areas where the line-type tube detectors might have insufficient tubing exposed to generate sufficient pressures to close the alarm contacts, air

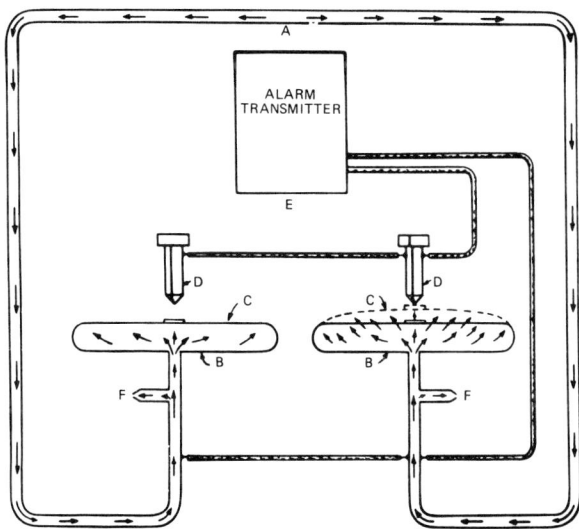

Fig. 9.8(d) Line-type rate-of-rise heat detector. The copper tubing A is fastened in a continuous loop to ceilings or walls and terminates at both ends in chambers B having flexible diaphragms C which control electrical contacts D. When air in the tubing expands under the influence of heat, pressure builds within the chambers, causing the diaphragms to move and close a circuit to alarm transmitter E. Vents F compensate for small pressure changes in the tubing brought about by small changes in temperature in the protected spaces. (American District Telegraph Co., and its associated companies.)

chambers or rosettes of tubing are often used. These units act like a spot-type detector by providing the volume of air required to meet the 5 percent or 25 ft (7.6 m) requirement. Since a line-type rate-of-rise detector is an integrating detector, it will actuate either when a rapid heat rise occurs in one area of exposed tubing, or when a slightly less rapid heat rise takes place in several areas where tubing on the same loop is exposed.

Spot-Type: The pneumatic principle is also used to close contacts within spot detectors. [See Figure 9.8(a).] The difference between the line- and spot-type detectors is that the spot-type contains all of the air in a single container rather than in a tube that extends from the detector assembly to the protected area(s).

Combination Detectors

Combination detectors contain more than one element which responds to a fire. These detectors may be designed to respond from either element, or from the combined partial or complete response of both elements. An example of the former is a heat detector that operates on both the rate-of-rise and fixed-temperature principles. Its advantage is that the rate-of-rise element will respond quickly to a rapidly developing fire, while the fixed-temperature element will respond to a slowly developing fire when the detecting element reaches its set point temperature. The most common combination detector uses a vented air chamber and a flexible diaphragm for the rate-of-rise function, while the fixed-temperature element is usually leaf-spring restrained by a eutectic metal. [See Figure 9.8(a).] When the fixed-temperature element reaches its design operating temperature, the eutectic metal fuses and releases the spring, which closes the contacts.

SMOKE DETECTORS

For the most complete protection of occupants within a dwelling, smoke detectors should be installed to protect each separate sleeping area and at the bottom of each basement stairway; smoke and heat detectors should be installed in all other major areas and rooms of the dwelling. If a detector is installed by the homeowner, the instructions furnished by the manufacturer concerning location, maintenance, and testing should be followed. The value of a household warning system is further enhanced if the family has an escape plan and has practiced the evacuation procedure to follow when the alarm is sounded.

Until the 1980s, attempts to persuade people to buy and use any type of alarm or extinguisher for the home met with appalling indifference. Now public interest in home fire warning devices has grown to the point that 75 percent of all U.S. homes have at least one smoke detector. As has been previously stated in this text, statistics show that the majority of fire deaths and injuries from fire occur in ordinary one- and two-family dwellings; thus, the more household fire warning devices installed, the better the fire casualty record will become.

Smoke detectors are identified by their operating principle. Two of the operating principles are ionization and photoelectric. As a class, smoke detectors using the ionization principle provide somewhat faster response to high-energy (open-flaming) fires, since these fires produce large numbers of the smaller smoke particles. As a class, smoke detectors operating on the photoelectric principle respond faster to the smoke generated by low-energy (smoldering) fires, as these fires generally produce more of the larger smoke particles. However, each type of smoke detector is subjected to, and must pass, the same test fires at testing laboratories in order to be listed.

Ionization Smoke Detectors

Smoke detectors utilizing the ionization principle are usually of the spot type. An ionization smoke detector has a small amount of radioactive material that ionizes the air in the sensing chamber, rendering the air conductive and permitting a current flow through the air between two charged electrodes. This gives the sensing chamber an effective electrical conductance. When smoke particles enter the ionization area, they decrease the conductance of the air by attaching themselves to the ions, causing a reduction in ion mobility. When the conductance is below a predetermined level, the detector responds. (See Figure 9.9.)

Photoelectric Smoke Detectors

The presence of suspended smoke particles generated during the combustion process affects the propogation of a light beam passing through

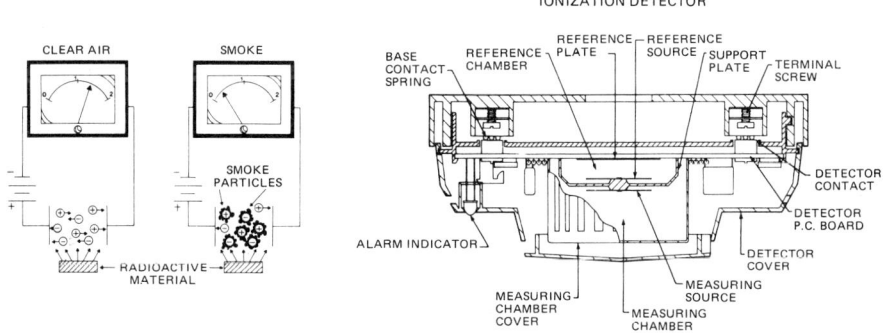

Fig. 9.9   (Left) Principle of operation for ionization smoke detector; (Right) Cross-section view of an ionization smoke detector.(Pyrotronics.)

the air. The effect can be utilized to detect the presence of a fire in two ways: (1) obscuration of light intensity over the beam path, or (2) scattering of the light beam.

Light-Obscuration Principle: Smoke detectors that operate on the principle of light obscuration consist of a light source, a light beam collimating system, and a photosensitive device. When dense smoke obscures part of the light beam, or less-dense smoke obscures more of the beam, the light reaching the photosensitive device is reduced, and this initiates the alarm. (See Figure 9.10.) The light source us usually a light-emitting diode (LED), a reliable long-life source of illumination having a low current requirement. Pulsed LEDs can generate sufficient light intensity for use in detection equipment while operating at even lower overall power levels.

Most light-obscuration smoke detectors are the beam type and are used to protect large open areas. They are installed with the light source at one end of the area to be protected, and the photosensitive device at the other. In some applications, mirrors direct the beam over the desired path, and this determines the area of coverage. For each mirror used, the rated beam length of the device must be progressively reduced by one-third. Projected beam detectors are generally installed close to the ceiling.

Light-Scattering Principle: When smoke particles enter a light path, scattering results. Smoke detectors utilizing the photoelectric light-scattering principle are usually of the spot type. They contain a light source and a photosensitive device arranged so the light rays normally do not fall onto the device. When smoke particles enter the light path, light strikes the particles and is scattered onto the photosensitive device, causing the detector to respond. (See Figure 9.11.) The photosensitive device used in scattering detectors usually is a photodiode or phototransistor.

Cloud Chamber Smoke Detection Principle: A smoke detector utilizing the cloud chamber principle is usually of the sampling type. An air pump draws a sample of air from the protected area(s) into a high-humidity chamber within the detector. After the air sample has been raised to a high humidity, the pressure is lowered slightly. If smoke particles are present, the

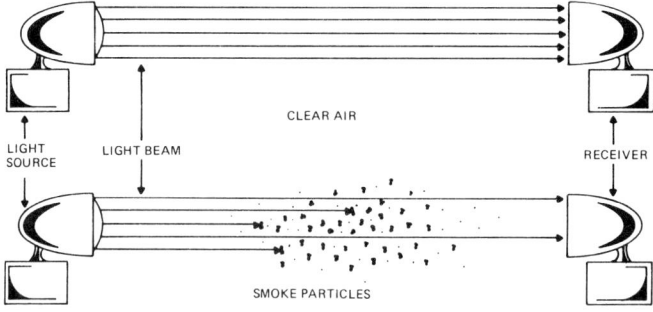

Fig. 9.10   Principle of operation for photoelectric obscuration smoke detector.

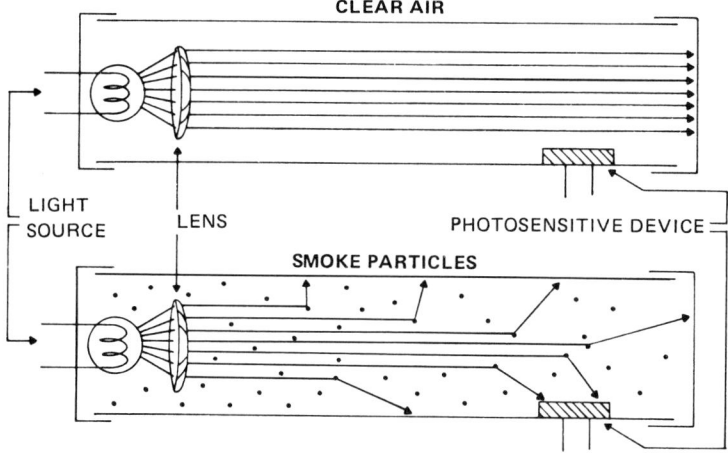

Fig. 9.11   Principle of operation for photoelectric scattering smoke detector.

moisture in the air condenses on them, forming a cloud in the chamber. The density of this cloud is then measured by a photoelectric principle. The detector responds when the density is greater than a predetermined level.

GAS-SENSING FIRE DETECTORS

Many changes occur in the gas content of the environment during a fire. In large-scale fire tests, it has been observed that detectable levels of gases are reached after detectable smoke levels and before detectable heat levels. One of two operating principles—semiconductor and catalytic element—may be used in a gas-sensing fire detector.

Semiconductor Principle

Fire-gas detectors of the semiconductor type respond to either oxidizing or reducing gases by creating electrical changes in the semiconductor. The subsequent conductivity change of the semiconductor causes actuation of the detector.

Catalytic Element Principle

Fire-gas detectors of the catalytic element type contain a material which in itself remains unchanged, but which accelerates the oxidation of combustible gases. The resulting temperature rise of the element causes detector actuation.

FLAME DETECTORS

A flame detector responds to radiant energy visible to the human eye (approximately 4,000 to 7,700 A) or outside the range of human vision. Such a detector is sensitive to glowing embers, coals, or flames which radiate energy of sufficient intensity and spectral quality to actuate the alarm.

Due to their fast detection capabilities, flame detectors are generally used only in high-hazard areas, such as fuel-loading platforms, industrial process areas, hyperbaric chambers, high-ceiling areas, and atmospheres in which explosions or very rapid fires may occur. Because flame detectors must be able to "see" the fire, they must not be blocked by objects placed in front of them. The infrared type of flame detector, however, has some capability for detecting radiation reflected from walls.

Infrared Flame Detectors

An infrared (IR) detector is basically composed of a ilter and lens system used to screen out unwanted wavelengths and focus the incoming energy on a photovoltaic or photoresistive cell sensitive to infrared energy. IR flame detectors can respond to the total IR component of the flame alone, or in combination with flame flicker in the frequency range of 5 to 30 Hz.

A major problem in the use of infrared detectors receiving total IR radiation is the possible interference of solar radiation in the infrared region. When detectors are located in places shielded from the sun, such as in vaults, filtering or shielding the unit from the sun's rays is unnecessary.

Ultraviolet Flame Detectors

Ultraviolet (UV) detectors generally use either a solid-state device, such as a silicone carbide or aluminum nitride, or a gas-filled tube as the sensing element. UV detectors are essentially insensitive to both sunlight and artificial light.

SPACING AND LOCATION OF DETECTION EQUIPMENT

Spacing and location of the previously mentioned heat, smoke, and flame detection devices is shown in diagrams in the appendices of NFPA 72E, *Standard on Automatic Fire Detectors.*[8] Generally, spot-type heat detectors should be located on the ceiling not less than 4 in. (102 mm) from the side wall, or on the side walls between 4 and 12 in. (102 mm and 325 mm) from the ceiling. Where complete coverage is required, strategically located detection devices should be installed throughout all parts of the building.

Factors to consider when spacing detection devices include ceiling construction, ceiling height, room volume, space subdivisions, the normal room temperature, possible abnormal room temperature conditions due to heat-producing appliances or manufacturing processes, and the draft conditions that might affect the normal operation of the device. Currently, heat devices are available that are rated for use with normal ceiling temperatures up to 300°F (149°C) and higher.

NFPA 72E[8] states that the location of smoke detectors should be based upon an engineering survey of the application of this form of protection to the area under consideration. Some conditions to consider are air velocity; ceiling shape, surfaces, and height; configuration of contents and burning characteristics of stored combustibles; the number of detectors required for complete coverage; and location of detectors in relation to ventilating and air conditioning facilities. Typical conditions of occupancy to be evaluated include possible obstruction of photoelectric light beams by the storage and movement of stock, and presence of dust or vapors that could interfere with operation of the smoke detection devices.

GENERAL UTILIZATION OF DETECTION DEVICES

Because single-station smoke and heat detection units commonly used in the home incorporate the detector, control equipment, and the alarm-sending device in one piece, they may be operated from the house current or by batteries. However powered, the homeowner or tenant must assume the responsibility of seeing that the power source is working and that the detector always is in operating condition. A neglected, inoperative warning device is worse than none because of the false sense of security it provides.

The various types of heat and smoke detection devices and systems are used to perform a number of helpful fire protection functions. They can: (1) cause fire doors to shut; (2) control dampers in air conditioning systems; (3) open valves to release water to sprinkler systems; (4) operate fixed extinguishing systems employing foam, water spray, carbon dioxide, dry chemical, halon, etc.; (5) open automatic drains; and (6) release dip tank covers. Detection systems can be arranged so that, if the fire is small, an alarm is sounded but the system will not turn on the extinguishing agent until the fire gets much larger. This gains time for operation of a hand extinguisher or hand hose, thus minimizing the possibility of damage from smoke or fire.

Summary

Fire alarm and detection systems and devices play an important function in saving life and property. Although most alarms of fire to the fire department are received by telephone, street fire alarm boxes are in use in many cities. There are three types of fire alarm boxes: (1) coded type, (2) voice type, and (3) code-voice type. Fire alarms are received at a communication center from which the fire department response is initiated. A communication center may serve one or more municipalities, an entire county, or other political jurisdictions.

There are six major types of automatic and manual protective signaling systems. These signaling systems are installed in buildings for fire alarm, fire prevention, and fire protection purposes. Generally, the more complex the property, the more sophisticated its signaling system.

Installation of smoke detectors for life safety in dwellings is becoming increasingly popular, as is the general use of the many types of heat and flame detection devices.

References

[1]NFPA 1221-1984. *Standard for the Installation, Maintenance, and Use of Public Fire Service Communication Systems*, National Fire Protection Association, Quincy, MA.

[2]NFPA 71-1985. *Standard for the Installation, Maintenance, and Use of Central Station Signaling Systems*, National Fire Protection Association, Quincy, MA.

[3]NFPA 72A-1985. *Standard for the Installation, Maintenance and Use of Local Protective Signaling Systems for Guard's Tour, Fire Alarm and Supervisory Service*, National Fire Protection Association, Quincy, MA.

[4]NFPA 72B-1986. *Standard for the Installation, Maintenance and Use of Auxiliary Protective Signaling Systems*, National Fire Protection Association, Quincy, MA.

[5]NFPA 72C-1986. *Standard for the Installation, Maintenance and Use of Remote Station Protective Signaling Systems*, National Fire Protection Association, Quincy, MA.

[6]NFPA 72D-1986. *Standard for the Installation, Maintenance and Use of Proprietary Protective Signaling Systems*, National Fire Protection Association, Quincy, MA.

[7]NFPA 74-1984. *Standard for the Installation, Maintenance, and Use of Household Fire Warning Equipment*, National Fire Protection Association, Quincy, MA.

[8]NFPA 72E-1984. *Standard on Automatic Fire Detectors*, National Fire Protection Association, Quincy, MA.

[9]NFPA 13-1987. *Standard for the Installation of Sprinkler Systems*, National Fire Protection Association, Quincy, MA.

[10]NFPA *101*-1985. *Life Safety Code*, National Fire Protection Association, Quincy, MA.

[11]NFPA 70-1987. *National Electrical Code*, National Fire Protection Association, Quincy, MA.

Additional Reading

Fire Protection Handbook 16th ed. 1986. National Fire Protection Association, Quincy, MA.

Fire Alarm Systems Handbook 1987. National Fire Protection Association, Quincy, MA.

Chapter 10

Municipal Fire Defenses

Evaluation and planning of fire protection must take into account many factors that influence the strength and effectiveness of a fire department. The ISO Fire Suppression Rating Schedule helps determine the level of fire protection for an area. Masterplanning for fire defenses helps establish good fire practices. Provision of necessary water supply systems ensures effective fire control.

PUBLIC FIRE PROTECTION EVALUATION AND PLANNING

Adequate cost-effective public fire protection is essential for a community. Public fire protection needs to be carefully planned and requires certain logical steps to achieve a comprehensive, acceptable, and workable plan. There are two important aspects to any good plan: (1) the plan itself, and (2) the process by which the plan is developed. The development process must ensure that all major goals are considered, and that each constituency to be affected by the plan is reasonably involved in the planning process.

Without adequate involvement of the various constituencies, implementation of the plan might fail due to a lack of cooperation. For a satisfactory plan to evolve, the planners must decide the end results they wish to achieve (goals), determine the status of the community in relation to those goals (evaluation), and calculate how much and what kinds of progress can take place over a certain period of time (objectives, tactics, time frame). Each of these three steps—setting goals, evaluation, and working out the details—requires the collection and analysis of relevant information, usually called data. Broad goals are achieved through planned strategies; precise objectives are achieved through implemented tactics. Each must be relevant to the other.

Because some degree of public fire protection almost always is in place, it is common for the entire planning process to begin with an evaluation of fire protection already available. Information obtained from the evaluation, when analyzed in terms of broad, generally recognized public protection goals, identifies needs and provides fire protection officials with the approximate parameters of the plan to be developed. As already noted, that plan needs the involvement of a wide variety of community groups. Even if the various constituencies seem willing to allow fire protection officials to develop and write the comprehensive plan without their input, the fire officials should be cautious; the citizens eventually must be willing to accommodate and pay for the implementation. Since a comprehensive plan envisions a larger system of integrated parts, a number of organizations and agencies outside the fire department will need to play important roles in implementation of the plan.

One basic concept of a comprehensive public fire protection plan is that it is infinitely better to prevent fires than for them to occur, even if they are extinguished quickly and at relatively low loss rates. The goal of reducing the incidence of fire involves all aspects of fire prevention. Historically, much more energy and many more resources have been devoted to evaluating, planning, and implementing fire fighting capabilities than to fire prevention programs. That focus is still exceedingly important, but additional emphasis on preventing fires is essential as well.

Planning groups have obvious difficulty, however, in evaluating the success of fire prevention programs. A reasonably effective method for conducting the evaluation involves two kinds of analysis. The first requires the community to maintain consistent, careful records concerning the number of fires which occur, and their cost in lives and dollars. When the human and physical costs are added to the expense of maintaining fire prevention and fire suppression programs, a standardized yearly total cost of fire to the community can be calculated and compared with that same cost the previous year, or over an average of three or more years. Necessary adjustments for inflation, for community growth or decline, and for other important variables can be made by local officials to permit a community to compare its present to its prior fire performance.

The second kind of analysis involves one community identifying others which are similar in ways important to fire protection, such as size, types of construction, hazards, and geography, and comparing its own total performance to the total performance of the similar communities. In the first analysis, the community uses an internal data source—itself—as a yardstick; in the second analysis, it uses an external yardstick—other communities. As more data are collected concerning the total cost of fires in various communities operating with various public fire protection plans, any given community will be able to benefit relatively quickly not only from its own experience, but from the experiences of others. The growing focus on reducing the incidence of arson is one example.

Important to this data collection, evaluation, and planning process is the degree of ability and willingness of the community to finance the total level of fire protection required to meet the plan's goals. The ability and willingness to implement segments of the plan which may be limited by existing legislation (or lack of it), or a lack of innovative approaches to solution identification, must be considered. While fire protection officials always must be concerned with reducing the total cost of fire (fire loss plus costs of prevention and suppression), citizens ultimately reserve the right to make decisions, or "tradeoffs," concerning the level of protection they wish their tax dollars to purchase.

To assist citizens in making decisions concerning community budgets, fire protection officials must accurately describe the effect on total cost if additional or fewer resources are applied to particular prevention or suppression efforts. Application of technical knowledge and analytical ability by officials provides a crucial element to comprehensive planning and evaluation and is one of the most important responsibilities of the public fire protection officer.

EVALUATION

In assessing the suppression capabilities of the fire department and related agencies, fire officials must take a number of factors into account. Such factors include known combustibles, life hazards, fire frequency, climatic conditions, demographic and geographic factors, and consideration of the specific role of a public fire department in providing fire protection to the community. Failure to adequately consider each of these factors can lead to a large-loss fire. A fire department's suppression capabilities can never be expected to compensate for deficiencies or the lack of built-in fire protection systems.

Urban Fire Protection

In urban areas, inadequate fire department response to initial alarms can be a major factor in fire losses due to high population densities and exposure of structures adjacent to the fire. The number of simultaneous fire fighting operations that may need to be conducted dictates the total amount of personnel and equipment required for *effective* fire fighting operations. In all but the smallest structural fire, several operations must be carried on simultaneously and the fire attack must be made from several points. This cannot be accomplished by the crew of a single fire apparatus. Multiple apparatus must be positioned properly, and adequate waterflow must be available to cope with the amount of fuel (fire load) involved or exposed.

In simplest terms, structural fire fighting involves simultaneous operation of three units under a chief officer: (1) a pumper company to undertake a fast initial fire attack; (2) a pumper to provide adequate water supply for a continuing operation; and (3) a company to handle rescue, ventilation, salvage, and various other activities not related to hose lines. At large-structure fires, additional fire fighting personnel are needed to cover the various points of fire attack. In some cases, certain functions can be handled most efficiently by crews with specific training, such as rescue companies and hazardous material teams operating from specially equipped apparatus.

In a light-hazard residential district, the minimum effective initial fire alarm response should consist of three pieces of apparatus. Two of these should be equipped to conduct pumping and water supply operations, with the remaining vehicle able to carry out the other operations. Each apparatus should carry the tools and appliances necessary to perform its designated operations. Twelve fire fighters and a chief officer are the personnel required for reasonably satisfactory operation of this equipment.

The rationale for a 12-person response team is based on this distribution: two fire fighters driving the two pumpers and remaining to operate the pumps, distribute equipment, and help other personnel; four fire fighters advancing and operating two hose lines; four fire fighters performing laddering, forcible entry, ventilation, rescue, salvage, and other "ladder company" duties; and two fire fighters handling hydrants and water supply lines, and advancing and operating a third hose line. One person also functions as direct supervisor (company officer) of the hose lines, and another as supervisor of laddering and other operations. The chief officer directs overall operations at the working fire. Response crews with fewer personnel and equipment risk being unable to perform rescue and extinguishment rapidly enough to do much good. While some case studies indicate that fewer personnel are sufficient to handle a percentage of fire calls, there is a dramatic history of small fires which have developed into large-loss fires—with possibly related fire fighter and civilian injuries or deaths—because of inadequate initial response.

Many times all of the initial response (first alarm) apparatus will not arrive at the fire scene simultaneously. When apparatus has to respond from more than one station, some may have longer travel times than others to the fire scene. In volunteer departments, personnel must travel varying distances to get to the fire station or the fire scene, making it unlikely that all apparatus can go into operation at the same time. Those fire fighters and vehicles which cannot arrive at the fire scene within the first critical time period do not support the initial attack, regardless of the department's response assignment ("running card"). Communities often have a false sense of security in this regard, until actual response times are tested and working initial attack personnel are counted.

Commercial, industrial, and mercantile areas generally require an additional piece of pumping apparatus in response to the initial alarm. (See Table 10.1.) If properties with considerable life hazard (schools, hospitals, nursing homes, etc.) are involved, at least four pumping apparatus, two aerial ladder trucks, and two chief officers should be considered a minimum response to initial alarms. Especially large numbers of personnel are needed for search and rescue operations in these properties.

The required fire fighting units should arrive on the scene soon enough after the initial alarm to operate as an effective fire fighting unit following planned tactical procedures. Often, application of the "task force" concept, where vehicles are housed together and respond together as a tactical unit, may prove to be the most efficient fire fighting tactic, even though a slight increase in response time is necessary in some areas of the first alarm district.

Table 10.1   Evaluation of Fire Department Response Capability

High-Hazard Occupancies (schools, hospitals, nursing homes, explosives plants, refineries, high-rise buildings, and other high life hazard or large fire potential occupancies):

At least four pumpers, two ladder trucks, two chief officers, and specialized apparatus as may be needed to cope with the combustible involved; not fewer than 24 fire fighters and two chief officers.

Medium-Hazard Occupancies (apartments, offices, mercantile, and industrial occupancies not normally requiring extensive rescue or fire fighting forces):

At least three pumpers, one ladder truck, one chief officer, and specialized apparatus as may be needed or available; not fewer than 16 fire fighters and one chief officer.

Low-Hazard Occupancies (one-, two- or three-family residences and scattered small business-es and industrial occupancies):

At least two pumpers, one ladder truck, one chief officer, and specialized apparatus as may be needed or available; not fewer than 12 fire fighters and one chief officer.

Rural Operations (scattered dwellings, small businesses, and farm buildings):

At least one pumper with a large water tank holding 500 gal (1.9m³) or more, one mobile water supply apparatus 1,000 gal (3.78 m³) or larger, and such specialized apparatus as may be necessary to perform effective initial fire fighting operations; at least six fire fighters and one chief officer.

Additional Alarms:

At least the equivalent of that required for rural operations for second alarms, and equipment as may be needed according to the type of emergency and capabilities of the fire department. This may involve the immediate use of mutual aid companies until local forces can be supplemented with additional off-duty personnel.

Improved efficiency often can outweigh this slight increase in response time. A minimum task force unit should consist of two pumping engines, one ladder truck, and a chief officer. Any plans for fire station consolidation or relocation should consider the possibility of a task force.

In evaluating the adequacy of fire protection in any given area, major consideration must be given to the ability of the fire department to efficiently handle any reasonably anticipated work load. This requires evaluation of the possibility of several simultaneous working fires, of weather factors that might contribute to the spread of fire, of delay in response or the possibility of slow operations at the fire scene, and of demographic or geographic conditions that could affect the frequency of fire occurrence and the response time of initial fire fighting units. Where fire frequency patterns lead any fire company to expect two or three working fires per day, or where structures to be protected require a heavy initial response, closer geographic spacing of or increased personnel assigned to individual fire companies may be necessary. The number of other fire fighting or related operations, such as at grass, brush, rubbish, and automobile fires, and emergency rescue operations also may require greater-than-normal staffing of equipment and closer spacing of fire companies. Staffing fire apparatus at a level far below minimum requirements usually will result in less effective fire fighting performance. Low staffing also has an adverse effect on the number of required fire companies for various alarms, since additional fire companies must be dispatched to the scene of an emergency to provide adequate coverage.

The highly desirable practice of assigning emergency medical responsibility to the fire department must be calculated into the staffing formula. Here it also is difficult to obtain effective teamwork and coordination with understaffed crews. Some fire departments have attempted to solve this problem by supplementing their crews with part-time or volunteer fire fighters, or by providing off-duty fire fighters with radio receivers and paying them for overtime when they respond to a fire. On-duty personnel perform the initial fire attack and holding action while off-duty personnel usually provide the additional assistance needed for continuing fire fighting operations, although off-duty personnel usually do not respond immediately. Efficiency definitely is lost, and increased fire losses can be expected with this arrangement. Such protection should not be relied upon as an adequate replacement for staffing and equipment needed immediately at the scene.

Personnel requirements are not merely a matter of numerical strength; they are based on maintenance of a well-trained and coordinated team necessary to utilize complicated and specialized equipment under the stress of emergency conditions. A general practice should be to avoid attempting to operate more fire companies than can be staffed effectively, even if some response distances must be increased somewhat. The effectiveness of pumper companies must be measured by their ability to get required hose streams into service quickly and efficiently. NFPA 1410, *A Training Standard on Initial Fire Attack*,[1] should be used as a guide in measuring this ability. Often a crew of fewer than four fire fighters may be unable to apply half as much water in a given time as a company of four or five fire fighters with the same equipment. Seriously understaffed fire companies generally are limited to using small hose streams until additional help arrives. Often this action may be totally ineffective in containing a small fire and in conducting effective rescue operations. Research indicates that a crew of four is only 65 percent as effective as a crew of five, and that a crew of three is only 38 percent as effective as a crew of five.

Consideration also must be given to maintaining additional forces to handle multiple alarms at the same fire, while still providing minimum fire protection coverage for the other areas under fire department protection. If available personnel prove adequate for routine fires but inadequate for major emergencies, arrangements should be made to supplement the fire protection coverage by calling back off-shift personnel and by promptly calling nearby fire departments for mutual aid. Off-shift personnel may operate reserve apparatus or relieve or supplement personnel on the fireground. Fire companies not dispatched or utilized on the fire scene should be repositioned throughout the remaining area of the jurisdiction to assure minimum response times to any other alarms.

In cases where several fire departments occupy contiguous territories, arrangements should be made for joint response along common boundaries to high-risk hazards and for assistance in covering vacant fire stations at times of major fires. However, mutual aid or mutual response should not be relied upon to provide assistance in all major emergencies, since there could be times when local commitments preclude the anticipated assistance. Mutual aid agreements do not reduce the responsibility of each jurisdiction to maintain adequate facilities to meet normal fire protection needs. It also must be assumed that teamwork and tactical efficiency at a fire will be somewhat less than expected of equal units from the same department under a united command. Often, however, specialized units (such as hazardous materials response teams) are organized to protect larger areas encompassing several fire departments.

Some may argue that it is not the public's responsibility to provide adequate fire protection to high-hazard risks that should have built-in fire protection systems. However, failure to attempt to provide fire protection for large taxable values on which the economy of a community may be based subject some municipal officials to extreme criticism in the event of a serious fire loss.

Time is another critical factor in the evaluation of public fire protection. It generally is felt that the first-arriving apparatus should be at the emergency scene within five minutes of the sounding of the alarm, with additional minutes needed to size up the situation, deploy hose lines, initiate search and rescue, etc. Delays in sounding an alarm obviously must be minimized or eliminated, along with delays in responding and in initiating rescue and attack. Time, however, must not become the all-important factor at the expense of safety.

In an increasing number of instances, specialized apparatus and equipment must be available to municipalities. One category of specialization concerns apparatus designed to handle hazardous materials, including spills of petroleum products and other chemicals. Such hazmat incidents require special extinguishing agents, such as foams or dry powders, and special equipment to apply these agents. Dangerous substances may be present because of manufacturing or storage facilities, or transportation routes in the district. Fire departments often conduct specialized functions such as extricating people from automobile wrecks, performing water and mountain rescue, and providing emergency medical services. These activities also may require special equipment and apparatus.

As with standard fire fighting equipment and apparatus, specialized tools cannot be used effectively and safely unless personnel are trained to use them under a wide variety of circumstances. Whether personnel are volunteer or career, in rural or urban areas, no plan can be implemented and no reasonable level of fire protection afforded to the community unless well-designed and well-managed training programs are carried out.

Fire Prevention

Fire prevention activities can be difficult to evaluate. In a real sense, if prevention efforts are effective, fires and fire-related tragedies occur with less frequency. There is a reduction or absence of fire activity, with results evident statistically although without dramatic newspaper articles and photographs. With careful and systematic long-term recordkeeping concerning the incidence of fires and fire losses, the effect of prevention programs can be documented. Inability of fire officials to demonstrate the value of committing community resources to the broad range of fire prevention activities may well result in a withdrawal from prevention programs and a subsequent increase in the need for a much larger suppression budget. Rational decisions and sound recommendations concerning evaluation and planning cannot be made unless fire officials learn to reallocate resources available for the total fire defense system.

Both evaluation and planning require recognition of the components and integrated parts of a fire prevention system. Modern approaches to fire prevention involve a comprehensive program that includes all organized activities, other than suppression, which reduce the incidence of fire and lowers fire-related losses. Ideally, these activities would be carried out in all communities, rural and urban, with adjustments made for community size, type, location, and fire history.

Prevention activities may be categorized in several ways, but it usually is helpful to group them as follows:

1. Activities which relate to construction, such as application of building codes, approval of building and facility plans, and occupancy certification.

2. Activities which relate to the enforcement of codes and regulations such as inspection of occupancies, licensing of hazardous facilities, development of regulations and codes, and promotion of legislation to adopt model codes.

3. Activities which relate to the reduction of arson, such as fire cause investigation and collection of information and data related to incendiary fires.

4. Activities which relate to the collection of data, such as standardized fire reporting, preparation of case histories, and fire research.

5. Activities which relate to public education and training, including fire prevention, evacuation and personal safety, fire protection training for industrial workers and other employee groups, hazardous materials and devices safeguards, and encouragement to install signaling and extinguishing systems.

An analysis of the community's fire history, conducted during the evaluation phase of the fire prevention plan, usually will indicate to fire experts and citizen groups which categories need strengthening. Comparing the number of fires and fire-related incidents, plus fire loss statistics, over several years as more prevention activities are phased in will provide an assessment of program effectiveness. Calculating the total cost of fire to the community (fire loss plus prevention and suppression costs) makes it possible to estimate the cost-effectiveness of a proposed fire prevention program.

Rural Fire Protection

One principal difference between rural and urban fire departments is that rural departments must pay more attention to water supplies, even though this is an important factor in any department's operations. Rural fire department operations and apparatus emphasize not only fire fighting requirements, but also the provision of water for fire fighting. Rural fire apparatus must have large water tanks to permit effective initial attack on fires while supplementary water supplies are being brought into action. Supplementary water supplies include suction sources—such as lakes and ponds—on or adjacent to rural properties, and mobile water tank vehicles for transporting water from more distant sources. Rural fire departments often are forced to use apparatus and hose to relay water from sources several thousand feet (1,000 ft equals 304.8 m) from the emergency. Initial response of pumpers, tankers, and auxiliary apparatus should be adequate for a quick attack on burning property. With good highways and well-designed apparatus, it is possible to bring substantial fire fighting forces to an emergency in rural areas in time for a properly planned and executed initial fire attack operation to be effective. Most rural properties are now located in areas which provide some level of fire protection. Some properties, however, still must depend entirely upon their own fire protection and whatever help they can obtain from forestry agencies or distant fire departments.

Minimum protection for a rural area should include a pumper with a large water tank, plus other apparatus necessary to provide a minimum of 100 gpm (378 L/m) at a fire scene. Since a larger flow often is required to provide adequate fire suppression services, additional resources should be available to supplement the apparatus responding to the initial alarm.

Equipment such as rescue and aerial ladder vehicles should be available, as needed. Generally, elevated master streams are not needed extensively in rural operations. Normal ladder truck equipment for rescue, forcible entry, ventilation, and salvage operations is carried on the pumper and equipment vehicles.

To be even minimally effective in controlling a fire, the initial responding apparatus must reach the emergency scene within approximately ten minutes of the sounding of the alarm, which itself is sometimes delayed. This apparatus should be capable of extinguishing small fires, preventing "flashover" or very rapid fire spread, and possibly preventing the extension of fires to exposed structures.

The provision of fire protection in a rural area should never be viewed as any less important or necessary than it is in a more populated region. The longer response distances in the country, coupled with often-limited water resources, increases the challenge to fire fighters to develop and maintain efficient fire suppression and fire prevention programs.

Many fire departments have shown that they can provide a level of fire protection in rural areas that is equal to the service they give in more developed areas. Much of what has been said about urban fire protection applies equally well to rural areas.

CLASSIFYING PUBLIC PROTECTION

The "Grading Schedule for Municipal Fire Protection" originally was developed by the National Board of Fire Underwriters, then continued by its successor, the American Insurance Association, and by the Insurance Services Office. The schedule is a guideline for municipalities working to classify their fire defenses and physical conditions. The gradings obtained under the schedule also were used in establishing base rates for fire insurance purposes. The schedule has been subject to amendment, with sweeping changes being made in the 1980 edition with development of a revised *Fire Suppression Rating Schedule* (FSRS).[2]

The *Fire Suppression Rating Schedule*[2] produces ten different Public Protection Classifications, with Class 1 receiving the most recognition for public fire protection services and Class 10 receiving no recognition. The FSRS defines different levels of public fire suppression capabilities which are credited in individual property fire insurance rates. This is accomplished by using the classification produced by the FSRS with the Commercial Fire Rating Schedule (CFRS) which correlates the construction, occupancy, exposures, and private fire protection to develop an applicable rate for an individual property.

The *Fire Suppression Rating Schedule*[2] is an insurance industry tool. It can be used to assist in an objective review of those aspects of public fire protection that have a significant influence on minimizing damage once a fire has occurred.

The *Fire Suppression Rating Schedule*[2] evaluates three areas of public fire protection: water supply, the fire department, and the receipt and handling of fire alarms. The water supply evaluation includes the supply works; fire flow delivery; distribution of hydrants; hydrant size, type, and installation; and the inspection and condition of hydrants. The water supply is assigned 40 percent of the relative weight of the FSRS. The fire department evaluation covers engine companies, truck companies, distribution of companies, pumper capacity, department manning, and training. These account for 50 percent of the relative weight of the FSRS. The remaining 10 percent of the relative weight is assigned to evaluation of the receipt and handling of fire alarms. This includes alarm receipt capability, operators, and alarm dispatch circuit facilities.

The *Fire Suppression Rating Schedule*[2] consists of two sections. Section I can be used to develop a public protection classification which reflects the community's ability to handle fires in buildings of small to moderate size; these are defined as having a needed fire flow of 3,500 gpm (13 m³/min) or less. Section II can be used to develop public protection classifications for large individual properties having needed fire flows greater than 3,500 gpm (13 m³/min).

Most communities design their fire protection needs around normally expected fires. This design is recognized in the differing concept between these two sections. The public protection classification in Section I is applicable to average buildings, with the influence of the fire protection demands for larger buildings being removed from that analysis. Section II can be applied individually to each large building to develop a public protection classification that reflects the protection available to that specific property.

PLANNING

Whenever a community—rural, suburban, or urban—considers its fire defenses, it must scrutinize the past and present and make forecasts for the future. Reviewing the past is called "data analysis" and depends upon good recordkeeping. "Evaluation," which is looking at the present, requires the ability to examine a situation objectively. The process of forecasting conditions requires that a "planning" process be followed, resulting in formulation and implementation of a plan so that future challenges to the community can be met. As the plan is implemented, the process must include establishment of a "feedback loop," providing a continuing assessment of how well the plan is contributing to successfully meeting goals and objectives, and feeding revised data into the plan so continuing redesign can occur.

Fire protection organizations—especially fire departments—need to develop several kinds of plans related to fire prevention and fire suppression. These plans are specific, directed at one clearly defined goal, and are operational over a relatively brief time period (usually from one to five years). Typically, the plans are internal to the department and do not involve outside broad-based planning groups. Examples of these plans, most often technical in nature, might involve apparatus replacement, training programs, revised initial response procedures, setting up a hazardous materials attack unit, and adjusting fireground procedures to incorporate the use of larger diameter hose. However, once department planning begins to consider aspects of fire protection which will have an impact on external groups, those groups will need to be consulted and incorporated into the planning process. Fire department planning must dovetail, for example, with new, broad-based, emergency management planning.

Examples of fire department planning which require the early involvement of other groups are station relocations or closings, building inspection programs, public education projects, and changes in the scheduling of work platoons. These plans, while involving some other groups, are still fairly narrow in scope and usually can be formulated over a relatively short time.

Another type of planning, called "comprehensive" or "master" planning, addresses the total community fire protection problem. Incorporating both prevention and suppression, masterplanning involves many community agencies and organizations, perhaps even county, state, and federal agencies. Masterplanning, a necessity for communities, is aimed at integrating all community efforts at prevention and suppression, and improving efficiency and cost-effectiveness of those activities. Better total community performance is the goal of this planning. Its degree of success must be measured by figures relating to the total cost of fire to the community—not just in gains for one subsystem. Masterplans often consist of a number of subplans from various agencies, developed at the same time as part of a larger, total process. The subplans fit together to make a comprehensive and integrated plan.

Comprehensive plans have clearly stated goals with agreed-upon ways of measuring their attainment. These overall goals are reached through strategies acceptable to all involved agencies and to the citizens who must pay for public fire protection. Each goal is comprised of various subgoals or objectives, and for each objective there is a tactic designed to reach it. All objectives must make sense in terms of the overall strategies. When the objectives and tactics are laid out on a time line, the total time required to implement the comprehensive plan becomes apparent, as does the timing for attaining each objective.

Fire protection has been largely a local responsibility, and for good reasons it seems destined to remain so. Each community has a unique set of conditions; it cannot be assumed that a system of fire protection that works well for one community will work equally well in other localities. To be adequate, the fire protection system must respond to local conditions, especially those that are changing. Planning is the key: without local-level planning, the fire protection system is apt to be poorly suited to the changing needs of the community.

Excellent fire protection (for example, in the form of automatic extinguishing systems such as residential sprinklers) is technically available and certainly can be provided with the resources of most communities. Even with considerable public support, however, it may require several years to attain this protection. Meanwhile, in every fire jurisdiction (whether a municipality, county, or region), standards aiming at significantly increasing fire protection must be set. The following sections discuss some of the concepts to be defined in setting these standards.

Adequate Level of Fire Protection: The question of "adequacy" is addressed not only in terms of day-to-day needs, but also for major contingencies that can be anticipated and for future needs. Definitions are needed of "optimal" protection—in contrast to "minimal" protection which fails to meet contingencies and future needs—and of "maximal" protection which is more expensive than most communities can afford.

Comprehensive planning must include contingencies drawn from an analysis of community hazards. A first step in emergency management, this process of hazard identification and analysis is crucial to fire department planning, since fire departments are called to respond to almost all types of emergencies and disasters.

Reasonable Community Costs: The costs of fire, both as threat and reality, include deaths, injuries, property losses, hospital bills, and lost tax revenues, plus the costs of maintaining fire departments, paying fire insurance premiums, and providing built-in fire protection. Each community must decide its appropriate level of investment in fire protection. Some costs which the public is unwilling to bear might be transferred to the private sector, as when automatic extinguishing systems are required in buildings over a certain size or height or with a certain occupancy. Service and use fees also should be considered.[3]

Acceptable Risk: A certain level of fire loss must be accepted as tolerable simply because no community has unlimited resources. Conditions that endanger the safety of citizens and fire fighters beyond the acceptable risk level must be identified as targets for reduction.

Consideration of these matters helps determine what functions and emphasis should be assigned now and in the future to the fire department, other municipal departments, and the private sector. It helps to define new policies, laws, or regulations that may be needed. Most importantly, consideration of these matters makes it clear that firesafety is a responsibility shared by the public and private sectors. Because the fire department cannot prevent all fire losses, formal obligations to have built-in fire protection fall on owners of certain kinds of buildings and occupancies. For the same reason, citizens have an obligation to exercise prudence with regard to fire in their daily lives. Such prudence requires education in firesafety, with the obligation to provide that education falling in the public sector—chiefly the fire department. The public sector (again, chiefly to the fire department) also has an obligation to enforce requirements for built-in protection in the private sector.

Masterplanning

Fire protection is only one of many community services. Not only must it compete for dollars with other municipal needs, such as the educational system and police protection, but in planning for growth the fire protection system must account for the ongoing changes in the community. For example, if a slum area is to be torn down and replaced with high-rise apartment buildings, the fire protection needs of that area will change. Revision in zoning also will alter the fire protection needs in different parts of the community.

To cope with the future, local administrators increasingly are turning to the concept of masterplanning of municipal functions. Such plans include examination of existing programs, projection of needs of the community, and determination of the means to fill those needs. Administrators must seek the most cost-effective allocations of resources to help assure that the needs will be met.

A major section of a community's general plan of land use should be a masterplan for fire protection. This plan should, first of all, be consistent with and reinforce the goals of the community's overall general plan and its time frame. For example, managers should plan the deployment of manpower and equipment according to the kinds and specific areas of growth that the community foresees. Goals and priorities for the fire department should be part of the masterplan. Not only is it important to set objectives in terms of lives and property to be saved, but also to decide allocations among fire prevention, inspection, firesafety education, and fire suppression to determine ways to accomplish the objectives. A masterplan must be an integrated part of the overall emergency management plan for the community.

Because fire departments exist in a real world where a variety of purposes must be served with a limited amount of money, it is important that every dollar be invested for maximum return. The fire protection masterplan should seek not only to provide the maximum cost-benefit ratio for fire protection expenditures, but should also establish a framework for measuring the effectiveness of these expenditures. Lastly, the plan should clarify the fire protection responsibility of other groups, both governmental and private, in the community.

Action in the following four phases can serve as guidelines to fire department administrators for developing and presenting a master fire protection plan.

Phase I

1. Identify the fire protection problems of the jurisdiction.
2. Identify the best combination of public resources and built-in protection required to manage the fire problem within acceptable limits:
   a. Specify current capabilities and future needs of public resources;
   b. Specify current capabilities and future requirements for built-in protection.
3. Develop alternative methods that can involve tradeoffs between benefits and risks.
4. Establish goals, programs, and cost estimates to implement the plan:
   a. Develop department goals and programs, including maximum possible participation of fire department personnel of all ranks;

b. Provide goals and objectives for all divisions to support overall goals of the department;
c. Design management development programs that increase accept-ance of authority and responsibility by all fire officers as they work to meet objectives and carry out programs.

Phase II

1. Define the roles of other government agencies in the fire protection process.
2. Present the proposed municipal fire protection system to the community administration for review.
3. Present the proposed system for adoption as the fire protection element of the jurisdiction's general plan. The standard process for development of a general plan provides the fire department administrator with an opportunity to inform the community leaders of the fire protection goals and system, and to obtain the leaders' support.

Phase III

In considering the fire protection element of the general plan, the governing body of the jurisdiction should pay special attention to:
1. Short- and long-range goals.
2. Long-range staffing and capital improvement plans.
3. Code revisions required for better fire loss management.

Phase IV

The fire loss management system must be reviewed and updated as budget allocations, capital improvement plans, and code revisions occur. Continuing review of results should concentrate on these areas:
1. Did fires remain within estimated limits?
2. Should limits be changed?
3. Did the level of losses prove to be acceptable?
4. Could resources be decreased, or should they be increased?

WATER SUPPLY REQUIREMENTS FOR FIRE PROTECTION

The amount of water needed and the economics of supplying it are basic considerations of good planning for fire suppression.

Most water supply systems serving a substantial number of customers are designed to supply water for normal domestic demands as well as for emergency use by fire departments. Domestic use covers drinking and sanitary purposes, plus processing and other industrial applications. Water for emergency use must supply fire hydrants, pumping engines, and fixed fire suppression systems, such as automatic sprinklers and standpipes.

Water systems meeting normal consumption demands and fire protection requirements must satisfy the design objective of providing simultaneous demand rates for both purposes with reliability. To meet this goal, it is necessary to focus upon the variations in normal consumption of water on the basis of the time of the year, the day of the week, and even the hour of the day. Obviously, in a given system, as more water is used for normal consumption demands, less remains for fire protection. Normal consumption demands usually are expressed in the following terms:

1. Average daily consumption—the average total amount of water used each day during a one-year period.
2. Maximum daily consumption—the maximum averaged total amount of water used during any 24-hour period within three years. (Unusual situations which may have caused an excessive use of water, such as refilling a reservoir after cleaning, should not be considered when determining this figure.)
3. Peak hourly consumption—the maximum amount of water used in any given hour of a day.

The maximum daily consumption normally is about 1.5 times the average daily consumption. The peak hourly rate varies from two to four times the normal hourly rate. The effect these varying consumption rates have on the ability of the system to deliver required fire flows varies with the system design. Both maximum daily consumption and peak hourly consumption should be considered to ensure that water supplies and pressures do not reach dangerously low levels during these high-use periods, and that adequate water will be available if there is a fire.

WATER FOR FIRE FIGHTING

Historically, water supply systems for cities and towns were developed primarily to provide water for drinking and for sanitary purposes rather than for fire protection. However, large cities requiring substantial amounts of water for domestic purposes usually had sufficient water for fire fighting purposes as well. This led to inquiries into the cost of waterworks that could provide water for fire fighting as well as other uses. A number of distinguished engineers associated with individual waterworks examined the problem and presented their findings in technical papers at engineering society meetings late in the 19th century. Papers by Shedd,[4] Fanning,[5] and Kuichling[6] give the details of discussions in which standards now followed in American and Canadian waterworks were first developed. (See Table 10.2.)

Table 10.2   Estimates of Fire Flow

Populations Thousands	Number of Fire Streams Required Simultaneously				
	Shedd 1889	Fanning 1892	Freeman 1892	Kuichling 1897	NBFU 1910
1			2–3	3	4
4		7		6	8
5	5		4–8	6	9
10	7	10	6–12	9	12
20	10		8–15	12	17
40	14		12–18	18	24
50		14		20	26
60	17		15–22	22	28
100	22	18	20–30	28	36
150		25		34	44
180	30			38	48
200			30–50	40	48

Sources (these authorities define streams slightly differently, but the streams were of the order of 200 gpm to 300 gpm):

Shedd, J. Herbert, discussion on a paper by Sherman, William, B., *Ratio of Pumping Capacity to Maximum Consumption* (Shedd 1889).[4]

Fanning, J. T., *Distribution Mains and the Fire Service* (Fanning 1892).[5]

Kuichling, E., *The Financial Management of Water Works* (Kuichling 1897).[6]

Freeman, John R., *The Arrangement of Hydrants and Water Pipes for the Protection of a City Against Fire* (Freeman 1892).[7]

Figures furnished by National Board of Fire Underwriters (Metcalf, Leonard, *et al* 1911).

Engineering: Distribution Network, Hydrant Spacing, Storage

Freeman noted (in 1892)[7] a fundamental difference between systems designed to meet ordinary water needs and those for fire protection. Fire draft required concentration of the water supply, whereas domestic draft was a matter of distribution.

Freeman[7] asserted that if a water system were to meet fire protection needs, the distribution system should be designed to concentrate the needed amounts of water. While small pipes were sufficient for distribution, larger pipes were needed for concentration of supply to fire streams. He suggested 6-in. diameter pipe as the minimum for residential districts, and noted that 8-in. diameter pipe was adequate only if it formed part of a network of distributing pipes whose intersections were not far apart.

Another important point Freeman[7] made was that hydrants should be placed where they could concentrate streams at specific blocks or groups of buildings, rather than placing the hydrants arbitrarily a uniform distance apart on the street mains. His work on hose streams showed how long hose lines reduced the water that could be delivered promptly to a fire. He therefore suggested a working rule for hydrant spacing of 250 ft (76 m) between hydrants in compact mercantile and manufacturing districts, and

400 (122 m) to 500 ft (152 m) in residential districts. These working rules can still be used as guides for good design.

Freeman[7] further insisted that fire water supply should be in addition to maximum domestic consumption and laid the foundation for eventual recognition of this principle. He also calculated how much water should be stored in standpipes or elevated reservoirs. He figured that flow for all of the hose streams required should be supplied from a reliable source, such as an elevated storage reservoir, for at least six hours when the system also was fulfilling maximum demands for domestic and other uses. Freeman calculated that to supply the combined fire and domestic needs in a system provided with reliable pump capacity, a one-hour supply in a standpipe or elevated reservoir would be acceptable.

FIRE PROTECTION REQUIREMENTS IN WATER SYSTEMS

The capacity of a water system is determined by the total amount of water it must furnish. This is the sum of water required for domestic or industrial uses, and water required for fire service. In small towns, the requirements for fire protection exceed other requirements.

In North American cities, a public water system is expected to furnish water for many uses. In some places there may be a heavy industrial demand. Water needs for such purposes as air conditioning and lawn sprinkling also can affect the required capacity of the system. Because the adequacy of a public water system for fire protection cannot be taken for granted, the other demands must be determined to estimate their effects on the system's capacity.

A joint report of committees of the American Society of Civil Engineers, the American Water Works Association, and others suggested that the maximum general service demand on a waterworks system be taken as the peak hourly demand during a test year.[9] This, the report noted, was the only figure which can fairly be compared with the maximum fire flow requirement.

Evaluating System Capacity

In most large cities, the peak hourly rate of water use exceeds the maximum daily consumption rate plus fire flow, and therefore is the controlling factor in the supply system design. In smaller communities, however, the reverse is true, with the maximum daily consumption rate plus fire flow being the controlling factor. For many years water consumption has been increasing in most municipalities, resulting in increased peak hourly rates. Consequently, there has been an increase in the number of municipalities in which the peak hourly rate controls design of the supply system.

In all areas served by the distribution system, fire flows are an important consideration and, in many instances, govern the size of pipe used. Every system should have a water supply sufficient to provide for automatic sprinklers and other automatic fire protection systems in addition to the normal demand rates. To meet water needs, for example, many communities restrict lawn watering to specified periods, usually two to four hours in the evening. In many water supply systems, the demand rates imposed by lawn watering are excessive, depleting storage facilities and reducing pressure throughout the system for many hours. In these situations, there might be little or no water available for fire protection systems, particularly at higher elevations.

Pressure Characteristics of Systems

The pressures for which systems normally are designed are based on several practical considerations which attempt to provide pressures adequate for water supplies both for domestic consumption and for fire protection. If either use demands special ranges of pressure, they can be provided. Selection of pipe and related fittings and methods of using them will allow almost any desired range.

San Francisco, for example, has a separate system, designated the "high-pressure system," controlled by the fire department. All of the pipe is heavy cast iron, tar coated and lined, and is tested on installation and repair to 450 psi (3103 kPa). Two steam-operated pump stations can pump water from San Francisco Bay into the system, and 20,000 gpm (75,700 L/min) at 250 psi (1724 kPa) can be delivered to most of the principal mercantile district. San Francisco maintains this system primarily because an earthquake could put the regular public water system out of service. A few other cities have similar high-pressure installations.

Modern fire department pumpers can create heavy streams and high pressures from ordinary water systems where adequate water volume is available. Cities that formerly had separate systems of fire mains, operating at so-called high pressures, now generally operate these systems at normal public water pressures. An advantage is retained because the second system, although not at high pressure, is still available.

Public water systems reflect a compromise on the question of pressures. Commonly in the range of 65 to 80 psi (448 to 552 kPa), these pressures are adequate for ordinary consumption in buildings up to about ten stories. This pressure range will provide a good supply of water for automatic sprinkler systems in buildings of about four stories in which occupancies are classified as "ordinary." Where pressures of this order are provided, it is reasonably easy to compensate for local fluctuations in draft.

A minimum residual pressure of least 20 psi (138 kPa) should be maintained at hydrants delivering the required fire flow. Pumpers can be operated, but with difficulty, where hydrant pressures are less. Where hydrants are well distributed and of the proper size and type (without excessive friction losses in the hydrant and suction line), it may be possible to set 10 psi (69 kPa) as the minimum pressure. Hydrant pressure should be sufficient to prevent a negative pressure from developing in the street mains; this might cause back siphonage of polluted water from some interconnected source. Using residual pressures less than 20 psi (138 kPa) is prohibited by most state health departments.

Pressures in a public water system may be considered excessive when they approach 150 psi (1034 kPa). As pressures increase, they tend to cause leaks in domestic plumbing. High pressures also require special attention to restrain pipelines in the ground. Although pipe and fittings used in ordinary public water systems are designed for maximum working pressures of 150 psi (1034 kPa), it is not good practice to operate with pressures that high. Pressure-reducing valves can be used in some sections of a system where variations in topography result in excessive pressures. Water services to individual buildings may require pressure-reducing valves to keep the pressure on domestic piping at safe levels.

Systems for Higher Elevations

When water must be supplied to a high elevation, a separate water distribution system usually is provided for the elevated section in order to maintain normal pressure. In such cases, the elevated area should have its own water storage facility as well as pumps to boost the water from other parts of the system. Likewise, the upper stories of a high building should be provided with water supply systems in the building itself. These systems have the same requirements as areas on a hill. High-rise structures normally are divided into a number of pressure zones, with zones of more than 12 stories tending to get outside the normal pressure ranges. In any case, each pressure zone must have storage of water in amounts needed for the sprinkler service or hose streams, and a system of pumps to supply each zone from the zone below. Care should be taken to ensure that pumps will be able to operate even during power failures.

ADEQUACY AND RELIABILITY OF SUPPLY

The adequacy of any given water supply system can be determined by engineering estimates. The water source, including storage facilities in the distribution system, must be sufficient to furnish all the water that combined fire and domestic needs may call for at any one time. Arrangement of the supply works and details of the pumping facilities can limit the adequacy of the supply or affect its reliability.

In a pumping system, a common arrangement is to have one set of pumps that takes suction from wells or from a river, lake, or other body of water. If the water does not have to be filtered, the pumps can discharge directly into the distribution system. Where filtration or other treatment is necessary, pumps take suction from the primary or raw water source and discharge to sedimentation basins or other facilities and then to filter beds. After processing, the water flows to clear water reservoirs from which a second set of pumps takes suction and discharges the water directly into the supply system. Unfortunately, failure of any part of the equipment may affect the entire system. This usually is taken care of by duplication of units and by arrangement of the system to facilitate repairs.

When assessing the reliability of the supply works, features which should be evaluated are: minimum yield; frequency and duration of droughts; condition of intakes; possibility of earthquakes, floods, and forest fires; ice formations, silting up, or shifting in river channels; and availability of guards to protect the facility from physical damage. Reservoirs out of service for cleaning, and the interdependence of parts of waterworks, also affect reliability. Other factors include the condition, arrangement, and dependability of individual units of plant equipment, such as pumps, engines, generators, electric motors, the fuel supply, and electric transmission facilities. Pumping stations of combustible construction are subject to destruction by fire unless protected by automatic sprinkler systems.

Duplication of pumping units and storage facilities, and arrangement of mains and distributors so that water may be supplied to them from more than one direction, are measures that can assure continuous operation. The importance of duplicate facilities is shown by the frequency of their use.

WATER SUPPLY AND DISTRIBUTION SYSTEMS

Water systems generally are subdivided into two divisions: supply systems and distribution systems. However, in small water systems there may be no way of differentiating between supply and distribution systems, since the functions of both may be carried out in a single element of the system.

Supply System

The supply portion of a water system usually is that part of the system where the source or sources of supply are found. It also includes the storage and transmission of that supply through large conduits and aqueducts, and, in some cases, includes the arterial feeders extending to the distribution system.

Distribution System

The distribution system is that portion of the works that actually delivers water to the individual consumer connections and to which fire hydrants are attached.

SOURCES OF WATER SUPPLY

Sources of water supply have two major divisions: ground-water supplies and surface-water supplies.

Ground-Water Supplies

Of the world's total quantity of fresh water, by far the most is available from ground-water supplies. Ground-water supply is water that percolates into the ground from precipitation, and is stored in underground strata called aquifers. The free surface of water in an underground stratum is referred to as the water table. The height of the water table varies throughout the year, depending upon variations in precipitation, water movement in the aquifer, and water withdrawn from the aquifer through springs or wells.

In some locations, now greatly diminished in number because of usage, water is stored in the underground strata under a positive head. When this aquifer is penetrated, water rises to an elevation greater than that in the aquifer, resulting in free flow at the surface.

The supply available from underground aquifers to wells is influenced by, among other factors, the precipitation falling on the area of recharge and the amount of water drawn from all of the wells penetrating the same aquifer. When the aquifer is of substantial size, static water levels usually lag the effects of drought and also lag the effects of increased precipitation. Many different types of wells can be constructed; however, high-capacity wells used for municipal water supplies usually are drilled and are equipped with deep-well turbine or submersible pumps.

NFPA 20, *Standard for the Installation of Centrifugal Fire Pumps*,[10] permits the installation of vertical turbine pumps in properly developed and tested wells under specific conditions.

Surface-Water Supplies

Surface-water supplies consist of rivers, lakes, streams, and impounded water supplies. As with underground supplies, the availability and reliability of the surface supply is dependent upon precipitation falling within the supply's drainage area or watershed. Usually, surface supplies respond relatively quickly to diminished precipitation or drought. Water levels may vary substantially between wet and dry periods. At certain times of the year, excessive runoff from the watershed can result in high water levels at intake structures and low-lift pumping stations. At other times, under drought conditions, water levels in surface supplies could be low enough to require installation of pumps in a reservoir discharging into normally gravity-supplied intake structures. Large-surface reservoirs supplied by substantial watersheds or runoff areas are reliable sources, provided water consumption demands do not increase beyond the recharge capabilities of the watershed involved.

Several factors can affect operation of the intake structures of surface supplies, particularly in colder climates. For example, ice formation is a hazard. Several kinds of ice can form and affect the functioning or threaten the stability of intake structures or cribs. Anchor ice, which forms on submerged dark metallic pipe, fittings, or gratings, can restrict flow into the intakes. Surface ice and the ice dams it forms can impose considerable thrust upon intake structures and cribs. Frazil ice, i.e., displaced anchor ice, or ice that has formed about small suspended particles, can clog intake parts.

When anchor or frazil ice begins to clog an intake, a relatively small change in temperature will free the obstruction. Some intakes are designed with a grid which is in essence a heating coil. Activation of the circuit will heat the intake element sufficiently to keep it free from ice. This method has been found economical in a number of locations, since it requires small amounts of electric power and is used only a relatively few times per year. When the intake is designed to permit continuous temperature recording, formation of anchor or frazil ice can be predicted.

A river also can be a reliable source of water supply if the flow rates during drought periods are not seriously affected. A river can be susceptible, however, to ice formation, scouring of the bottom, changing of the channel, and silting. Before a river intake is constructed, a careful study must be made of the stream bottom, the degree of scour, the extent of formation of surface ice, and the likelihood of the formation of ice jams. An intake can be destroyed by an ice jam, or the entire flow of a river may be stopped by ice. Therefore, the intake must be designed so it can withstand the forces which will act upon it during times of flooding, heavy silting, or ice conditions.

GRAVITY AND PUMPING SYSTEMS

The two basic types of water systems—supply and distribution—are based on gravity and direct pumping action. Most water systems are a combination of gravity and direct pumping.

Gravity Systems

A true gravity system is one which delivers water from the source directly to the distribution system without the use of pumping equipment. This type of system usually is ideal for a fire supply, providing pressures are adequate to meet fire demands and normal consumption rates. A gravity system is extremely reliable because the supply is not dependent upon the operation of mechanical equipment; however, a well-designed and safeguarded pumping system can be developed to the extent that no distinction is made between the reliability of gravity and pumping systems.

Pumping Systems

When water cannot sufficiently be obtained at an elevation to provide working pressures from the elevation head, it is necessary to provide pumps on the system. These pumps normally are located at the source of supply. They are used to develop the pressure needed to overcome friction loss in the supply system, and to provide satisfactory working pressures in the distribution system. Public systems sometimes have water treatment facilities associated with the pumping station. Although it is not the intent to discuss water treatment here, treatment facilities will affect flow rates and quantities of water available to the distribution system. Because many times the limiting features of supply are due to some element of water treatment, it is imperative that the effects of water treatment on the availability of supply be thoroughly understood and considered.

Combination Systems

Often associated with pumping systems are water distribution storage facilities. These provide for storage of water during times of least demand, and then supply water when demand is at its peak.

Storage can be located so that pumps directly supply the storage facility from which water flows to the distribution system. Storage also can be provided at a remote location within the distribution system; under this arrangement, water can be pumped directly into the distribution system, with any excess automatically dumping into the storage facility. The more water that can be maintained in elevated storage, the more reliable a system can be considered because water flowing from a storage facility is the same as a gravity system. Any failure of pumping equipment will not prevent this water from being available for fire protection purposes.

Distribution storage also can consist of large water tanks located at surface elevations equal to or even somewhat lower than the areas of the distribution system they serve. These tanks are filled during periods of relatively low consumption in the system. When demand rates are high, pumps deliver water from these storage facilities to the distribution system. Duplication of pumping equipment and proper design and operation can enable tank storage facilities to be nearly as reliable as elevated storage.

PIPELINES

Pipelines are designed to withstand pressure and to distribute water to the point of use. Three classes of pipelines, or distribution mains, in a large system are:

1. *Primary feeders* consisting of large pipes with relatively wide spacing. They convey large quantities of water to various points of the system for local distribution to the smaller mains.
2. *Secondary feeders* forming a network of pipes of intermediate size. They reinforce the distribution grid within the various panels of the primary feeder system and aid the concentration of the required fire flow at any point.
3. *Distribution lines* consisting of a gridiron arrangement of small mains. They serve the individual fire hydrants and blocks of consumers.

To provide for reliability, two or more primary feeders should extend by separate routes from the source of supply to the high-value districts of a city. Similarly, secondary feeders should be arranged in loops as much as possible to give two directions of supply to any point. This practice increases the water capacity at any given point and prevents a break in a feeder main from completely cutting off the supply. Secondary feeders in built-up areas generally should be installed not more than 3,000 ft (914 m) apart.

Where water systems are divided into pressure zones, water can be transferred from one zone to another in two ways: by operating valves, or by using fire department pumpers to pump from the hydrants in one zone to hydrants in the other. The same action can be taken between the water systems of adjoining communities or between a private system and the public system. However, great care must be taken to prevent the damage that can occur when parts of the system are subjected to excessive pressures and possible contamination. Usually, zone transfer is not a good practice and should not be attempted without advance planning, adequate controls, and specific written approval from health authorities.

The Size of Pipe

Pipe less than 6 in. in diameter is not recommended for fire service. Pipes of this size should be used only when looped in a gridiron where no leg is longer than 600 ft (183 m). In congested districts, it is recommended that distributors be not less than 8 in. in diameter and interconnected within every 600 ft (183 m). On principal streets and for all long lines, the distributors should be 12 in. in diameter or larger.

The cost of a line of pipe includes expenses for trenching (sometimes with piling), laying the pipe, backfilling, and testing, as well as the price of the pipe delivered to the job. All of these costs apply, regardless of the size of the pipe being used. It usually is good practice to install pipe for fire protection which is one or more sizes larger than the bare minimum that might be required. Increasing the pipe diameter only one size often will nearly double the possible flow. The figures in Table 10.3 show the relative capacity of pipe obtained by increasing diameter sizes above 6 in.

Table 10.3   Comparison of Pipe Capacity

Size of Pipe, Inches*	Relative Capacity
6	1.0
8	2.1
10	3.8
12	6.2
14	9.3
16	13.2

*Nominal pipe sizes are not directly convertible into metric sizes. For comparison, one inch equals 25.4 mm.

In designing a system, it also is important to consider the probable development of the area under consideration and, in a general way at least, to plan protection for its ultimate development. Then install that part of the system for which there is immediate need.

The size of pipe needed is based on the rate of water flow required (domestic consumption plus fire flow) and the hydraulic gradient in the area.

Arrangement of Pipe Systems

As previously stated, pipe systems should be arranged in loops wherever possible. This allows hydrants and other connections to be fed from at least two directions and makes it much more possible to deliver water without excessive friction loss.

Internal Condition of Pipe Systems

In time, the internal cross-section areas of unlined cast-iron pipe might be reduced or the pipe's interior surface roughened because of tuberculation, incrustation, or sedimentation. Incrustations may be due to: (1) tubercular growth, (2) deposits of chemical constituents normally in solution in the water, or (3) growth of biological or living organisms. Deposits in all kinds of pipe may be due to: (1) sediments, such as mud, clay, leaves, or vegetable decay, or (2) foreign matter other than sediment.

Serious trouble generally can be detected by careful flushing tests. Flushing of the system will remove ordinary sediment. Operation of valves sometimes will indicate sediment or corrosion. Local water conditions must be taken into account in establishing a regular procedure of flushing and testing.

Pipes can also be cleaned by use of a scraper or rotating auger, pulled through the pipe by a cable or forced through by water pressure. When pipe has been cleaned in this manner, the rate of tuberculation following cleaning usually will be very rapid, resulting in quickly decreasing carrying capacities. The addition of cement lining will retard or prevent further deterioration of capacity. A cost analysis should be made to determine if the pipe should be lined or should be cleaned periodically throughout its life.

FIRE APPARATUS

Engine Company

Engine companies normally comprise the largest number of companies within any fire department. The engine company is considered the basic unit of a fire department and is usually supplemented by other types of companies. The basic unit of apparatus is the pumper, which carries hose, nozzles, an onboard water tank, and a pump. The engine company's basic role in tactical operations is to deliver water through hose lines to control fires, although a variety of additional functions and equipment generally is assigned to engine personnel. In most cases, at least one engine company is based at each fire station to respond quickly and initiate fire control operations at a fire scene.

Ladder Company

The basic apparatus of a ladder company is an aerial ladder or elevating platform device which provides access at higher levels or directs elevated master streams on a fire. Ladder trucks also carry a complement of ground ladders and a selection of hand and power tools. Ladder companies perform a supporting role in fireground operations, which includes search and rescue, forcible entry, ventilation, salvage, overhaul, and the use of ladders to gain access and to rescue people above ground level.

Ladder companies are established according to the degree of urban development and the need for aerial apparatus. In a densely developed city, one ladder company might be provided for every two or three engine companies, while a rural area might have no ladder companies. Where there are no ladder companies, their supporting functions must be assumed by other types of companies.

Rescue Company

Many fire departments utilize separate rescue companies for fire fighting rescues and nonfire-related rescue incidents. Such rescue companies often primarily are involved in the delivery of emergency medical services and in physical rescues, such as extricating victims from vehicle accidents, removing injured people from perilous locations, and assisting victims of industrial accidents.

Rescue company vehicles range from small vehicles designed for emergency medical service delivery to heavy squad vehicles carrying a large variety of tools and equipment.

In fire fighting operations, rescue companies usually are assigned search and rescue along with responsibility for medical treatment. Additional duties often are similar to those of a ladder company, particularly forcible entry, ventilation, and the use of power tools.

Squad Companies

In many fire departments, "squads" supplement engine and ladder companies with additional personnel or highly specialized apparatus. The term "squad" occasionally is used interchangeably with "rescue company."

Special Apparatus

In addition to the apparatus normally assigned to engine and ladder companies, fire departments often maintain a selection of specialized vehicles, including off-road vehicles for brush fires, water tankers, hose wagons, foam pumpers, hazardous materials units, lighting trucks, breathing air supply trucks, and command vehicles. Staffing can be organized as individual companies or assigned to regular companies.

Fire apparatus can be ordered with a variety of options and configurations to suit the needs of a particular community or fire department. These options include aerial devices, water towers, and foam systems on engine company apparatus; high-volume pumps on ladder trucks; remote-control nozzles; and large electrical generators and air supply systems.

NFPA 1901, *Standard on Automotive Fire Apparatus*,[11] includes listings of equipment and appliances needed with various categories of fire apparatus. The standard specifies equipment for each type of apparatus which the manufacturer is required to furnish, as well as equipment required to be carried on each type of apparatus.

Apparatus must be equipped with the tools necessary to accomplish fireground operations. Where apparatus is delivered with only the minimum items of equipment, other equipment must be supplied as needed. The latest edition of NFPA 1901[11] should be consulted for up-to-date equipment lists. These lists generally are used when evaluating fire department equipment.

FIRE SERVICE SAFETY

Fire fighting has been recognized as the most hazardous occupation in North America, in terms of occupational death and injury statistics. In recent years, more than 100 line-of-duty deaths have been recorded annually among career and volunteer fire fighters in the United States alone. The statistics on fire service deaths and injuries compiled by NFPA are more than sufficient evidence to demonstrate the need for increased efforts to reduce this toll.

In addition to the direct line-of-duty deaths, there is growing concern with the number of fire fighters who suffer disabling injuries. Others develop occupational diseases and conditions that force them to discontinue their fire service activities and often have debilitating or fatal consequences. The link between respiratory and heart diseases and fire service careers has been well documented and established. There is growing evidence of a similar link to cancer and related diseases, derived from occupational exposure to carcinogens, toxic products of combustion, and hazardous materials.

The fire service is involved not only in fire suppression activities, but plays an increasing role in delivery of emergency medical and rescue services and in response to incidents involving hazardous materials. The fire fighter may be exposed to a wide range of dangers arising from these nontraditional activities that present another complex set of occupational health and safety hazards; there is a growing concern with the aspects of fire department activities and functions that are directly related to stress and to the emotional and psychological consequences of providing emergency services.

NFPA 1500, *Standard on Fire Department Occupational Safety and Health Program*,[12] became effective in August 1987. Before that time, there was no consensus standard for an occupational safety and health program in the fire service. Depending on governmental authority and legislative actions, a fire service organization may or may not be subject to mandatory occupational health and safety requirements.

The intent of NFPA 1500[12] is to provide the framework for a safety and health program for a fire department or any type of organization providing similar services. The standard addresses the basic organizational components that must be in place for an appropriate program approach to safety and health management, including the stated organizational commitment to provide a safe and healthy work environment. It defines basic roles and responsibilities, including those of member organizations, that are intended to focus efforts toward health and safety concerns and to facilitate cooperative action. The need for a fire department safety officer and the role of the occupational health and safety committee are established in the standard.

NFPA 1500[12] also addresses training requirements for individuals who may provide emergency services, as well as precautions to be taken during training. The standard requires that all personnel receive basic training before engaging in emergency scene operations.

Due to the number of reported deaths and injuries resulting from emergency vehicle accidents, NFPA 1500[12] establishes requirements for driver/operator training and for the operation of emergency vehicles. The standard prohibits riding in exposed positions and requires all personnel to ride in seats with seat belts or safety harnesses. In the case of new fire apparatus, it specifies that all seating must be provided in enclosed areas. Requirements for maintenance, inspection, and repair of vehicles, tools, and equipment also are included in NFPA 1500.

NFPA 1500[12] also mandates the provision and use of appropriate protective clothing and equipment, as well as what equipment must be used under various conditions. Included in the standard is a requirement for the mandatory use of positive pressure, self-contained breathing apparatus when personnel are exposed to respiratory hazards. These provisions interface with the NFPA standards that establish design and performance criteria for clothing and equipment.

NFPA 1500[12] requires that emergency operations be conducted in a standard manner that recognizes and provides for the inherent dangers of fire suppression and related activities. Roles and responsibilities are defined for incident command and safety supervision, including provisions that limit operations to those that can be performed safely with the available personnel and equipment. Where special dangers or hazards exist, the standard requires supervision of operations by qualified personnel, and backup crews to provide immediate assistance.

NFPA 1500[12] addresses areas of concern that have been recognized through analysis of fire fighter injury and disability statistics and attempts to establish realistic and necessary approaches to issues, including eye and face protection and hearing conservation. It further requires that all fire department facilities be maintained and inspected for conformance with all applicable health and safety standards, to provide for health and safety on par with any other workplace. An additional aspect of this standard is the delineation of medical, physical fitness, and employee assistance components that are directly related to health and safety in the fire fighter's work environment.

Provisions of NFPA 1500[12] apply to career, volunteer, mixed career and volunteer, part-time, private, military, and public sector organizations that engage in the activities normally associated with a fire department. This standard is meant to be appropriate for voluntary compliance, as a state-of-the-art document, whether or not it is adopted as a mandatory requirement by an authority having jurisdiction over a particular organization.

Protective Clothing and Protective Equipment

Each year, one of every two fire fighters is injured, and many die, in the line of duty. Often these mishaps occur because fire fighters are not sufficiently equipped with, or do not properly utilize, protective clothing and equipment. No equipment can guarantee a fire fighter's safety, but self-contained breathing apparatus (SCBA) and full protective clothing improve fire fighters' chances to carry out their jobs without suffering injury or death. The safety odds are even better if that equipment is carefully selected, properly maintained, and used at all appropriate times. In addition to protecting fire fighters, SCBA and protective equipment will make fireground operations more effective.

Injuries to fire fighters are costly to the community. Although the initial cost of good protective clothing may seem high, it is small compared with the expense of hospitalization or replacement of an injured fire fighter. Recruitment and training, medical treatment, disability payment, and early retirements are costly. The loss of a volunteer fire fighter also can disrupt a community's fire service effort.

The protective equipment that fire fighters wear while combating structural fires must be viewed as a system that includes SCBA and full protective clothing. Together, these must protect fire fighters from toxic fumes and gases, heat, moisture, puncture, impact, and electrical shock.

NFPA publishes several standards on fire fighter protective clothing and equipment, including:

NFPA 1971, *Standard on Protective Clothing for Structural Fire Fighting.*[13] As far as practicable, the standard was written as a performance standard rather than a manufacturing specification. The performance requirements cover garment and textiles, outer shell, moisture barrier, thermal barrier, thread, visibility, hardware, and labeling. The testing requirements include thermal protective performance (TPP); thermal shrinkage resistance; heat, char, and ignition resistance; tear resistance; and retroreflectivity.

NFPA 1972, *Standard on Structural Fire Fighters' Helmets.*[14] Intended to provide minimum performance criteria and test methods, NFPA 1972 looks at the helmet as part of a protection system. It requires all renewals, repairs, or additions of accessories to use parts approved by the helmet manufacturer whose product complies with this standard.

NFPA 1972[14] lists ten major performance requirements and test procedures dealing with impact force, impact acceleration, penetration resistance, heat resistance, flame resistance, resistance to electric current, effectiveness of the chin strap and suspension system, flammability of ear flaps, resistance of the face shield to heat and flame, and, finally, a test of the brightness and surface area of fluorescent markings. The tests are intended to be as realistic and representative of actual fireground conditions as could be duplicated in a laboratory.

It should be noted that the face shield does not provide complete face and eye protection against flying particles, splashes, gases, and vapors. For known eye hazards, such as those accompanying the use of power tools, additional specific protection to be used might be goggles or the SCBA facepiece.

NFPA 1973, *Standard on Gloves for Structural Fire Fighters.*[15] Like the other standards in this series, NFPA 1973 is a performance standard. It requires that gloves for structural fire fighters be made of durable outer material designed to withstand the effects of flame, heat, vapor, liquids, sharp objects, and other hazards that are encountered during structural fire fighting. The gloves also must be designed for minimum interference with physical movement and the use of tools, and to provide close-fitting wrist protection to prevent burns and injury in that area.

Major portions of the standard deal with performance requirements and test procedures. The performance requirements address resistance to conductive and radiant heat, water penetration, cut resistance, puncture resistance, dexterity, and grip. Labeling requirements for gloves that meet provisions of NFPA 1973[15] also are specified.

NFPA 1974, *Standard for Protective Footwear for Structural Fire Fighting.*[16] To comply with this standard, footwear must meet certain performance requirements dealing with resistance to puncture, flame, heat, abrasion, and electrical current.

Chapter 1 of this standard provides a picture of a combat-type boot to aid in understanding the nomenclature. This is not meant to exclude a rubber-type boot. Protective footwear that meets the criteria outlined must contain specific labeling, including the name and country of the manufacturer, size and width, model or stock number, and the lot or serial number.

NFPA 1975, *Standard on Station/Work Uniforms for Fire Fighters.*[17] This standard concerns station/work uniforms which, when worn under protective clothing, will not contribute to fire fighter injury and will not cause any degradation of the performance features of the protective clothing. The requirements also focus on the resistance of the uniform's material to flame and heat.

Dress uniforms are not intended to be worn under the fire fighter's protective clothing, but, if they are, they should meet the performance requirements of this standard. Likewise, undergarments should be of materials which do not compromise the intent of this standard.

NFPA 1981, *Standard on Self-Contained Breathing Apparatus for Fire Fighters.*[18] This standard sets performance requirements and test procedures designed to simulate various environmental conditions that SCBA can be exposed to during stowage and use. These performance criteria and test methods are in addition to the NIOSH/MSHA certification.

Performance requirements and testing includes:

- Positive pressure operation.
- Minimum rated service life of 30 minutes.
- Airflow of 100 liters per minute.
- Resistance to temperatures that the SCBA might be exposed to during normal fire fighting operations.
- Resistance to particulate matter, such as dust, commonly present during fire fighting operations.
- Resistance of the facepiece lens to scratches that could reduce the visibility of the fire fighter.
- Evidence that the SCBA does not significantly reduce the fire fighter's normal voice communications.
- Resistance to corrosion that would interfere with the performance and function of the unit.
- Resistance to damage from vibration and shock associated with being carried on fire apparatus.
- Resistance to damage of the harness assembly as a result of exposure to flame and heat.

The standard requires SCBA that meet all of the requirements to be labeled accordingly.

The fact that a SCBA meets the provisions of NFPA 1981[18] is no guarantee that it will not fail or that a fire fighter will not be injured. Even the best-designed SCBA cannot compensate for equipment abuse, lack of training, or improper maintenance. When using SCBA, fire fighters always should assume that the atmosphere is immediately dangerous to life or health. There is no way to predetermine hazardous conditions, concentrations of toxic materials, or percentages of oxygen in air in a fire environment. This applies during fire fighting, overhaul operations, or under other emergency conditions which may involve spills or releases of hazardous materials. Thus, SCBA are required at all times during operations involving fire fighting, hazardous materials, or overhaul.

NFPA 1982, *Standard on Personal Alert Safety Systems (PASS) for Fire Fighters*.[19] The purpose of the PASS device is to provide an automatic audible warning signal should a fire fighter become incapacitated or need assistance while operating at an emergency. This standard specifies minimum performance criteria, functioning, and test methods for PASS devices to be used by fire fighters engaged in rescue, fire fighting, and other hazardous duties.

The performance requirements include motion detection, signal types and strength, retention system, and intrinsic safety while the testing covers environmental, radiant heat, flame, impact, retention, and sound level pressure.

NFPA 1983, *Standard on Fire Service Life Safety Rope, Harnesses, and Hardware*.[20] This standard specifies the design and construction, performance requirements, and testing for rope, and the associated hardware and harness, used by the fire service to support fire service personnel and civilians during rescue, fire fighting, or other emergencies or during training.

All of the above standards set forth minimum requirements for the particular item described. The standards are not intended to serve as detailed manufacturing or purchasing specifications. NFPA does not "approve" specific products, but publishes standards for these products, developed in accordance with the established Regulations Governing Committee Projects.

Summary

Fire protection is largely a local responsibility responding to the specific considerations and needs of the community the fire department serves. A fire defense system that works well in a rural area may not be appropriate to fire protection needs in an urban area.

For many years, improvements in fire protection service have followed the recommendations of the ISO *Fire Suppression Rating Schedule*, formerly the *ISO Grading Schedule*, used to assign community insurance ratings. The municipal *ISO Grading Schedule* and the *Fire Suppression Rating Schedule* used by insurance companies to develop proper base rates for the community have greatly influenced the strength of fire departments and water supplies in our cities, and has been largely responsible for upgrading municipal water supplies.

Every community should have a masterplan for fire protection. The concept of masterplanning for present and future needs involves many additional factors and considerations not addressed by the *ISO Fire Suppression Rating Schedule*. Fire departments that have been relying solely on the *ISO Fire Suppression Rating Schedule* will find that the masterplan involves attention to many more factors and calls for custom-tailoring future priorities to meet local needs.

As communities undertake a basic reassessment of their fire services, they will have to find solutions best suited to their conditions. Some communities are at an early stage of growth where they can consider a number of alternatives to their present system of fire protection. Others have a heavy investment in their present system and can consider only a gradual shift of priorities. For most communities, improving the effectiveness of the fire service calls for gradual changes within the present structure: a shift of priorities toward fire prevention, better deployment systems, and revised and improved management practices. Still other communities will want to consider a major shift from their present system.

Water systems and water supplies needed for fire protection are major considerations to be evaluated when assessing the level of fire protection service available in a community. A community water system must be designed to provide adequate water for fire fighting while at the same time meeting the maximum anticipated consumption needs for all domestic and industrial purposes. An ideal water supply would come from a good watershed into a large impounded lake where it would flow by gravity into the community's distribution system. In many systems, however, pumps must be used in order to provide proper working pressures in the mains.

References

[1]NFPA 1410-1979. *A Training Standard on Initial Fire Attack*, National Fire Protection Association, Quincy, MA.

[2]*Fire Suppression Rating Schedule* 1980. Insurance Services Organization, NY.

[3]Coleman, R.J. 1980. "Service and Use Fees Place the Burden on Users," *The International Fire Chief*, Vol. 46, No. 7, July.

[4]Shedd, J. Herbert, 1889. Discussion on a paper by William B. Sherman, "Ratio of Pumping Capacity to Maximum Consumption," *Journal of New England Water Works Association*, Vol. 3.

[5]Fanning, J.T. 1892. Distribution Mains and the Fire Service. *Proceedings of the American Water Works Association*, Vol. 12.

[6]Kuichling, E. 1897. "The Financial Management of Water Works," *Transactions of the American Society of Civil Engineers*, Vol. 38.

[7]Freeman, John R. 1892. "The Arrangement of Hydrants and Water Pipes for the Protection of a City Against Fire," *Journal of the New England Water Works Association*, Vol. 7.

[8]Metcalf, L., *et al* 1911. "Some Fundamental Considerations in the Determination of a Reasonable Return for Public Fire Hydrant Service," *Proceedings of the American Water Works Association*, Vol. 31.

[9]ASCE 1951. "Fundamental Considerations in Rates and Rate Structures for Water and Sewage Works: A Joint Report of Committees of the American Society of Civil Engineers and the Section of Municipal Law of the American Bar Association and of Representatives of the American Water Works

Association, National Association of Railroad and Utilities Commissioners, Municipal Finance Officers Association, Federation of Sewage Works Association, American Public Works Association, and Investment Bankers Association of America." ASCE Bulletin No. 2. American Society of Civil Engineers, NY.

[10]NFPA 20-1987. *Standard for the Installation of Centrifugal Fire Pumps*, National Fire Protection Association, Quincy, MA.

[11]NFPA 1901-1985. *Standard on Automotive Fire Apparatus*, National Fire Protection Association, Quincy, MA.

[12]NFPA 1500-1987. *Standard of Fire Department Occupational Safety and Health Program*, National Fire Protection Association, Quincy, MA.

[13]NFPA 1971-1986. *Standard on Protective Clothing for Structural Fire Fighting*, National Fire Protection Association, Quincy, MA.

[14]NFPA 1972-1985. *Standard on Structural Fire Fighters' Helmets*, National Fire Protection Association, Quincy, MA.

[15]NFPA 1973-1983. *Standard on Gloves for Structural Fire Fighters*, National Fire Protection Association, Quincy, MA.

[16]NFPA 1974-1987. *Standard for Protective Footwear for Structural Fire Fighting*, National Fire Protection Association, Quincy, MA.

[17]NFPA 1975-1985. *Standard on Station/Work Uniforms for Fire Fighters*, National Fire Protection Association, Quincy, MA.

[18]NFPA 1981-1981. *Standard on Self-Contained Breathing Apparatus for Fire Fighters*, National Fire Protection Association, Quincy, MA.

[19]NFPA 1982-1983. *Standard on Personal Alert Safety Systems (PASS) for Fire Fighters*, National Fire Protection Association, Quincy, MA.

[20]NFPA 1983-1985. *Standard on Fire Service Life Safety Rope, Harnesses, and Hardware*, National Fire Protection Association, Quincy, MA.

Additional Reading

Bosewell, C.R. 1978. *Standards for Rural Fire Protection*, Department of Extension Forestry, Kansas State University, Manhattan, KA.

Didactic Systems, Inc. 1977. "Fire Prevention Activities," *Management in the Fire Service*," National Fire Protection Association, Quincy, MA.

Federal Emergency Management Agency 1984. *The Integrated Emergency Management System: Process Overview*, U.S. Government Printing Office, 1984-445 004/18624, Washington, DC.

Granito, J.A. 1983. "Trends in Fire Service Management," *Fire Service Today*, August.

Hickey, Harry E. 1973. *Public Fire Safety Organization*, National Fire Protection Association, Quincy, MA.

Houlihan, John C. 1977. *Alternatives to Traditional Public Safety Delivery Systems: A Tale of Two Cities*, Institute for Local Self-Government, Berkeley, CA.

290 PRINCIPLES OF FIRE PROTECTION

International City Management Association 1979. *Managing Fire Services*, ICMA, Washington, DC.

ISO 1974. *Grading Schedule for Municipal Fire Protection*, Insurance Services Office, NY.

Master Planning for Fire Protection 1980. Federal Emergency Management Agency, U.S. Fire Administration, Washington, DC, March.

"National Association of Counties, Multi-Jurisdictional Fire Protection Planning," prepared for the Federal Emergency Management Agency, U.S. Fire Administration, Washington, DC.

National Fire Prevention and Control Administration (now U.S. Fire Administration) 1978. *A Basic Guide for Fire Prevention and Central Master Planning*, U.S. Dept. of Commerce, Washington, DC.

NCFPC 1973. "America Burning" The Report of the National Commission on Fire Prevention and Control, U.S. Government Printing Office, Washington, DC.

Research Triangle Institute, *et al* (no date). "Evaluating the Organization of Service Delivery," Fire, Research Triangle Institute, Center for Population and Urban-Rural Studies, Durham, NC.

Research Triangle Institute, *et al* (no date). *Municipal Fire Service Workbook*, Government Printing Office, Washington, DC.

Chapter 11

Fire Department Organization, Administration, and Operation

Fire departments protect a large number of people and properties, and their organization and operation play a vital role in fire waste control. Fire department administration and management is responsible for maintaining highly trained and efficient operational units for the performance of effective tactical and nontactical operations.

FIRE DEPARTMENT ORGANIZATION

There are approximately 30,000 fire departments in the United States. About 4500 of these fire departments protect cities and towns having a population of 10,000 or more. About 2 percent of the fire departments protect larger communities (50,000 population or greater). The remaining departments operate in cities and towns with populations of less than 10,000, with nearly 6 percent of all fire departments protecting populations of less than 2,500.

It is estimated that there are about 238,500 paid fire fighters in American fire departments and 839,500 volunteer fire fighters. Fire departments comprise a large operation that involves great numbers of people and large expenditures of funds.

The history of fire departments is long and colorful. Up until a hundred years ago fire fighting, even in our larger cities, was accomplished by volunteers. Records indicate that the first big city fire department with fully paid fire fighters was established in Cincinnati in 1853. The apparatus for this department was a horse-drawn steamer. Self-propelled steamers and aerial ladders appeared about 1870, and the automotive fire department pumper dates from 1910. The first NFPA standard on automotive fire apparatus was adopted in 1914.

Fire Department Objectives

The four traditional objectives commonly accepted by most modern fire departments are:

291

1. To prevent fires from starting.
2. To prevent loss of life and property when fire starts.
3. To confine the fire to the place where it started.
4. To extinguish fires.

Whether documented or implied, these are likely the only goals of many fire departments. All four goals are presented here as broad, general statements that are not definitive in terms of achievement or performance. To be more meaningful, performance-oriented objectives should be developed to support the stated goals.

Each fire department, regardless of size, should develop performance objectives that specify the results expected and the time required for achievement. The following material explains a system for developing such objectives and their related enabling objectives. This procedure and the examples given are a brief overview for illustrative purposes. Those desiring to use this concept are advised to consult specific texts on the subject of management by objectives.

Developing Performance Objectives

The first step in developing performance objectives is to determine the purpose of the organization, i.e., why it exists. This should be done not only for the department as a whole, but for each subordinate operating division or section. When determined, these items should show the relationship between the department and each of its operating divisions, and the relationship among the divisions. For example, lists with standard headings such as: "The fire department exists for the purpose of ...," "The fire suppression operations divisions exists for the purpose of...," "The fire prevention division exists for the purpose of...," etc., can be used.

The next step is to develop a list of responsibilities. Each operating division should be asked to develop a list of its current responsibilities and activities. The responsibilities listed should be detailed and specific. Once complete, these lists should be examined to determine where there is overlap or deficiencies in areas of responsibility.

The third step is to write a series of statements that describe desired goals. These statements should describe in detail the definite and measurable goals that a department or division would like to achieve. It is important that the statements be realistic and achievable. Statements that are beyond the current scope or resources of the department may be meaningless unless they are properly labeled as long-range goals. Most departments or operating divisions should prepare not one, but a series of statements. Following are two examples of such statements:

1. *Fire suppression operations division*—Each company will conduct an inspection of all target hazards in their first-due area.
2. *Fire prevention division*—Work with the board of education to implement the NFPA *Learn Not to Burn Curriculum*[1] at the elementary school level.

Once the divisions have determined goals, an evaluation should be made by comparing what is currently being done with what has been proposed. This comparison might indicate the need for a decision concerning present status and desired goals. Perhaps the extra effort needed to achieve the desired goals may not be productive, and in many cases there might be gaps between present and desired levels of performance. It is in these areas that priorities should be assessed to establish realistic objectives.

In the example for the fire suppression operations division, the stated goal is that all companies will conduct inspections of all target hazards in their first-due area. Are the companies currently conducting inspections? If not, and if it has been decided that all companies will accomplish this goal, then enabling objectives that explain how to achieve this goal must be established.

Enabling objectives must be measurable, and have a standard of performance established, and a definite time period for accomplishment. If the enabling objectives are not expressed in quantitative terms, they will be of little value since there will be no means of measuring the performance needed to achieve the overall objectives or goals.

Management by Objectives

Once specific objectives have been established, consideration should be given to the possibility of developing a more functional operating system. This requires that the specific steps necessary to achieve results be identified, problem areas determined, details of activity planned, and a time sequence programmed. It is important that time deadlines be set at various points so that programs and procedures may be monitored and schedules met.

The establishment of a more efficient operating system necessitates a determination of whether or not the present system is adequate. Based upon this decision, changes in the present system may be necessary. Using the previous fire suppression operations division as an example, if it has been decided that all companies are to conduct inspections of target hazards in their first-due area, the following questions might be asked:

1. Does the present system have the capacity required?
2. Have the fire fighters been trained in inspection techniques? If not, what training is required?
3. Does the training division have the time and resources to conduct this training?
4. How long will the training take?

The answers to the preceding questions will help fill any existing gaps in the present system, and identify key areas that need improvement.

Next, specific enabling objectives for any changes and improvements should be written. The enabling objectives should be expressed in terms of how these changes and improvements are to be accomplished, and should be used to revise the present operating system or establish a new system.

Before the overall program can be completed, objectives for individual positions within the system must be developed. These objectives must also be specified in terms that are performance oriented and measurable, with a time limit established for their completion. In this way, each individual will know what is expected and the amount of time allowed for completion of each activity.

Examples of two individual objectives for the fire suppression inspection project (involving more than one division) are as follows:

1. *Fire suppression*—Each company officer will arrange for and ensure that the target hazards in each first-due area are inspected starting in January of the next calendar year. Each company will submit weekly progress reports on the inspections completed, and the completed inspection forms. All inspections will be completed prior to the last day of March.

2. *Training*—The training officer will design and develop a training course for suppression personnel on the techniques of inspecting target hazards. The training course will include all items necessary for the identification of potential hazards and recommendations for their elimination based on the local fire prevention code and nationally recognized good practices. Progress reports on course development will be submitted weekly to the chief of training. All work will be completed prior to the last day of October.

Finally, it is necessary to monitor results at timed intervals as specified in the objectives. Monitoring provides an opportunity for evaluation, and documents activity. There may be instances during evaluation when it becomes evident that some of the established objectives are unrealistic (depending upon the department's capacity) and may require change.

Managing an operating system by the use of performance objectives is not a static process. If good objectives are established, the process is dynamic and continuously changing. Constant feedback and realignment, change, and addition of objectives are required.

Fire Department Structure

The organization of fire departments takes many forms. The purely volunteer department often operates independently and raises its own funds at fairs, carnivals, and by public subscriptions. Many volunteer fire departments now receive contributions of funds and equipment from the local government of the community. In the larger cities, the fire department is one division of the local government, usually with the fire chief directly responsible to the chief administrative officer of the city. There are a growing number of regional fire departments (such as fire districts, fire protection districts, and county fire departments). Fire districts are organized under special provisions of state laws with their own governing bodies, and are supported by a district tax levy. Fire protection districts are set up to contract for fire protection from a nearby fire department, and are supported by taxes and property owners in the area.

County fire departments are now found in a number of metropolitan areas. Numerous small communities in the county combine to maintain a large, professionally administered public fire department with countywide communications and fire prevention services. The kind, size, and makeup of the local fire department will, of course, depend on the financial resources available, the frequency of fires to be expected, the area to be protected, and so on.

Larger fire departments are typically staffed by all career personnel. Departments that protect smaller populations may be staffed by a combination of career and noncareer personnel. The noncareer personnel may receive some compensation for their time or may truly volunteer their time. The more rural areas and small population communities generally have totally noncareer personnel.

Principles of Organization: Much has been written about fire department organization and organizational principles. The following principles are generally considered to be universal and mandatory.

One of the most basic principles is that work should be assigned to individuals and units based on a careful, well-arranged plan. Another important principle relates to the size of the department. As the department grows as a result of more complex community needs, careful consideration must be given to coordination of the various departmental and divisional functions.

Another well-known principle involves lines of authority. Individuals should be aware of their relationship to the total organization. In the same manner, each operational unit should also be aware of its place and function in the total organization. Additionally, it is important that when responsibility is given to an individual or to a unit, the authority to carry out that responsibility must also be clearly given. In this way, when authority is necessary to complete and carry out a given responsibility, it is there.

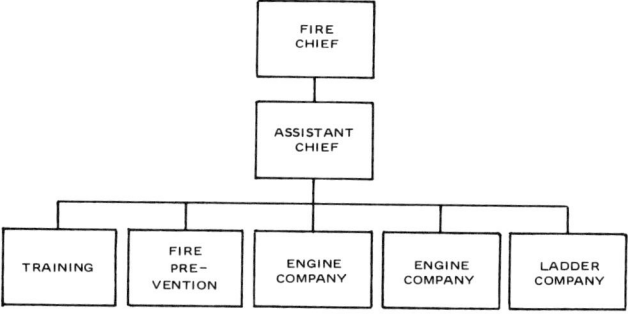

Fig. 11.1   Typical organizational structure of a small-sized fire department.

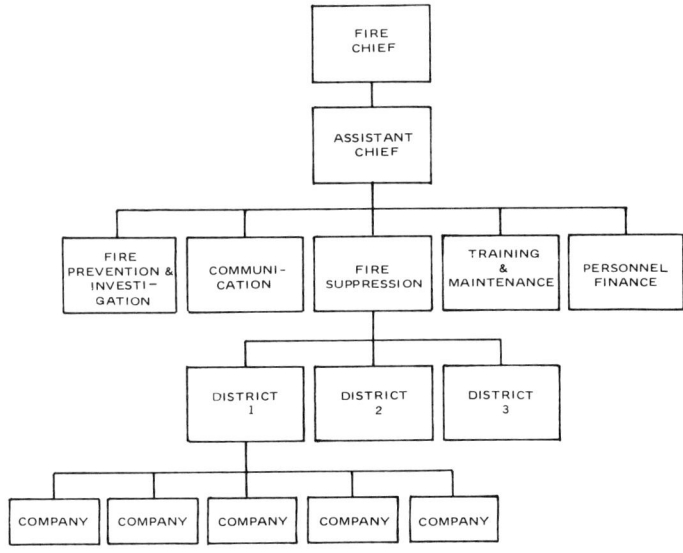

Fig. 11.2   Typical organizational structure of a medium-sized fire department.

Unity and clarity of command and command levels are important principles of fire department organization. When an individual receives conflicting orders from several superiors, or when a superior has too many individuals and functions to coordinate and supervise, poor organization is reflected in the general confusion that usually results.

Typical organization charts for small, medium, and large fire departments are shown in Figures 11.1, 11.2, and 11.3. The evolution from a small organization to a larger organization follows the growth and congestion of the area. There is no set pattern, however, as to when the changeover takes place. Local circumstances and fire experience of the area will dictate the changeover.

Line Functions: Line functions in fire departments normally refer to the activities directly involved with fire suppression operations. Officers directing fire suppression are primarily considered line officers. This does not mean, however, that these officers do not have other functions. As these officers are promoted to higher levels within the department, their line responsibilities may be equally divided with staff responsibilities. At the highest officer levels within the department, line responsibilities diminish, while staff responsibilities increase.

Staff Functions: Staff functions are those activities that do not involve fire fighting. When the department is divided into divisions or bureaus (such as fire prevention, training, communications, maintenance, and personnel), a

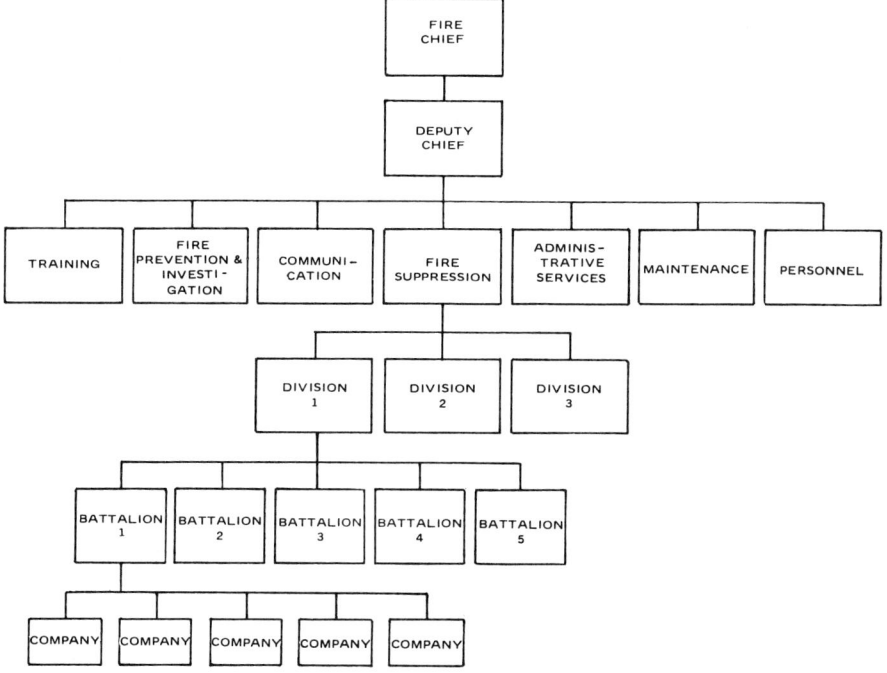

Fig. 11.3 Typical organizational structure of a large-sized fire department.

staff officer will usually be assigned to supervise each such division. Staff officers are normally not involved in line functions.

Organizational Plans: The manner in which fire departments are organized is dependent upon the size of the department and the scope of its operations. Organizational plans are designed to illustrate or show the relationship of each operating division to the total organization. It is essential that each fire department have available an organizational plan that reflects the current status and functions of the department.

A list of responsibilities or a job description for each position should accompany the organizational plan. In small departments, a single individual may have responsibility for more than one function. For example, a single officer may be responsible for both training and maintenance. This should be detailed in the job description.

The organization chart should show how the various functions that may demand time and support from other personnel groups will be coordinated within the fire department. Both the personnel within the ranks of the fire department and the public need to see a clear coordinated effort of providing fire protection to the community.

Rules and Regulations

As with any organization, rules and regulations are needed to govern the operations of that organization. This is especially true in the fire service due to the hazardous nature of much of the activity, and the need for a clear understanding of expected performance.

Every fire department should have a set of rules and regulations which outline performance expectations for its members, the standard operating procedures for the department, and disciplinary action which can be taken against personnel not following the regulations. These rules and regulations can be, and often are, supplemented by orders from the fire chief who may supplement or clarify the rules or change them for a special event or specific purpose. Both the rules and regulations and subsequent orders from the chief should be written and distributed in such a manner as to ensure all involved persons are properly made aware of them.

ADMINISTRATION AND MANAGEMENT

The management of a fire department should be much the same as that of any well-run business establishment. The governing body is normally that of the municipality with the power to levy taxes for the support of the fire service, to own property such as fire stations, and to pay the personnel. The fire chief is the manager of the department and is usually appointed by the mayor, city manager, or chief administrative officer of the city or town.

Function of Management

Fire department management is responsible for maintaining highly trained and efficient operational units to perform assigned tasks both in the prevention and suppression of fires. Fire department management is generally the responsibility of the local government that supports the service. As with any governmental or business operation, fire department management involves three major areas of responsibility: (1) fiscal management, (2) personnel management, and (3) productivity.

In general, fiscal management practices follow those used by the government agency supporting the department and includes budgeting, cost accounting, personnel costs (including payroll), and purchasing or procurement costs. The degree varies to which these factors are a direct responsibility of fire department management, depending upon the practices of local government.

Fire departments utilize persons with specialized skills who are organized into various operational and staff units. Fire department management is involved to some degree in the recruitment, selection, and promotion of personnel needed to fill various positions in the organization. Largely, these matters are governed by local and/or state law; by personnel agencies, including civil service authorities; and by direct decisions of the governmental agency operating the fire department. The assignment of available personnel to positions provided in the budgeted organizational structure, and supervision of personnel performance, are normally the direct responsibility of the fire department management, although certain assignments are frequently governed by work contract agreements.

Productivity in the fire service is the most difficult ingredient for management to measure. The basic objective of the fire service is the protection of life and property. Modern fire service practice involves two major activities: (1) control of hazards to minimize fire losses and to prevent fires, and (2) dealing with actual fires and emergencies to minimize suffering and losses. It is difficult to assess the number of fires and suffering that have been prevented by fire department activities; however, experience has demonstrated that lack of effective fire prevention and control measures invites disastrous experiences. Likewise, the fact that most fires are suppressed with minimum losses and injuries does not indicate conclusively that an adequate level of fire department service has been provided. Experience shows that major fires and emergencies often arise from combinations of circumstances beyond the immediate control of fire department management, but which must be dealt with effectively to protect the public. It is imperative that fire department management be concerned with maintenance of reasonable standards of organization and highly trained and efficient operational units to perform assigned tasks in both the prevention and suppression of fires.

PERSONNEL

The recruitment of personnel is seldom the responsibility of a municipal fire department, except in those cases where there is no local governmental personnel agency. Fire districts and volunteer departments recruit their own members.

It is the responsibility of fire department management to notify the personnel agency of existing vacancies in the organization and to request the number of persons needed to fill these vacancies. In connection with recruitment, fire department management has three responsibilities. The first is to recommend appropriate recruitment standards to the personnel agency. The second is to provide the basic training necessary for the new personnel so they can properly perform their assigned duties. The third is to certify, after providing the basic training, that the new members are ready for appointment as permanent fire fighters or, where individuals prove unable to perform satisfactorily, to recommend that their services be terminated before permanent appointment.

Selection of personnel must meet local, state, and federal standards. U.S.

courts have ruled previously that there must be no discrimination in hiring practices. Some rulings prohibit residency requirements for recruitment, although fire department rules of employment may stipulate that because of the emergency nature of the work employees must reside within a reasonable distance of the community. One court decision has ruled out examinations that require knowledge of fire department practices and equipment prior to appointment and in-service probationary training.

Fire Fighter Qualifications

It is imperative, for obvious reasons, that all persons functioning in the fire service be fully qualified and capable of efficiently performing the wide range of services necessary to protect life and property. Many states have enacted legislation establishing commissions on fire fighter standards which require that all personnel employed by fire departments must satisfactorily complete required basic training before being given permanent employment.

In 1970 the Joint Council of National Fire Service Organizations (JCNFSO), consisting of ten national organizations directly involved in various aspects of the fire service, recommended that national fire service professional qualification standards for fire fighters be developed through NFPA technical committee procedures. The Joint Council established a National Professional Qualifications Board (NPQB) to supervise a nationally coordinated continuing professional development program for the U.S. fire service. The Board has nine members appointed by the Joint Council. The Board agreed that the desired professional qualifications standards should be developed through the NFPA standards-making system, and that the secretariat for the committees and the Board would be provided by the NFPA staff. The Board reviews all draft standards before these are submitted to NFPA for final adoption.

Four technical committees have been organized to develop professional qualification standards. These are the Fire Fighter Qualifications Committee, the Fire Inspectors and Investigators Qualifications Committee, the Fire Service Instructors Qualifications Committee, and the Fire Service Officers Qualifications Committee. Standards have been developed by these committees covering the following positions in a fire department:

- Fire fighter I, II, and III
- Fire apparatus driver/operator
- Airport fire fighter
- Fire fighter medical technician
- Fire officer I through VI
- Fire inspector I, II, and III
- Fire investigator I, II, and III

- Public fire educator I, II, and III
- Fire instructor I through IV

All of the standards are expressed in measurable performance or behavioral objectives covering both required knowledge and demonstrated skills. The standards are prepared for use as a basis for nationally standardized examinations by authorized agencies, and are available for adoption by federal, state, and local authorities in the U.S.

The establishment of standards and testing procedures does not ensure that all personnel will achieve the required level of competency. Training programs are necessary to prepare members of the fire service to acquire the skills and knowledge necessary to achieve the terminal performance objectives set forth for each position.

Recruitment: Paid fire fighters are not, as a rule, recruited directly by the fire department but by a municipal personnel or civil service agency. Recruitment is subject to federal, state, and local employment standards. Local regulations with regard to age, physical fitness, and education can vary widely.

In many states the Civil Service Commission maintains a roster for individuals who wish to join a paid fire department. To become eligible, a candidate usually must pass both a written examination and a physical examination. Those who qualify are usually entered on the roster in order of their test results. Several names from the top of the list are submitted to a fire department for specific selection. Either a personnel officer, the company officer, or a small selection team must then choose from these candidates the one who is best qualified to fill the position.

The U.S. Supreme Court decisions based upon constitutional prohibition of sex discrimination in employment now has enabled women to apply for fire fighter jobs. In connection with both racial and sexual discrimination, courts have ruled that height and weight requirements are discriminatory. However, candidates may be required to demonstrate their ability to perform the required duties.

Fire fighting requires a major degree of physical strength, and an important factor is the interdependence of fire fighters on each other in fire suppression and rescue operations. If there is a great disparity of physical strength and endurance among crew members, then an unreasonable burden and strain is placed upon those with the most strength and stamina. This is not only dangerous to the fire fighter, but also adversely effects the protection of the public.

In some jurisdictions, the services of the fire department training division may be utilized in testing recruits and conducting promotional examinations. In all such cases, recognized standards, such as NFPA 1001,[2] *Standard for Fire Fighter Professional Qualifications*, should be followed carefully so that the results will not be subject to valid charges of discrimination, and qualifications essential to the work will be tested adequately.

Promotional Practices

In the vast majority of fire departments, promotions to various officer ranks are made from personnel serving in the next lower rank or ranks, although more fire departments are recognizing the potential benefits of allowing lateral entry, transfers, and promotions of well-qualified personnel from other areas and departments. Promotional procedures are designed to take into account technical qualifications for the particular rank and fire department experience. It is essential that examination procedures in the civil service be fully competitive and nondiscriminatory. In general, as with initial appointments, promotional procedures are administered by personnel departments or by state or local civil service authorities.

Some promotional practices include an oral interview as part of the process. This can easily become a vehicle for discrimination on the basis of race, color, or mannerisms, and few such oral interviews are scientifically designed or professionally administered. Therefore, increasing emphasis is being given to requiring candidates for promotion to take a written exam regarding the required technical knowledge.

In recent years some fire departments have relied upon assessment centers as a means of selecting candidates for promotion. The assessment center requires the candidate to demonstrate certain abilities through the use of problem-solving exercises, role playing, and other simulated exercises. Each person is observed by a trained assessment team and scored according to performance. Some feel that the assessment center is the most realistic means of determining a candidate's suitability for a particular position.

NFPA 1021, *Standard for Fire Officer Professional Qualifications*,[3] specifies the levels of performance for fire officers. NFPA 1021, written in performance terms, requires an individual to demonstrate competence by knowledge and performance. More fire departments are including educational requirements, such as a community college fire science certificate for all officers, or a bachelors degree for chief officers. It is not uncommon to find fire departments that may seek individuals with graduate-level degrees in public administration/management for fire chief positions.

The fire department administration, as an arm of the municipal administration, does have an important role to play in the promotional process. First, it must advise the personnel agency as to the qualifications required in any job or rank to be filled, where such qualifications have not been previously established. Second, when a list of successful candidates for promotion is received from the personnel agency, the agency should advise the promoting authority regarding the promotions to be made. The usual practice is to fill vacancies from the top of the promotional list, except where the head of the fire department specifies in writing valid reasons for rejecting an individual.

Such reasons might be a record of serious disciplinary problems, including disobedience of written orders, frequent bad judgment when performing assigned duties, a record of conflicts with other employees, and other major personality problems. While such problems might not be serious enough to warrant severance from the present level of employment, they do indicate that the particular candidate, although technically qualified, would be less preferable than another candidate on the list. Personnel records should be available to substantiate any such reasons for rejection.

STAFFING PRACTICES

Staffing levels for fire departments vary considerably and are influenced by such things as population protected, population density, hours per work week of fire fighters, response distances, and fire fighter safety. Generally, fire departments utilize a three- or four-person platoon system which will accommodate a 42- to 56-hour work week, respectively. A four-platoon system requires about 25 percent more personnel than a three-platoon system. Staffing levels for major metropolitan cities within the U.S. range from 1 to 3 fire fighters per thousand population with an average of 1.5 per thousand. Communities must assess their needs to determine the level of staffing that meets their requirements. It has been demonstrated, however, that when staffing falls below 4 fire fighters per company, critical fireground operations are not carried out when needed. Tests conducted in 1984 with the Dallas, Texas Fire Department indicated that staffing below a crew size of four can overtax the operating force and lead to higher losses.

When fire departments operate emergency medical service and rescue squads, additional personnel are needed in order to maintain basic fire company strength. In some of the smaller communities, the staffing ratio per population protected may be relatively high because of the need for sufficient on-duty personnel for effective initial attack and rescue operations, especially in "bedroom communities" where call personnel are not readily available during the work day and where there are high-value properties to be protected that may be more significant than population ratios in determining the number of fire fighters to be provided.

In many core cities, as well as suburbs, the fire departments must protect substantial concentrations of values that exceed the average values related to populations. For example, a core city of 80,000 persons may be the business center for an area of 500,000 persons, and house a high percentage of the low-income groups. The number of high-rise and large-area structures to be protected and the frequency of alarms for fires and emergencies should be considered in determining on-duty fire department staffing.

Some very large fire departments may operate with a lower relative strength per 1,000 population than cities of a more average size, because with high population densities these departments have sufficient companies to provide needed coverage while handling working fires. For example, a large city fire department may operate one engine company per 15,000 to 20,000 population and still have a large number of well-distributed fire companies, whereas a city of 30,000 persons could not be properly protected with only two engine companies.

Mutual aid plays an important role in providing additional resources. Almost all jurisdictions rely to some extent on mutual aid from surrounding areas to provide fire fighting resources on a routine or major emergency basis. Some departments use automatic mutual aid on initial response. Even large cities are making increased use of both regularly assigned and automatic mutual aid. Often this is practical because companies from neighboring fire departments may be much nearer to a fire location than some of the assigned fire companies.

Frequently it is impossible for small cities to fully staff all of the fire companies needed for proper distribution of companies throughout the community to handle working fires. In many cases, the population density and the values protected per square mile are relatively low. In such communities, some engine companies may respond with only three persons on duty, and ladder companies with only two. Such low levels of staffing should be backed up promptly by off-shift or call personnel, or by multiple-alarm response to ensure adequate personnel. Combination fire departments which utilize a mix of career and either paid on-call or volunteer fire fighters are found throughout the U.S. In some cases, additional apparatus may be assigned to respond, offsetting deficient company strength. In communities with a large geographic area having relatively low concentrations of value, this may be an acceptable arrangement. However, in general, a minimum of four fire fighters on duty, including an officer, should be provided for each engine company where there is no assigned off-shift or call personnel on first alarms.

INTERGOVERNMENTAL RELATIONS

Fire departments are but one agency of local government, and much of their success depends upon their working relationships with other local, state, and federal agencies. Some of the more important contacts are mentioned here.

Building Department

Proper construction and arrangement of buildings is essential to a sound fire protection program.

State laws and local ordinances, or agreements between fire and building departments, increasingly require written approval by the head of the fire department concerning specified fire protection features before building permits can be issued. Also, close cooperation is needed between these departments to control serious fire hazards which commonly are present while buildings are under construction, before the required fire resistance or protection features have been installed. In small communities the fire chief generally must handle this assignment, but in many fire departments it is one of the responsibilities delegated to the fire prevention bureau.

Law Enforcement

Cooperation between the fire department and law enforcement officials is essential. Regular law enforcement response to fire alarms is necessary to control traffic and crowds. It is also important for fire and law enforcement agencies to develop coordinated plans in the event of an incident which requires the evacuation or closure of areas. Law enforcement and fire officials also need to develop working relationships to deal with fire investigations. In many areas a combination of fire/law enforcement fire investigation teams has been developed. In these cases, both agencies work together and bring their special expertise to a coordinated effort in cases where arson is suspected.

Water Department

Adequate water supplies, including hydrant service, are essential for fire fighting. A knowledgeable fire officer should be assigned to maintain liaison with the water authority and to develop water flow requirements of the fire department for various areas and types of property. In some communities the fire department is responsible, by ordinance, for determining the location and setting of hydrants. Where such ordinances do not exist, the fire department should still be involved in locating fire hydrants. It is important that all hydrants be serviced regularly and after each use, especially in cold weather to prevent ice formation; the fire department should promptly report all hydrants it has used in a particular incident to the water department. Alert fire departments maintain a list of hydrants in each fire company inspection district with flow data on each hydrant. Hydrants should be properly marked for flows and painted for nighttime visibility.

Personnel Department

Members of career fire departments are public employees, and as such their recruitment and promotion may involve cooperation with the personnel agency, which in some jurisdictions is responsible for conducting entrance and promotional examinations. A fire department officer should be assigned to maintain liaison with the personnel office.

Finance Department

Fire departments need to work closely with finance officials when developing and administering budgets. Budgeting is generally very complex and it is important that fire officials prepare budget documents correctly. After budgets have been prepared and adopted, it is necessary to assure that proper records of expenditures are maintained and that spending is kept within the adopted budget. Although some fire departments may have a finance officer, liaison with the finance department may be handled by designated staff officers.

Purchasing

All purchases exceeding stipulated amounts must be made according to specifications, and usually with competitive bidding by the purchasing department. Close liaison is necessary to assure that specifications are properly drawn to meet fire department needs, and that when bids are opened any proposals that deviate from specifications are rejected. This liaison may be handled directly by a staff officer.

Data Processing

Increasingly, fire departments are utilizing electronic data processing for keeping fire records, payroll records, and for statistical analysis. Each fire department should have persons knowledgeable in the use of data processing. It may be desirable to appoint one officer as coordinator of this activity, but commonly there would be an administrative committee in the fire department which would include representatives of plans and research (where provided), administration, fire prevention, and fire suppression. This same committee may also be involved in long-range planning for the department.

Planning

Fire departments, particularly those in rapidly growing areas, should maintain a close working relationship with local and regional planning groups. These agencies can provide valuable information on growth patterns that will affect the resources of the department. The plans, studies, and reports prepared by the planning agencies can be used to determine the need to increase personnel, add equipment, and expand station facilities.

RESOURCE UTILIZATION

The principal resource of a fire department is its highly trained personnel. The vast majority of personnel are assigned to the fire fighting division, and possibly two percent to three percent of the personnel are assigned full time to the fire prevention bureau. Thus, for the most effective resource allocation, maximum use must be made of the fire fighting personnel through careful time utilization schedules assigning appropriate allocations of work periods to apparatus and equipment maintenance, fire service training, and scheduled inspections. Such programs require close coordination among the chiefs of the fire fighting, training, and fire prevention bureaus or divisions. In well-managed fire departments, most of the routine fire prevention inspections are conducted by fire companies in their assigned inspection districts. Time utilization studies in some fire departments have shown that undue amounts of time were being wasted in common janitorial duties that could be done more cheaply under contract or by less-skilled labor. In many cases only the fire chief is available on a regular basis to handle all staff duties, and, accordingly, many of the management functions outlined in this text tend to be neglected. Personnel are not fully utilized when they are not given management responsibilities for the various programs within the fire department.

Considering that 20 percent or more of all fire department personnel are officers, there should be ample, technically qualified managerial help available even in a small fire department who can contribute to maximum efficiency and productivity.

FIRE DEPARTMENT OPERATIONS

The basic organization and orientation of almost all public fire departments is primarily directed toward fire suppression and emergency service delivery. The most realistic appraisals of such orientation, in almost every case, reaffirms the need for the fire department to respond to and control fires as they occur. In many cases this role has been extended to include the delivery of additional emergency services to deal with situations presenting an immediate threat to lives and property in the community. While the fire service may place an emphasis on fire prevention, public education, risk reduction, and hazard abatement programs, its ability to respond and to control fires is an overriding operational priority.

Fire suppression is generally organized around a system of decentralized fire stations, providing the capability to respond quickly with personnel and equipment to control and extinguish fires. This organization may be staffed by career, part-time, or volunteer personnel; may reflect a variety of characteristics derived from local needs, structure, and tradition; and may be involved in the delivery of other emergency and nonemergency services. The ability to respond to the life safety and property protection needs of the local community is the common denominator in fire department emergency operations.

ORGANIZATION FOR FIRE SUPPRESSION

The company is the basic organizational unit of the fire department involved in fire suppression. A company is a complement of personnel operating one or more pieces of apparatus under the supervision of a company officer. A number of different types of tactical companies are used by fire departments, depending upon local needs. Engine and ladder companies are the most numerous, although a variety of others are often provided to perform specialized functions, including rescue companies and personnel squads, as well as companies operating special tactical or support function apparatus.

FIREGROUND OPERATIONS

Fire suppression operations involve the utilization of a fire department's resources to combat a fire. The success of a fire fighting operation depends upon the ability of a fire department to effectively and efficiently use the available resources to protect lives and property. The fireground commander is responsible for managing the available personnel and equipment to achieve maximum results, depending upon the situation, associated conditions, and resources available. The same principles apply to other types of incidents in addition to fires which the department may be expected to handle.

A fireground commander is responsible for the direction and control of operations in every incident. From the arrival of the first unit at the fire scene, there should be one identified person in command, with the responsibility and authority to direct all phases of the operation. The officer in charge of the first-arriving company assumes the role of fireground commander until relieved by a chief officer. In complex situations, command may be transferred one or more times as higher-ranking chief officers arrive and assume command.

The fireground commander is responsible for strategic decision making and the translation of strategic goals into tactical objectives and task assignments. (See Figure 11.4.) Intermediate levels of command are responsible to the fireground commander for geographical portions of the operation (sectors) or for the supervision of particular functions. These sector officers coordinate the operations of a group of companies under the overall command of the fireground commander.

A major incident may require a fairly complex command staff with a number of sector officers reporting to the fireground commander. Several officers may be assigned to directly support the fireground commander at a command post, providing information or further subdividing the span of control into manageable units.

Standard operating procedures are an important aspect of fireground command. Every fire department should have a set of procedures that outlines the basic operating principles to be employed for any situation, from the most simple to the most complex. The procedures should be flexible

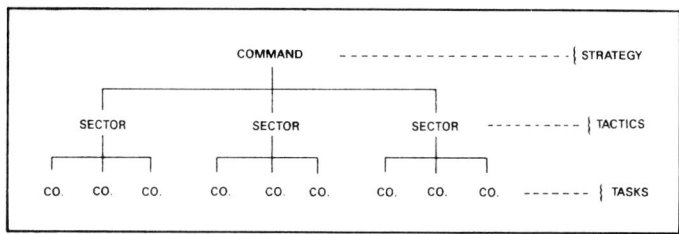

Fig. 11.4   A typical fireground organization. The fireground commander is involved in developing a strategic plan. The sector officers use tactical operations to achieve the goals of the plan, assigning specific tasks to individual fire companies within each sector.

enough to allow fire fighters to react to different situations and incremental to permit adjustment to the scale of the incident. Standard operating procedures provide a menu of basic functions that can be employed as needed and establish consistent approaches to fire control situations.

The fireground commander must utilize strategy and tactics in the management of a fire suppression or similar incident.

Strategy

Strategy involves the development of a basic plan to most effectively deal with a situation. The plan must identify major goals and prioritize objectives for the tactical elements. Strategic decisions are based upon an evaluation of the situation, the risk potential, and the capabilities of the available resources.

The strategic options available to the fireground commander involve some very important, but basic decisions. The most fundamental decision is the choice between offensive and defensive modes of operation, based on the capability of available resources and the risk to fire personnel. For offensive operations, companies extend hose lines into the interior of an involved fire area and extinguish the fire where they find it. In defensive operations, heavy streams are applied from the exterior to confine or control a fire, conceding the loss of the involved area. Offensive and defensive modes of operation must not be mixed in the same place at the same time. The fireground commander must make a conscious decision to identify what can and what cannot feasibly be saved without undue risk to personnel.

Strategic decisions also identify the priorities for committing resources to various tactical positions, and activities based on standard approaches and prevailing conditions. The generally accepted priorities for strategic decisions are rescue, fire control, and property conservation. The strategic plan identifies where and when the forces will attempt to control the fire and how their activities will be combined and prioritized.

Tactics

Tactics are the methods selected by the fireground commander to implement the strategic plan. The tactical objectives define specific functions that are assigned to groups of companies operating under sector officers. The achievement of these objectives contributes to the strategic goals and must be compatible with the overall strategic plan.

Tactical activities must be conducted in relation to three distinct phases or priorities. The first priority is to provide for the safety of the public by extending search and rescue efforts to all areas where potential victims may be located. The second priority is to control the fire, and the third priority is to conserve property. Fireground tactics usually involve a coordinated mixture of tasks directed toward these objectives in the above order.

Tasks

The translation of tactical objectives to task assignments results in assignments to individual companies. A company generally is involved in the actions relating to one or two specific tasks at any particular time. These tasks must be coordinated and combined to achieve tactical objectives.

Tactical Functions

Several tactical operations may be employed at each fire incident, and several tactical operations may be carried out simultaneously during multi-company operations. Every company must be trained to carry out all basic operations and be prepared to contribute to tactical objectives when possible.

Search and Rescue: Rescue is the first and most important considera-tion at any fire incident and until completed, may preclude any fire control efforts. The fireground commander may have to initiate fire control activities to protect the rescue operation or to keep the fire away from potential victims. Rescue operations may be simple, requiring only one or two fire fighters, or may require resources beyond the capabilities of the entire first-alarm assignment. All involved or threatened occupancies should be thoroughly searched for occupants without delay by companies specifically assigned to this task. Every company must be prepared to perform search and rescue as a first priority. Rescue operation difficulties may be compounded by the time of the incident, the occupancy, and the height and construction of the structure. Rescue is the only acceptable reason for exposing fire fighters to otherwise unnecessary risks.

Exposure Protection: The second fire suppression priority is to control the fire. This begins with confinement of the fire to the property initially involved. The most basic responsibility of fire departments, with respect to property, is to protect the community from large-loss fires. Failure to adequately protect exposed structures may allow a fire to extend beyond the building of fire origin. The problem of exposure protection may be com-pounded by closely spaced buildings, combustible construction, the type of occupancy, the lack of fire department access to the fire, and the lack of fire department resources. Exposure protection is a vital and necessary tactical consideration, and should be the major objective in defensive fire control situations.

Confinement: The confinement of a fire to its area of origin is often a complex problem. Fire control is achieved when the fire is successfully confined to a manageable area. All avenues of possible fire travel must be secured. The concept of surrounding the fire (over, under, and around) is necessary for successful confinement. Additional factors that influence the success or failure of confinement operations are the type of fuel involved, the location of the fire, building construction features, the presence of built-in fire suppression systems, and the availability of fire department resources.

Extinguishment: Offensive fire control strategies are aimed at controlling and extinguishing the fire, where encountered, by attack forces. Successful offensive operations are regulated by the type of fuel involved, the location of the fire and the degree of involvement, and the ability of fire control forces to apply sufficient extinguishing agents directly on the fire. In some instances, the use of special extinguishing agents may be required.

In defensive operations, final extinguishment may be achieved only when the fire burns down to a size which can be extinguished by the fire department. Defensive tactics depend upon the department's capability to apply large volumes of water or other agents to confine and eventually extinguish the fire.

Ventilation: Ventilation operations are the planned and systematic removal of heat, smoke, and fire gases from the structure. In some cases, it may be necessary to initiate ventilation with rescue in order to protect occupants from combustion products and heat and to provide visibility and tenability during rescue operations. Ventilation is also necessary during confinement and extinguishment to aid in locating the fire and to provide safer working conditions for fire suppression personnel, as well as to reduce overall damage to structure contents.

Property Conservation: Salvage operations are conducted by fire suppression personnel to conserve property and minimize damage to the structure and contents due to heat, smoke, and water. This involves, in part, covering contents and removing excess water. Salvage is an integral part of tactical operations and should commence as soon as possible to prevent additional damage to the structure and its contents.

Overhaul: Overhaul operations are required to completely extinguish the fire, place the structure in a safe condition, and aid in determining the fire ignition sequence. Overhaul may involve only a few personnel for a short period of time, or large numbers of personnel over an extended period. Extensive overhaul may require the use of special pieces of equipment beyond those normally provided by the fire department. It is important that extensive overhaul not be commenced prior to a thorough investigation to determine the cause of the fire. Once the investigation is complete, overhaul should continue to ensure that the premises are as safe as possible, and that all fires are extinguished.

FIRE FIGHTING SAFETY

Fire fighting is considered to be the most dangerous occupation in North America. Fire fighter deaths and injuries occur at a rate exceeding that for all other categorized labor activities. Though inherently dangerous, it is possible and essential to improve the safety of fire department operations. Safety should be a primary concern for all personnel, involving everyone from the fireground commander to the individual fire fighters.

All fire fighters must be provided with complete turnout clothing and protective equipment, including self-contained breathing apparatus, which comply with the appropriate NFPA standards. Regulations must be implemented and enforced to require use of the proper protective equipment in every situation. All fire departments should implement a complete safety program, including a physical fitness program for personnel, health monitoring and medical support, the provision and use of protective equipment, ongoing training, and the management of situations which present an unusual hazard to personnel. A comprehensive breathing apparatus program for all fire incidents is an essential element of a successful fire department safety program. NFPA 1500, *Standard on Fire Department Occupational Safety and Health Program*,[4] provides a good basis for such a program.

Safety must be a primary concern of the fireground commander at all times. In complex situations, safety officers should be assigned specifically to monitor conditions and advise the fireground commander and sector officers on hazardous conditions and safety concerns. Sector officers must be particularly concerned with the safety of personnel under their direction. Every tactical option must be evaluated in terms of its risk to fire fighters.

TRANSPORTATION INCIDENTS

All public fire departments need to be as well prepared for transportation incidents as they are for structure fires. These incidents may range from automobile fires to aircraft accidents and train derailments, and may occur anywhere and at any time. Preincident planning is essential for large-scale incidents, such as those involving mass transit systems, airports, and railways.

Fires and accidents that occur at such locations as bridges, freeways, railroads, tunnels, mass transit systems, etc., must be a part of the planning process for every fire department, since these incidents combine difficult access, limited water supplies, and special requirements for rescue and fire suppression. These situations may require specialized apparatus, equipment, training, and procedures so personnel can deal effectively with the variety of problems that can occur. Incidents ranging from a gasoline tanker burning on a bridge to a subway train collision and fire in an underground tunnel must be considered, and effective plans must be in place to deal with them. Some of the most disastrous fire incidents have occurred where unanticipated events and circumstances exceeded the capabilities of emergency response organizations.

Large quantities of flammable liquids and other hazardous materials are constantly in transit by road and rail. Any incident involving trucks or freight trains should be considered as a potential hazardous materials incident. Personnel must be familiar with labeling and placarding systems for hazardous materials, the use of cargo manifests and waybills to identify cargoes, and resources such as CHEMTREC.

MUTUAL AID AND MAJOR EMERGENCIES

The possibility of fire and disaster problems that exceed the capacity of the local fire fighting forces must always be considered. For this reason, most fire departments traditionally have rendered mutual assistance to other departments in times of need. Mutual aid plans establish procedures for requesting and dispatching help between fire departments so that each party will know what is expected. Mutual aid plans may include the following functions: (1) immediate joint response of several fire departments to high-risk properties, (2) joint response to alarms adjacent to the boundaries between fire department areas (automatic aid), (3) coverage of vacated territories by outside departments when the resources of the local department are engaged, (4) provision of additional units to assist at major fires that may be too large for the local department to handle, and (5) provision of specialized types of fire fighting equipment not available locally in adequate quantity for the particular incident.

Mutual aid plans also should include provisions for standard operating procedures, interdepartmental communications, common terminology, maps, adaptors, and other considerations that directly affect the department's ability to operate effectively. Command responsibility, jurisdictional questions, insurance coverage, and legal constraints should be covered in written agreements supported by enabling legislation to properly establish mutual aid systems for the participating departments.

Some jurisdictions have extended the mutual aid concept to multijurisdictional agreements in which fire department resources are pooled or merged into an integrated system, with standardized training procedures and communications. These networks may include shared facilities, joint purchase of specialized apparatus and equipment, and a coordinated approach to long-range planning.

Experience with natural disasters and large-scale incidents has focused attention upon the importance of plans and organizational procedures for systematically mobilizing fire forces for large-scale operations. Some states and provinces have established large-area disaster plans involving all emergency services, coordinated under standard procedures such as the Integrated Command System (ICS). For successful disaster operations, it is imperative that such plans integrate with the normal organizational and command procedures used by fire departments. These large-scale mutual aid networks form a natural basis for smaller-scale, more routine mutual assistance plans.

FIRE PREVENTION

Fire prevention encompasses all the means used by fire departments to decrease the incidents of uncontrolled fire. The fire prevention methods employed by fire department personnel involve engineering, education, and enforcement. Good engineering practices can do much to provide built-in safeguards to help prevent fires from starting and to limit the spread of fire should it occur. Education can instruct and inform groups and individuals of the dangers of fire and its possible effects, and of appropriate behaviors and reactions when faced with a fire situation. Enforcement is the legal means of correcting deficiencies that pose a threat to life and property. In addition, fire investigation aids fire prevention efforts by indicating problem areas that may require additional educational efforts or legislation to correct deficiencies.

In the United States, fire prevention has historically taken a secondary role to fire suppression as the principal activity of the public fire department. In 1973, fire prevention received its greatest endorsement when the National Commission of Fire Prevention and Control reported on the fire problem in America. Throughout the report, top priority was given to the necessity for increased fire prevention activities to reduce fire loss. The results of such efforts are being more clearly defined every year as progressive departments initiate more effective fire prevention efforts, in addition to maintaining their fire fighting forces.

Organizations for Fire Prevention

Most states have a state fire marshal charged by law with the responsibility for the administration and enforcement of state laws relating to safety to life and property from fire. All of the Canadian provinces have a provincial fire marshal or fire commissioner who performs similar functions.

All of the larger fire departments and many smaller ones have a fire prevention bureau in the fire department engaged primarily in the prevention of fires through property inspections and enforcement of a fire prevention code or other fire laws and regulations. These bureaus also carry on public education programs in fire prevention and investigate the cause and origin of fires.

The following defines the various fire prevention departments or offices typically present in government:[5]

> *State Fire Marshal*: The makeup of state fire marshal offices differs from state to state. Most receive their authority from the state legislature and are answerable to the governor, a high state officer,

or a commission created for that purpose. In some states the fire marshal's office may be a division of the state insurance department, state police, state building department, state commerce division, or other state agency. Few are organized as separate agencies.

State or provincial agencies normally function in those areas that go beyond the scope of the municipal, county, or fire district organizations. Local fire protection organizations are sometimes granted the authority to act as agents for the state in stipulated areas of inspection, enforcement, and investigation.

Chief of Fire Prevention or Local Fire Marshal: The laws of the county, municipal, or fire districts delegate the responsibility and authority of fire prevention to the fire chief or fire department head. Provision is then made for that officer to delegate this authority to an individual or division, depending on the size of the department. The individual or head of the division should be a high-ranking chief officer and should also function as a staff officer to the fire chief. This division of the fire service is normally called the fire prevention bureau, and its top officer is chief of fire prevention or the local fire marshal. Where size permits, the bureau is divided into subdepartments of inspections, investigations, and public education. These subdepartments are then headed by subordinate chiefs.

Fire Inspector or Fire Prevention Officer: The terms fire inspector or fire prevention officer have different meanings, depending on department classification. Sometimes the two titles are the same and denote the position responsible for conducting fire inspections assigned to the fire prevention bureau. In bureaus not large enough for three subdepartments, the fire inspector is also responsible for fire investigations and public education duties.

Inspections

Inspections by the fire department are extremely important if fires are to be prevented. Inspections are also made by the fire companies, not only to uncover and correct fire hazards but also to familiarize the fire fighters with the buildings in their district before a fire starts in any of them. Any code violations or special hazards discovered by company inspections must be referred to the authorized bureau of fire prevention for action.

The bureau of fire prevention makes regular inspections of target risks—e.g., hospitals, nursing homes, homes for the aged, hotels, theaters, halls, and other places where the public may congregate, and of all important manufacturing and business establishments—to make sure no code violations exist in these occupancies.

A key provision of the ordinance establishing a bureau of fire prevention gives the fire marshal or chief of the fire prevention bureau the following power: whenever any inspector finds in any building or upon the premises or other places combustible or explosive matter, or dangerous accumulation of rubbish, or unnecessary accumulation of waste paper, boxes, shavings or any highly flammable materials especially liable to fire, which is so situated as to endanger property; or finds obstructions to or on fire escapes, stairs, passageways, doors or windows liable to interfere with the operations of the fire department or egress of occupants in case of fire, the inspector must order the same to be removed or remedied, and such order must forthwith be complied with by the owner or occupant of such premises or buildings.

Home Inspections: Home inspections are not required by law, but are of immense value to the public and to the fire department. If carefully planned and publicized in advance, such inspections are welcomed by the great majority of families and not only eliminate many hazards, but engender goodwill toward the fire department.

The purpose of home inspections by fire departments is to point out such common fire hazards as needless accumulations of rubbish, overloaded electric circuits, improper storage of paints and flammable liquids, space heaters too close to flammable material, matches kept within reach of children, etc.

Home inspections do not include multifamily residential properties where the common areas are often covered by inspections for local code compliance.

Business Inspections: Most larger manufacturing and mercantile establishments recognize the disruption that a fire would cause to their business, and not only spend money on fire protection but conduct their own inspections and organize private fire brigades. These and smaller business properties are also subject to inspections by public fire forces, insurance carriers, and the Occupational Safety and Health Administration. Good inspection procedures from whatever source are the best way to keep fire losses down. The policy and attitude of the management of the business toward fire prevention and protection is of prime importance. As a rule, plants with obvious fire hazards (such as oil refineries, chemical and explosive manufacturing, and plants manufacturing plastics) usually have a good fire record because of their acute awareness of the fire hazards of their processes and products. A large number of small plants and stores generally depend on inspections by their insurance carriers or by the fire department to bring potential fire hazards to their attention.

Enforcement of Codes

Most communities have adopted a fire prevention code or at least some ordinances covering storage and handling of flammable liquids, installation of electrical wiring, heating appliances, etc. It is the duty of the fire marshal or fire prevention bureau to see that such codes and ordinances are enforced.

A good local fire prevention code will include provisions for control of not only the common hazards found in any community, but any special hazards that the fire department may face in industrial or mercantile operations. The code should also cover the servicing and maintenance of interior fire protection required in various occupancies (such as automatic sprinklers, fire detection and alarm systems, portable fire extinguishers, and standpipes). The installation requirements are generally covered in a building code. Many of the national standards referred to in building and fire prevention codes are primarily for purposes of preventing fires. Typical of these are the many storage standards covering indoor and outdoor storage of combustibles, rack storage, and storage of tires, records, explosives and flammable liquids, pyroxylin plastics, and forest products.

Other well-known fire prevention standards include those for oil-burning equipment, fuel gas, cutting and welding processes, prevention of dust explosions, and the safe installation of chimneys and fireplaces.

INVESTIGATION OF FIRES

Another important function of the fire marshal or fire prevention bureau is the investigation of fires to determine cause and origin. At first glance, it may seem that fire investigation is not related to fire prevention. However, careful investigation to discover all the factors that influence the start and spread of a fire will provide the data required to develop the codes and inspection procedures needed to prevent future fires. For this reason, all fires should be properly investigated, not just those thought to be "suspicious." The fire prevention bureau should maintain records to show not only the number of fires by location and occupancy, but fire cause, time of day, room or floor in which the fire occurred, and other data basic to an understanding of the problems that the department faces.

The importance of fire investigation is recognized by the courts. Fire marshals are usually given broad powers to investigate fires, and to subpoena persons and records. Arson can often be discouraged if the potential firesetter knows that fires are carefully investigated to determine any suspicious circumstances.

It is extremely important that anyone investigating a fire approach the investigation with an open mind and not jump to conclusions. Premature judgements will often lead to a wrong conclusion which could jeopardize a criminal investigation or result in wrongly assigning responsibility.

PUBLIC EDUCATION

The fire prevention arm of the fire department is usually given the responsibility for a continuing program of public education. There are many ways to motivate people to do the things that will save them and their families from the threat of fire. The fire department should work with the school authorities to ensure that all school children are taught at least the basics that they need to know about what to do when fire strikes and how to eliminate the common fire hazards in their own home. The *Learn Not to Burn Curriculum*[1] published by the NFPA provides an excellent means of accomplishing this. In addition, the fire department should send personnel to the schools to talk about the danger of false alarms, fire hazards, etc. Many departments invite children and the public to visit the fire station to see how the department works and to learn how they may cooperate to reduce fire losses. Another excellent procedure is the distribution of home inspection forms to pupils. Many parents have been alerted to fire hazards in the home by having to sign their child's home inspection form.

The fire prevention division should also use television, radio, and the press to inform the public about hazardous toys, garments, or home appliances; special fire department problems in times of large fires, winter storms, floods, and other disasters; and special programs directed at particular fire hazards.

Seasonal Activities: Fire departments often sponsor or participate in Fire Prevention Week, spring clean-up week, Christmas holiday warnings, Fourth of July safety from fireworks, and similar seasonal campaigns. Many fire departments sponsor junior fire department programs, such as Sparky's Junior Fire Department, and encourage fire prevention activities by Boy Scouts, Girl Scouts, etc. One excellent activity, endorsed by the Fire Marshals Association of North America, is Operation EDITH (Exit Drills In The Home). Homeowners are urged to conduct a fire drill in their homes at a certain day and time during Fire Prevention Week. For most effective results, public education in fire prevention has to be constantly updated and upgraded to maintain public interest and support.

Summary

The United States is protected by approximately 30,000 fire departments. The majority of these departments are small volunteer fire departments. About 2 percent of the American fire departments protect large communities (50,000 population or greater). Although fire departments have evolved to fit local needs, much can be done to improve performance in areas of both fire protection and fire prevention. Through the development of specific fire department goals and objectives that can be measured and evaluated by performance standards, the organization and management of fire departments can be made increasingly more efficient and effective.

In addition to more effective utilization of the resources of the fire department itself, effective coordination between other public agencies and the fire department is important so that the needs and requirements of the fire department will be clearly communicated and implemented. Fire department effectiveness is further ensured by careful prefire planning and training.

The importance of fire prevention activities, as a fire department responsibility, is becoming an area of increased emphasis. Most fires start from such common causes as poorly maintained heaters and furnaces, careless smoking, frayed electric cords, and improper storage of gasoline and paint. Every school should teach fire safety, and fire departments should regularly conduct public education programs for all age groups. Most states have a state fire marshal, and most cities have a fire marshal or a fire prevention bureau in the fire department. A fire prevention code regulating the principal fire hazards is now law in many states and cities, with the result that inspections for fire hazards should be regularly carried out in industries and commercial establishments by state and city fire inspection staff. Such inspections are particularly important in places such as hospitals, homes for the aged, nursing homes, hotels, theaters, halls, and other places of public assembly. Home inspections by fire departments are of great value and should be universal.

References

[1]*Learn Not to Burn Curriculum* 1981. National Fire Protection Association, Quincy, MA.

[2]NFPA 1001-1987. *Standard for Fire Fighter Professional Qualifications*, National Fire Protection Association, Quincy, MA.

[3]NFPA 1021-1983. *Standard for Fire Officer Professional Qualifications*, National Fire Protection Association, Quincy, MA.

[4]NFPA 1500-1987. *Fire Department Occupational Safety and Health Program*, National Fire Protection Association, Quincy, MA.

[5]*Fire Protection Handbook*, 16th ed. 1986. National Fire Protection Association, Quincy, MA.

Additional Reading

Brunacini, Alan V. 1985. *Fire Command*, National Fire Protection Association, Quincy, MA.

Bryan, J.L., and Picard, R.C., eds. 1979. *Managing Fire Services*, International City Management Association, Washington, DC.

Coleman, R.J. 1978. *Management of Fire Service Operations*, Bretton Publishers, North Scituate, MA.

Conducting Fire Inspections 1982. National Fire Protection Association, Quincy, MA.

Deltaan, John D. 1983. *Kirk's Fire Investigation*, 2nd ed., John Wiley & Sons, Inc., New York.

Didactic Systems, Inc. 1977. *Management in the Fire Service*, National Fire Protection Association, Quincy, MA.

Fire Safety Educator's Handbook 1983. National Fire Protection Association, Quincy, MA.

Hall, J.R., *et al* 1979. *Fire Code Inspections and Fire Prevention: What Methods Lend to Success*, National Fire Protection Association, Quincy, MA.

Managing People: Fire Service Personnel Strategies 1984. National Fire Protection Association, Quincy, MA.

O'Hagan, John T. 1984 through 1985. "Staffing Levels: A Major New Study," *Fire Command* (published in six parts), Vol. 51, Nos. 11 and 12; Vol. 52, Nos. 1,2,3, and 5 (November 1984 through May 1985).

Small Community Fire Departments: Organization and Operation 1982. National Fire Protection Association, Quincy, MA.

Thomas, Peter, *Management by Objectives: A Handbook for Governmental Managers and Supervisors* 1978. Masterco Press, Ann Arbor, MI.

Chapter **12**

Codes and Standards

Throughout history there have been building regulations for preventing fire and restricting its spread. Over the years these regulations have evolved into the codes and standards developed by committees concerned with fire protection. In many cases, a particular code dealing with a hazard of paramount importance may be enacted into law.

THE HISTORY AND DEVELOPMENT OF FIRE PROTECTION REGULATIONS

King Hammurabi, the famous law-making Babylonian ruler who reigned from approximately 1955 to 1913 B.C., is probably best remembered for the *Code of Hammurabi*, a statute primarily based on retaliation. The following decree is from the *Code of Hammurabi*:

> In the case of collapse of a defective building, the architect is to be put to death if the owner is killed by accident; and the architect's son if the son of the owner loses his life.

Today, we no longer endorse Hammurabi's ancient law of retaliation but seek, rather, to prevent accidents and loss of life and property. From these objectives have evolved the rules and regulations that represent today's building codes and current standards for fire prevention, fire protection, and fire suppression.

Early Building and Fire Laws

The earliest recorded building laws apparently were concerned with the prevention of collapse. During the rapid growth of the Roman Empire under the reigns of Julius and Augustus Caesar, the city of Rome became the site of a large number of hastily constructed apartment buildings—many of which were erected to considerable heights. Because building collapse due to structural failure was frequent, laws were passed that limited the heights of buildings—first to 70 ft (21.3 m) and then to 60 ft (18.2 m).

Later in history there evolved many building regulations for preventing fire and restricting its spread. In London during the 14th century an ordinance was issued requiring that chimneys be built of tile, stone, or plaster; the ordinance prohibited the use of wood for this purpose. Among the first

321

building ordinances of New York City was a similar provision, and among the first legislative acts of Boston was one requiring that dwellings be constructed of brick or stone and roofed with slate or tile (rather than being built of wood and having thatched roofs with wooden chimneys covered with mud and clay similar to those to which the early settlers had been accustomed in Europe). Obviously, the intention of these building ordinances was to restrict the spread of fire from building to building in order to prevent conflagrations. As an inducement for helping to prevent fires, a fine of ten shillings was imposed on any householders who had chimney fires. This encouraged the citizenry to keep its chimneys free from soot and creosote. Thus was the first fire code in America established and enforced.

In colonial America, the need for laws that offered protection from the ravages of fire developed simultaneously with the growth of the colonies. The laws outlined the fire protection responsibilities of both homeowners and authorities. Some of these new laws were planned to punish people who put themselves and others at risk of fire. For example, in Boston no person was allowed to build a fire within "three rods" of any building, or in ships that were docked in Boston Harbor. It was illegal to carry "burning brands" for lighting fires except in covered containers, and arson was punishable by death. Regardless of such precautions, in Boston and in other emerging communities, fires were everyday occurrences. It therefore became necessary to enact more laws with which to govern building construction and to make further provisions for public fire protection. There emerged a growing body of rules and regulations concerning fire prevention, protection, and control. From these small beginnings, various codes and types of codes have evolved in this country, ranging from the most meager of ordinances to comprehensive handbooks and volumes of codes and standards on building construction and firesafety.

Development of Building and Fire Codes

The rapid growth of early American cities inspired much speculative building, and the structures usually were built close to one another. Construction often was started before adequate building codes had been enacted. For example, the year before the great Chicago fire of 1871, Lloyd's of London stopped writing policies in Chicago because of the haphazard manner in which construction was proceeding. Other insurance companies had difficulty selling policies at the high rates they had to charge. Despite these excessively high rates, many insurance companies suffered great losses when fires spread out of control.

The National Board of Fire Underwriters (now the American Insurance Association), organized in 1866, realized that the adjustment and standardization of rates were merely temporary solutions to a serious technical problem. This group began to emphasize safe building construction, control of fire hazards, and improvements in both water supplies and fire departments. As a result, the new tall buildings constructed of concrete and steel conformed to specifications that helped limit the risk of fire. These buildings were called Class A buildings. In 1905 the National Board of Fire Underwriters published the first edition of its *Recommended Building Code* (later the *National Building Code*).[1] This was a first and very useful attempt to show the way to uniformity.

In San Francisco in 1906, although there were some new Class A concrete and steel buildings in the downtown section, most of the city consisted of fire-prone wood shacks. Concerned with such conditions, the National Board of Fire Underwriters wrote that "San Francisco has violated all underwriting traditions and precedents by not burning up."

On April 18 of that same year, the city of San Francisco experienced a conflagration—started by an earthquake—that killed 452 people and destroyed some 28,000 buildings. Although the contents of many of the new Class A buildings were destroyed in the San Francisco fire, most of the walls, frames, and floors remained intact and could be renovated. (See Figure 12.1.)

Fig. 12.1   A great earthquake and ensuing conflagration devastated San Francisco in 1906. As many as 452 people were killed, and 28,000 buildings were damaged or destroyed. Total financial loss was $350 million.

Following analysis of the fire damage caused by the San Francisco disaster and other major fires, the National Board of Fire Underwriters became convinced of the need for more comprehensive standards and codes relating to the design, construction, and maintenance of buildings. With this increasing recognition of the importance of fire protection came more knowledge about the subject. Engineers started to accumulate information about fire hazards in building construction and in manufacturing processes, and much of this information became the basis for the early standards and codes.

BUILDING CODE DEFINED

A building code is a law that sets forth minimum requirements for design and construction of buildings and structures. These minimum requirements, established to protect the health and safety of society, generally represent a compromise between optimum safety and economic feasibility. Although builders and building owners often establish their own requirements, the minimum code requirements of a jurisdiction must be met. Features covered include structural design, fire protection, means of egress, light, sanitation, and interior finish.[2]

There are two types of building codes. Specification codes spell out in detail what materials can be used, the building size, and how components should be assembled. Performance codes detail the objective to be met and establish criteria for determining if the objective has been reached; thus, the designer and builder are free to select construction methods and materials as long as it can be shown that the performance criteria can be met. Performance-oriented building codes still embody a fair amount of specification-type requirements, but provisions exist for substitution of alternate methods and materials ("tradeoffs") if they can be proven adequate.

BUILDING CODES AND FIRE PROTECTION

The requirements contained in building codes generally are based upon the known properties of materials, the hazards presented by various occupancies, and the lessons learned from previous experiences, such as fire and natural disasters.[2] As previously noted, the promulgation of modern building codes in the U.S. began with the disastrous conflagrations and earthquakes which occurred at the turn of the century.

For a number of years, building codes dealt mainly with structural safety under fire or earthquake conditions. Since then, codes have grown into documents prescribing minimum requirements for structural stability, fire resistance, means of egress, sanitation, lighting, ventilation, and built-in safety equipment. More than 50 percent of a modern building code usually refers in some way or another to fire protection.

Building codes usually establish fire limits or fire districts in certain areas of a municipality. Only specific types of construction are allowed within the fire limits. Such a restriction is said to reduce the conflagration potential of the more densely populated areas. Use of a given type of building construction alone, however, is not necessarily a deterrent to conflagration. Outside the fire limits, the restriction of certain construction types is relaxed, due to such factors as decreased building density (increased spacing between buildings). Unfortunately, as areas outside the fire limits are developed, building density increases and the fire limits frequently must be extended. In addition, without construction restrictions, areas outside the fire limits invite the erection of large buildings despite public protection which is weak or lacking.

Another example of the impact of building codes on fire protection and prevention is the establishment of height and area criteria. The criteria establish height and size of a particular building based on its intended use. Unfortunately, these requirements vary considerably from one area to the next; there is no nationally recognized standard for setting height and area limitations. The types of building construction are important factors in establishing height and area limitations.

Other requirements found in building codes that directly relate to fire protection include: (1) enclosure of vertical openings such as stair shafts, elevator shafts, and pipe chases; (2) provision of exits for evacuation of occupants; (3) requirements for flame spread of interior finish; and (4) provisions for automatic fire suppression systems. Exit requirements found in most building codes are based on specifications in NFPA *101*, the *Life Safety Code*.[3]

Inasmuch as a building code is actually a law, many state and local jurisdictions write their own codes. Because of the complexities of modern building code development, several organizations develop model building codes for use by jurisdictions which can then adopt the model codes into law.

Relationships Between Building and Fire Codes

It often is difficult to differentiate between items that should go into a fire prevention code and those best included in a building code. Generally, whatever requirements deal specifically with construction of a building are part of a building code administered by the building department. A fire prevention code, on the other hand, includes information on fire hazards in a building, and usually is regulated by the fire official. Requirements for exits and fire extinguishing equipment generally are found in building codes, while the maintenance of such items is covered in fire prevention codes.

MODEL BUILDING CODE GROUPS

Three groups develop model building codes: International Conference of Building Officials (ICBO), Southern Building Code Congress International

(SBCCI), and Building Officials and Code Administrators (BOCA). Each organization and its model code are discussed in the following paragraphs.

International Conference of Building Officials (ICBO)

ICBO first published the *Uniform Building Code* (UBC)[4] in 1927. The UBC is principally used in the western U.S., but has been adopted in municipalities as far east as Michigan.

In addition to the building code, the Conference publishes a *Uniform Mechanical Code and Uniform Plumbing Code* in association with the International Association of Plumbing and Mechanical Officials (IAPMO), and a *Uniform Fire Code* in association with the Western Fire Chiefs Association. Code changes are made each year and amended versions of the codes are published every three years. Changes made between major reprintings are issued as supplements.

The Conference's stated objectives relate to the development and publishing of regulations and educational materials designed to increase the standardization and enforcement of building construction regulations. To encourage standardization, the Conference maintains a staff of architects and engineers in Whittier, California. This staff provides governmental bodies with technical assistance in the administration and enforcement of the ICBO codes.

Southern Building Code Congress International (SBCCI)

SBCCI was organized in 1940 and first published the *Southern Standard Building Code* in 1945. Now called the *Standard Building Code* (SBC),[5] this code is used principally throughout the southern portion of the U.S.

Like ICBO, SBCCI also publishes mechanical, plumbing, fire prevention, and gas codes to be used in conjunction with the SBC. The code is amended and reprinted every three years, with changes made yearly and printed in supplements.

The stated purpose of SBCCI is to develop, maintain, and promote the use of its series of codes. In addition, SBCCI intends to promote standardization in building regulation and enforcement of those regulations. SBCCI maintains a technical staff headquartered in Birmingham, Alabama.

Building Officials and Code Administrators (BOCA)

As the Building Officials Conference of America, BOCA published the first edition of the *Basic/Building Code* (BBC) in 1950. (For a period of time, this code was called the *Basic/National Building Code*.) The organization has changed its name to Building Officials and Code Administrators, International, and the BBC is now published as the *National Building Code*.[6] It is used principally in the midwest and northeast portions of the U.S.

Like the other code groups, BOCA also publishes mechanical, plumbing, and fire prevention codes. Each of these codes is revised annually and reprinted every three years. Changes approved between reprintings are published in supplements. The stated objective of the organization is to develop and maintain the *National Building Code*.[6] BOCA maintains a technical staff in Country Club Hills, Illinois.

American Insurance Association (AIA)

As previously noted, the National Board of Fire Underwriters (NBFU), now the AIA, first published the *National Building Code* (NBC)[1] in 1905. The code was used as a model for adoption by cities, as well as a basis to evaluate the building regulations of towns and cities for town grading purposes. The code was periodically reviewed by the NBFU staff, revised as necessary, and republished. The last code revision was in the 1976 edition. Since then, the AIA has discontinued updating or publishing the NBC, and Building Officials and Code Administrators (BOCA) has acquired the right to use the name *National Building Code*.[6] AIA also published the first model fire prevention code, which was most recently issued in 1976, but also has discontinued the updating and publishing of this document.

National Fire Protection Association (NFPA)

NFPA does not publish a model building code *per se*; however, NFPA *101*, the *Life Safety Code*,[3] establishes minimum requirements necessary for providing a reasonable degree of safety from fire. The *Life Safety Code*[3] addresses those construction, protection, and occupancy features necessary to minimize danger to life from fire, smoke, fumes, or panic. It also identifies the minimum criteria for design of egress facilities to permit prompt escape of occupants from buildings or, where preferable, evacuation into safe areas within buildings. The *Life Safety Code*[3] does not attempt to address those general fire prevention or building construction features that normally are functions of fire prevention and building codes. The NFPA Committee on Safety to Life was found in 1913 and initially developed a number of pamphlets on exits from various occupancies. The first edition of the NFPA *Building Exits Code* (later the *Life Safety Code*), developed by the Safety to Life Committee, was published in 1927.

The *Life Safety Code*[3] is one of more than 250 codes, standards, recommended practices, manuals, guides, and model laws developed and published by NFPA. The complete compendium of these documents, which encompass the entire scope of fire protection and prevention, practices, and safeguards comprises the NFPA *National Fire Codes®* (*NFC*). Included within the *NFC* are a number of individual firesafety codes, including NFPA 70, the *National Electrical Code*,[7] believed to be the most widely adopted code in the U.S.; NFPA 30, the *Flammable and Combustible Liquids Code*;[8] and NFPA 54, the *National Fuel Gas Code*[9]—just to name a few. NFPA also has developed its own *Fire Prevention Code*,[10] NFPA 1, which also is included in the *National Fire Codes*.

NFPA standards and codes are developed by more than 170 NFPA technical committees, each representing a balance of affected interests. More than 3,500 individuals serve on the Association's committees, all on a voluntary, unpaid basis. Committees operate according to detailed Regulations Governing Committee Projects[11] and are administered by the Standards Council which reports to the Association's Board of Directors.

Built into the standards development and adoption process is the publication of calls for proposals to amend existing documents, or to shape the content of new documents. These public proposals, together with the committee action on each proposal and committee-generated proposals, are published in the *Technical Committee Reports* for public review and comment. Public comments, together with the committee action on each document, are then published in the *Technical Committee Documentation*. Only after this public review and comment cycle has been completed is the final committee report brought before the membership for action. This democratic legislative procedure allows proponents and opponents to be freely heard. Once adopted by the NFPA membership at either an annual meeting or a fall meeting, and issued by the Standards Council, the documents are published and made available for voluntary adoption.

TOWARD UNIFORMITY OF BUILDING CODES

Although model codes originally were known as regional codes, boundary lines for their adoption are much less significant today. In fact, much effort was expended toward elimination of differences among the model codes through such organizations as the former Joint Committee on Building Codes. This group was composed of representatives of the organizations sponsoring the model codes and others concerned with the development of codes and standards. Formed in 1949, the Joint Committee later became the Model Code Standardization Council and is now the Board for the Coordination of Model Codes (BCMC), operating as a committee of the Council of American Building Officials (CABO). Code uniformity continues to be the Board's main objective.

OTHER CODE-RELATED ORGANIZATIONS

As noted earlier, the primary objective of the model code groups is to provide standardization and promote enforcement of construction regulations. Other organizations with similar goals also have been formed.

American National Standards Institute (ANSI)

ANSI was founded in 1918 to prevent economic losses caused by lack of nationally recognized standards and by uncoordinated standards development. ANSI coordinates and harmonizes private-sector standards activity in the U.S. In order for a document to be designated an American National Standard, the principles of openess and due process must have been followed in its development, and consensus among those directly and materially affected by the standard must have been achieved. ANSI also represents the U.S. interests in activities of the International Organization for Standardization (ISO) and the International Electrotechnical Commission (IEC).

Council of American Building Officials (CABO)

CABO is an organization formed by the three major model code groups—ICBO, SBCCI, and BOCA—in 1972. The intent of CABO is to promote the model code process and to represent it on a national level. One of the committees formed by CABO is the Board for the Coordination of the Model Codes (BCMC). Representatives of BOCA, ICBO, SBCCI, and, most recently (1981), NFPA participate in this effort.

The purpose of this committee is to develop uniform code language to be included in each of the model codes. Proposals resulting from BCMC are processed in accordance with the code change procedures of each code organization.

Two of the major accomplishments of CABO have been to organize the National Research Board and publish the *One- and Two-Family Dwelling Code.*[12] The National Research Board coordinates the research and evaluation programs of the three model code groups to eliminate the need for a manufacturer to work with three different organizations. The *One- and Two-Family Dwelling Code* contains regulations for detached housing which meet the requirements of the three model codes.

National Conference of States on Building Codes and Standards (NCSBCS)

NCSBCS is a nonprofit corporation founded in 1967 as a result of Congressional interest in reform of building codes. The Conference attempts to foster increased interstate cooperation in the area of building codes and standards and coordinates intergovernmental code administration reforms. NCSBCS is an executive-branch organization of the National Governors Association and includes as members governor-appointed representatives of each state and territorial government. It has a working relationship with the National Conference of State Legislatures and the Council of State Community Affairs Agencies.

Association of Major City Building Officials

The Association of Major City Building Officials was formed in 1974. This group recognizes code problems that are peculiar to large cities, defines problem areas, and seeks solutions.

National Institute of Building Sciences (NIBS)

NIBS was authorized by Congress in 1974 under Public Law 93-383 as a nongovernmental, nonprofit organization governed by a 21-member board of directors. Fifteen of the board members are elected and six are appointed by the President of the United States, with the advice and consent of the U.S. Senate. In establishing the objectives, powers, and structure of NIBS, Congress also outlined certain functions relating to building regulations in four general areas. These are:

1. Development, promulgation, and maintenance of nationally recognized performance criteria, standards, and other technical provisions for maintenance of life, safety, health, and public welfare that are suitable for adoption by building regulating jurisdictions and agencies. This also includes test methods and other evaluative techniques relating to building systems, subsystems, components, products, and materials, with due regard for consumer interests.
2. Evaluation and prequalification of existing and new building technology.
3. Conduct of needed investigations.
4. Assembly, storage, and dissemination of technical data and other related information.

Working under its very broad mandate, NIBS has established a Consultative Council with membership available to representatives of all appropriate private trade, professional, and labor organizations; private and public standards, codes, and testing bodies; public regulatory agencies; and consumer groups. The Council's purpose is to ensure a direct line of communication between such groups and the Institute and to serve as a vehicle for representative hearings on matters before the Institute.

The Institute is a core organization that serves primarily as an investigative body, offering its findings and recommendations to government and to responsible private-sector organizations for voluntary implementation. It carries out its mandated mission essentially by identifying and investigating national problems confronting the building community, and proposing courses of action to bring about solutions to the problems. NIBS activities are board-based and center around regulatory concerns, technology for the built environment, and distribution of technical and other useful information.

World Organization of Building Officials (WOBO)

WOBO was founded in 1984 with the primary objective of advancing education through worldwide dissemination of knowledge in building science, technology, and construction. WOBO was established because of increased participation of nations in the global marketplace; rapid development of new international building technologies and products; and development of international standards which now make it impossible for building officials to confine their concern to activities within their own national boundaries.

ENFORCEMENT OF CODES AND STANDARDS

Today the life and property of every citizen is safeguarded to at least some extent by fire legislation enacted by the Congress of the United States, state legislatures, city councils, town meetings, and many other jurisdictions and levels of government. The implementation and enforcement of this legislation is in the hands of administrative agencies of government, such as federal departments and agencies, state fire marshal offices and other appropriate state agencies, and local fire departments, building departments, electrical inspectors, etc.

In this country's earlier days, the protection of citizens from fire was solely the concern of the local community. Fire fighting is now, and always has been, carried on by local fire departments. While most communities have had some type of building code since the beginning of the 20th century, they did not have fire prevention codes until more recently.

With the need for more detailed, comprehensive standards and codes relating to the construction, design, and maintenance of buildings came the knowledge that regulations based on such codes certainly could prevent most fires and reduce losses in the fires that did occur.

Regulations relating to firesafety are determined and enforced by different levels of government. While some functions overlap, federal and state laws generally govern those areas that cannot be regulated at the local level.

Federal Firesafety Regulations

There is a substantial amount of federal regulation pertaining to firesafety. Under the Constitution, Congress has the power to regulate interstate commerce; this power has been interpreted to permit Congress to pass laws authorizing various federal departments and agencies to adopt and enforce regulations to protect the public from fire hazards.

Any federal department or agency can promulgate firesafety regulations only if authority to do so is granted by a specific act of Congress. These regulations have the force of law, and violations can result in legal action. In general, such federal laws can be enacted to provide: (1) that all state laws on the same subject are superseded by the federal law, (2) that state laws not conflicting with the federal law remain valid, or (3) that any state law will prevail if it is more stringent than the federal law. Among the federal agencies that have the authority to promulgate firesafety regulations are the Occupational Safety and Health Administration (OSHA), the Department of Health and Human Services (HHS), and the Consumer Product Safety Commission (CPSC).

FIRESAFETY REGULATIONS OF STATE AND LOCAL GOVERNMENT

Within the scope of the police power of state government is the regulation of building construction for the health and safety of the public—a power usually delegated to local governments by the state.

Building code requirements usually apply to new construction or to major alterations to buildings. Retroactive application of code requirements is very rare. Building code applicability usually ends with the issuance of an occupancy permit. The basic premise that fire legislation should regulate for the safety of current occupants and for current risk is not generally the province of building codes once a structure is occupied. Then after-occupancy fire codes—or, more precisely, fire prevention codes—apply. This also usually is the point at which the authority of the building official ends and the fire official begins.

This division of authority, however, does not preclude interaction between the two officials during both a building's development and its subsequent use. In practice, many jurisdictions assign responsibilities to building and/or fire officials in both building and fire code application. The division of authority varies considerably among communities.

In most states, the principal fire official is the state fire marshal. For the most part, the state fire marshal is the statutory official charged by law with responsibility for the administration and enforcement of state laws relating to safety to life and property from fire. Usually the state fire marshal also has the power to investigate fires and to suppress arson.

The manner in which each state handles the promulgation of building and fire regulations varies widely. In some states each local government may have its own code, while in others the local authority has the option of adopting the state building code. In still other states, the state code establishes the minimum below which the local regulations cannot go. Finally, in some states the local government has no choice and must adopt the state code.

These situations have resulted in a plethora of local building and fire codes. Some of the local governments adopt one of the model codes or a code based upon one of them. Others draft their own local code. This lack of uniformity has been criticized by materials producers, building designers, builders, and others, and some years ago prompted the appointment of federal commissions to study the situation and make recommendations to the administration.[13,14,15]

The legal procedure for adopting codes and standards into law can also vary from one enforcing jurisdiction to another. Usually, the simplest and best way is to adopt by reference. This method, applicable to public authorities as well as to private entities, requires that the text of the law or rule cite the standard by its title and give adequate publishing information to permit its exact identification. The standard itself is not reprinted in the law. All deletions, additions, or changes made by the adopting authority are noted separately in the text of the law. Adoption of a current edition of a standard obviates outdated editions maintained as law until a new law referencing a new edition is adopted.

Where local laws do not permit adoption by reference, a standard can be adopted by transcription. This requires that the text of the adopted standard be transcribed into the law. Existing material can be deleted and new material added only if such material does not change the meaning or intent of the existing or remaining material. Under adoption by transcription, the standard cannot be rewritten, although changes can be made for administrative provisions. Because the text of the standard is transcribed into the law, due notice of the copyright of the standard's developer is required. As a result, most code groups copyright their standards to prevent misuse and unlawful use.

THE ROLE OF STANDARDS IN BUILDING CODES

Many requirements found in building codes are excerpts from, or based on, the standards published by nationally recognized organizations. The most extensive use of standards is their adoption into building codes by reference, thus keeping the building codes to a workable size and eliminating much duplication of effort. Such standards also are used by specification writers in the design stage of a building to provide guidelines for the bidders or contractors.

Numerous NFPA standards are referenced by the model building codes and thus obtain legal status where these model codes are adopted. Notable examples of such referenced NFPA standards are those that deal with electrical systems, extinguishing systems, flammable liquids, hazardous processes, combustible dusts, liquefied petroleum gas, and fire tests.

Two of the model building codes—BOCA and SBCCI—contain appendices that list standards published by many recognized groups, including standards-making organizations, professional engineering societies, building materials trade associations, federal agencies, and testing agencies. The appendices are prefaced with a statement indicating that the standards are to be used where required by the provisions of the code or where referenced by the code. The other model building code organization—ICBO—publishes a book of standards to accompany its code. Some of the standards in the book have been written by ICBO; others are standards of other organizations adapted for ICBO use.

NATIONAL STANDARDS

A 1978 Federal Trade Commission (FTC) report on standards and certification quoted a Department of Commerce estimate that more than $500 million dollars are spent annually by participants in national standards activities. It is difficult for newcomers to the standards development business to conceive the enormity of the total enterprise because they see only those segments of the system in which they first become involved.

Currently, nearly 500 organizations in the U.S. write and maintain national standards. They can be divided into four principal groups:

1. Organizations structured to deal exclusively with the preparation, publication, and distribution of a wide variety of specifications intended for voluntary use in the private and public sectors.

    As an example, the American Society for Testing and Materials (ASTM), a private-sector organization, falls into this category. In the public sector, the Standards Development Services Section of the National Bureau of Standards and the U.S. Department of Commerce have similar functions.

2. Professional societies and technical organizations that deal with design standards for construction and equipment. The American Concrete Institute (ACI), the American Society of Mechanical Engineers (ASME), and the American Society of Heating, Refrigeration and Air Conditioning Engineers (ASHRAE) are examples in this category of activity.

    Prominent standards developed by these organizations include ACI 318, *Building Code Requirements for Reinforced Concrete*; the boiler and pressure vessel specifications by ASME; and the ASHRAE Standard 90A, *Energy Conservation in New Building Design*.

3. Industry Associations and Manufacturers, a group that produces proprietary standards for materials and test methods applicable to their particular industry products. Although such specifications are not developed under national consensus proceedings, their broad voluntary use achieves national acceptance.

Three examples of proprietary standards are the American Institute of Steel Construction (AISC) specifications for *Design, Fabrication and Erection of Structural Steel for Buildings*; the specifications for *Aluminum Siding and High Performance Coatings for Aluminum Siding* by the Architectural Aluminum Manufacturers Association (AAMA); and the American Plywood Association (APA) document *Design and Fabrication Specifications for Plywood Lumber Components*.

4. A mixture of standards-developing organizations which cannot properly be classified in any of the categories described above. For instance, the National Fire Protection Association (NFPA) focuses its standards development program on a singular interest: firesafety. Underwriters Laboratories Inc. (UL), an independent testing and certification organization, develops a variety of standards needed in the conduct of its work. UL specifications become nationally recognized standards and many receive approval as consensus standards through the ANSI process.

Similar to UL is the Factory Mutual Research Corporation (FMRC) which develops safety standards and test methods widely used by insurance and other interests.

Another organization in this category is the National Conference of States on Building Codes and Standards (NCSBCS). This group, whose principal members are state officials, develops certain mobile home standards and also is heavily involved in the development of standards dealing with energy conservation.

THE SYSTEM FOR CODE AND STANDARD DEVELOPMENT

Each organization within the heterogeneous community of codes and standards developers conducts its business as an independent body, and all are generally committed to different goals and purposes. Despite the diversified character of the industry, there is a national system in place by virtue of technology transfer from one organization to another.

The entire reticulation of national codes and standard activity can be diagrammed. (See Figure 12.2.)

Summary

Early building codes were concerned primarily with the prevention of building collapse. As civilization progressed and cities became crowded, regulations were formed to limit the types, numbers, and heights of buildings that could be constructed, and also to prevent the start and spread of fire in

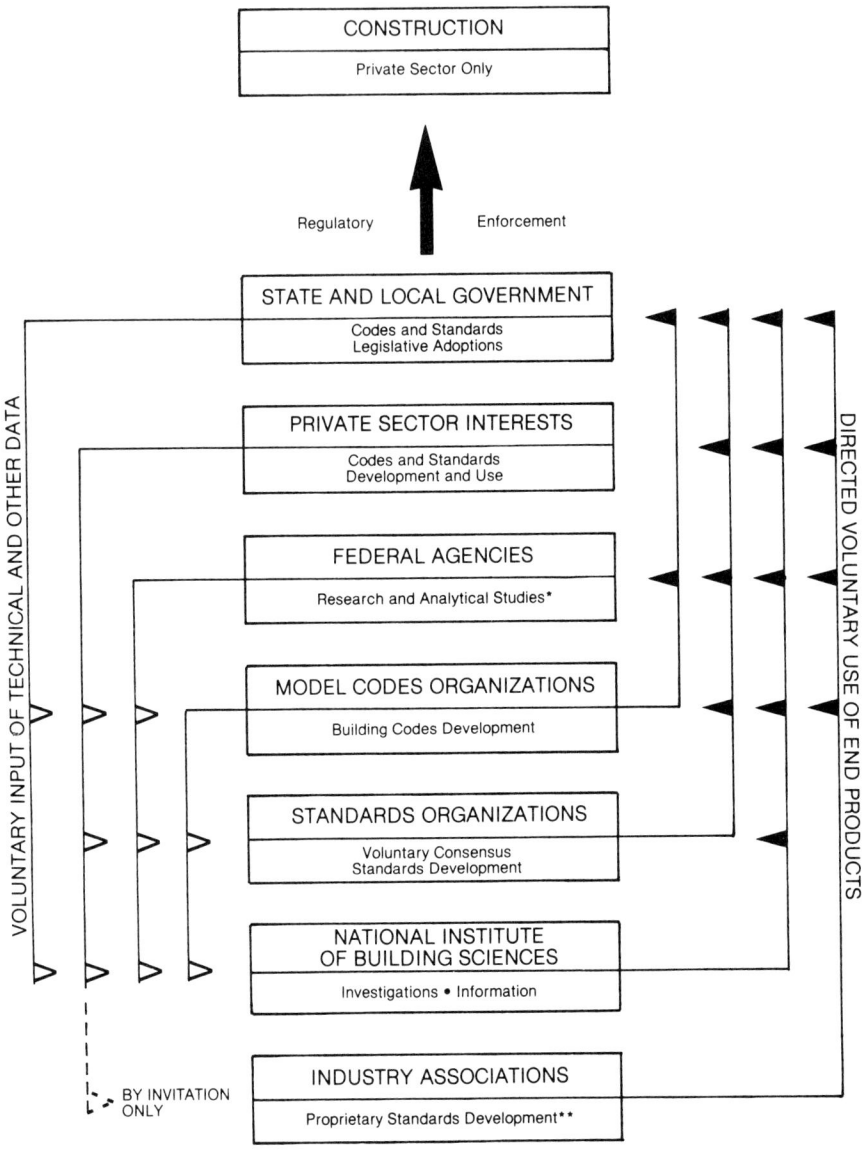

Fig. 12.2   *Network of Codes and Standards Activity.*

those buildings. As building and fire regulations developed throughout the United States, many were incorporated into the law at federal, state, and local levels of government.

Today's model building codes establish minimum requirements for construction and design, and fire protection codes and standards play an important part in community development.

References

[1]*National Building Code* 1976. American Insurance Association, Engineering and Safety Services, New York.

[2]Sanderson, R.L. 1969. *Codes and Code Administration*, Building Officials Conference of America, Inc., Chicago, IL.

[3]NFPA *101*-1988. *Life Safety Code*, National Fire Protection Association, Quincy, MA.

[4]*Uniform Building Code* 1985. International Conference of Building Officials, Whittier, CA.

[5]*Standard Building Code* 1985. Southern Building Code Congress International, Inc., Birmingham, AL.

[6]*National Building Code* 1984. Building Officials and Code Administrators, International, Country Club Hills, IL.

[7]NFPA 70-1987. *National Electrical Code*, National Fire Protection Assocation, Quincy, MA.

[8]NFPA 30-1987. *Flammable and Combustible Liquids Code*, National Fire Protection Association, Quincy, MA.

[9]NFPA 54-1984. *National Fuel Gas Code*, National Fire Protection Association, Quincy, MA.

[10]NFPA 1-1982. *Fire Prevention Code*, National Fire Protection Association, Quincy, MA.

[11]*NFPA Directory* (Annual). National Fire Protection Association, Quincy, MA.

[12]*One- and Two-Family Dwelling Code* 1987, Council of American Building Officials, Country Club Hills, IL.

[13]"Building the American City" 1968. Report of the National Commission on Urban Problems (NCUP), U. S. Government Printing Office, Washington, DC.

[14]*Building Codes: A Program for Intergovernmental Reform* 1966. Advisory Commission on Intergovernmental Relations (ACIR), U.S. Government Printing Office, Washington, DC.

[15]"Report of the President's Commission on Housing" 1982. U.S. Government Printing Office, Washington, DC.

Additional Reading

NFPA 220-1985. *Standard on Types of Building Construction*, National Fire Protection Association, Quincy, MA.

ASTM Standards in Building Codes 1978. American Society for Testing and

Materials, Philadelphia, PA.

Bihr, J. E. 1977. "Building and Fire Codes: The Regulatory Process, Fire Standards, and Safety," *ASTM STP 614*, American Society for Testing and Materials, Philadelphia, PA.

Curless, M. 1969. *Codes, Standards, and Fire Protection Engineering*, MP 69-1, National Fire Protection Association, Boston, MA.

Fire Protection Through Modern Building Codes, 5th ed. 1981. American Iron and Steel Institute, Washington, DC.

Hansen, A. T. 1984. "Applying Building Codes to Existing Buildings," *CBD 230*, Division of Building Research, National Research Council, Ottawa, Canada.

Harter, Philip J. 1979. "Regulatory Use of Standards: The Implications for Standards Writers," *NBS-GCR-79-171*, National Bureau of Standards, Gaithersburg, MD.

Kouba, Dennis, ed. 1982. *Code Administration and Enforcement: Trends and Perspectives*, International City Management Association, Washington, DC.

McConnaughey, John S. 1978. "An Economic Analysis of Building Code Impacts: A Suggested Approach," *NBSIR 78-1528*, National Bureau of Standards, Gaithersburg, MD.

Rawie, Carol C. 1981. "Estimating Benefits and Costs of Building Regulations: A Step-by-Step Guide," *NBSIR 2223*, National Bureau of Standards, Gaithersburg, MD.

Sanderson, R. L., *Readings in Code Administration*, 3 vols. 1975. Building Officials and Code Administrators International, Inc., Chicago, IL.

Smyrl, Elmira S. 1980. "Literature Review: The Building Regulatory System in the United States," *NBS-GCR-80-286*, National Bureau of Standards, Gaithersburg, MD.

Standards Activities of Organizations in the United States 1984. National Bureau of Standards, U.S. Department of Commerce, Washington, DC.

Sullivan, Charles D. 1983. *Standards and Standardization: Basic Principles and Applications*, Marcel Dekker, New York.

Taylor, D. M., *A Guide for Codes Adoption and Codes Enforcement* 1974. U.S. Department of Housing and Urban Development, Regional Office 4, Atlanta, GA.

Teague, Paul, E. 1977. "Mini-Max Building Codes and their Effect on the Fire Service," *Fire Journal*, Vol. 71, No. 3.

Vogel, Bertram M. 1976. *Standards Referenced in Selected Building Codes, Office of Building Standards, and Codes Services*, Center for Building Technology, Institute of Applied Technology, National Bureau of Standards, Washington, DC.

A

Absolute pressure, 81
Acetylene, 79, 80, 84
Acetate, 71, 72
Acrolein, 58
Additives
  flame-retardant, 70
  in plastics, 69, 70
Air shipments, of hazardous materials, 93
Alarms
  multiple, 260–261
  receipt of, 231, 234, 235
  on sprinkler systems, 173, 241–242
Alarm systems. *See* Municipal fire alarm systems; Protective signaling systems
Altruistic behavior, 21, 24, 26
Aluminum Siding and High Performance Coatings for Aluminum Siding (AISC), 335
"America Burning," 136–137, 148, 314
American Concrete Institute (ACI), 334
American Institute of Steel Construction (AISC), 335
American Insurance Association (AIA), 322, 327
American National Standards Institute (ANSI), 329
American Plywood Association (APA), 335
American Society for Testing and Materials (ASTM), 334
American Society of Heating, Refrigeration and Air Conditioning Engineers (ASHRAE), 334
American Society of Mechanical Engineers (ASME), 84, 334
Ammonia, 79, 80
Antifreeze solution, 178–179
Architectural Aluminum Manufacturers Association (AAMA), 335
Arcing, 52
Areas of refuge, 26
Argon, 79, 80
Arson, 16, 20, 116, 126, 129

Arundel Park Hall fire, 24–25
Asphalt, 77
Assembly occupancies, 33, 73
Association of Major City Building Officials, 330
Atkins, Thomas, 4
Atlanta, GA, 9
Atomic explosions, 55
Automatic sprinkler heads. *See* Sprinklers
Automatic Sprinkler Protection (Dana), 10
Automatic sprinkler systems, 9–10, 70–71, 142, 169–174
  combined dry pipe and preaction, 177, 182
  combined standpipe and, 192
  deluge, 177, 181–182
  design of, 175
  development of, 169–170
  economics of, 173–174
  fundamentals of, 174–175
  installation standards, 174–177
  nonstandard, 177, 182–184
  outside, 183
  partial, 171, 172, 183
  performance of, 171–172
  preaction, 177, 180–181
  regular dry pipe, 177, 179–180
  residential, 175–177, 187
  value of, 142, 160, 170–173
  water damage from, 172–173
  wet pipe, 177, 178–179
  see also Sprinklers

B

"Babcock" chemical engines, 8
Baltimore fire, 7
Basic/Building Code (BBC) (BOCA), 326
Bed key, 5
Beverly Hills Supper Club fire, 13, 24, 26
Blackstone, G.V., 6
BLEVEs (Boiling Liquid Expanding Vapor Explosions), 55, 82–83, 84